MACHINE VISION

McGRAW-HILL SERIES IN COMPUTER SCIENCE

SENIOR CONSULTING EDITOR
C. L. Liu, *University of Illinois at Urbana-Champaign*
CONSULTING EDITOR
Allen B. Tucker, *Bowdoin College*

Fundamentals of Computing and Programming

Computer Organization and Architecture

Computers in Society/Ethics

Systems and Languages

Theoretical Foundations

Software Engineering and Database

Artificial Intelligence

Networks, Parallel and Distributed Computing

Graphics and Visualization

The MIT Electrical Engineering and Computer Science Series

ARTIFICIAL INTELLIGENCE

Bowen: Prolog and Expert Systems Programming
Fu: Neural Networks in Computer Intelligence
*Horn: Robot Vision
Jain, Kasturi, and Schunck: Machine Vision
Levine: Vision in Man and Machine
Rich and Knight: Artificial Intelligence

*Co-published by the MIT Press and McGraw-Hill, Inc.

MACHINE VISION

Ramesh Jain
University of California, San Diego

Rangachar Kasturi
Pennsylvania State University

Brian G. Schunck
Adept Technology

McGraw-Hill, Inc.
New York   St. Louis   San Francisco   Auckland   Bogotá
Caracas   Lisbon   London   Madrid   Mexico City   Milan
Montreal   New Delhi   San Juan   Singapore   Sydney
Tokyo   Toronto

Machine Vision

Copyright ©1995 by McGraw-Hill, Inc. All rights reserved. Printed in the
United States of America. Except as permitted under the United States
Copyright Act of 1976, no part of this publication may be reproduced
or distributed in any form or by any means, or stored in a data base or
retrieval system, without the prior written permission of the publisher.

This book is printed on acid-free paper.

5 6 7 8 9 0 FGR FGR 9 0 9

ISBN 0-07-032018-7

The editor was Eric M. Munson;
the production supervisor was Kathryn Porzio.
The cover was designed by BC Graphics.
Project supervision was done by Publication Services, Inc.
Quebecor Printing/Fairfield was printer and binder.

Library of Congress Cataloging-in-Publication Data

Jain, Ramesh.
    Machine vision / Ramesh Jain, Rangachar Kasturi, Brian G. Schunck.
        p.   cm.
    Includes index.
    ISBN 0-07-032018-7
        1. Computer vision.   I. Kasturi, Rangachar, (date).
    II. Schunck, Brian G.   III. Title.
    TA1634.J35   1995
    006.4'2—dc20
                                                            95-3771

About the Authors

Ramesh Jain is currently a Professor of Electrical and Computer Engineering, and Computer Science and Engineering at the University of California at San Diego. Before joining UCSD, he was a Professor of Electrical Engineering and Computer Science and the founding Director of the Artificial Intelligence Laboratory at the University of Michigan, Ann Arbor. He was also the founder and chairman of Imageware Inc. His current research interests are in multimedia information systems, image databases, machine vision, and intelligent systems.

Ramesh is a Fellow of IEEE, AAAI, and Society of Photo-Optical Instrumentation Engineers, and member of ACM, Pattern Recognition Society, Cognitive Science Society, Optical Society of America, and Society of Manufacturing Engineers. He is currently the Editor-in-Chief of *IEEE Multimedia,* and is on the editorial boards of *Machine Vision and Applications, Pattern Recognition,* and *Image and Vision Computing.* He received his Ph.D. from IIT, Kharagpur, in 1975 and his B.E. from Nagpur University in 1969.

Rangachar Kasturi joined Penn State University in 1982 after completing his graduate studies at Texas Tech University (Ph.D. 1982 and M.S.E.E. 1980). He received a B.E. (Electrical) degree from Bangalore University in 1968. His primary research focus in recent years has been in the area of Document Image Analysis (DIA). His group's main contribution has been the design of efficient algorithms to generate intelligent interpretations of engineering drawings and maps to facilitate automatic conversion from paper medium to computer databases. He is the Editor-in-Chief of *IEEE Transactions on Pattern Analysis and Machine Intelligence.* He was the managing editor of *Machine Vision and Applications* during 1993–94. He is a coauthor of the tutorial texts, *Computer Vision: Principles and Applications* and *Document Image Analysis,* both published by IEEE CS Press, and a coeditor of the book *Image Analysis Applications* (Marcel Dekker, 1990). During 1987–90 he delivered lectures at many chapters of the IEEE Computer Society through its Distinguished Visitor Program. He has served the International Association for Pattern Recognition in various capacities.

Brian G. Schunck has worked for several years on the development of systems for machine vision and image processing. He was educated in computer science at the University of California, Irvine, where he received the B.S. magna cum laude in 1976. He studied electrical engineering, systems theory, and artificial intelligence at M.I.T., where he received the Master's and E.E. degrees in 1979 for work on control algorithms for robotic manipulators and the doctorate in 1983 for research on image flow. He was an assistant professor in the Department of Electrical Engineering and Computer Science and a member of the Artificial Intelligence Laboratory at the University of Michigan, Ann Arbor. Currently he is the Director of Vision Software at Adept Technology.

Brian's current interests include statistical methods for machine vision and industrial inspection; contour, surface, and volume models for computer vision and medical image processing; structure and motion estimation for mobile robots; reverse engineering part models from range data; computer graphics; user interfaces; and marine navigation.

Brian Schunck is a member of the IEEE, ACM, the Society for Industrial and Applied Mathematics, the American Statistical Association, the American Society for Photogrammetry and Remote Sensing, the Society for Manufacturing Engineers, and the Society for Automotive Engineers.

To

Sudha
    —*Ramesh Jain*

My Grandmother
    —*Rangachar Kasturi*

Gizmo
    —*Brian G. Schunck*

Contents

Preface

This book grew out of our efforts to provide a balanced coverage of essential elements of machine vision systems to students in our undergraduate and early graduate classes. The field of machine vision, or computer vision, has been growing at a fast pace. The growth in this field, unlike most established fields, has been both in breadth and depth of concepts and techniques. To make the situation more confusing, the number of new applications has also been growing. Machine vision techniques are being applied in areas ranging from medical imaging to remote sensing, industrial inspection to document processing, and nanotechnology to multimedia databases.

As in most developing fields, not all aspects of machine vision that are of interest to active researchers are useful to the designers of a vision system for a specific application. A designer needs to know basic concepts and techniques to be successful in designing or evaluating a vision system for a particular application. It may not be necessary to know the latest, often controversial, results from leading research centers. On the other hand, the techniques learned by a designer should not be ephemeral.

This text is intended to provide a practical introduction to machine vision. We made efforts to provide all of the details to allow vision algorithms to be used in practical applications. Intentionally omitted are theories of machine vision that do not appear to have sufficient practical applications at this time. We want this to be a useful introduction to machine vision rather than a state-of-the-art collection of research on machine vision.

The text is intended to be used in an introductory course in machine vision at the undergraduate or early graduate level and should be suitable for students with no prior knowledge of computer graphics or signal processing. Students should have a working knowledge of mathematics through calculus of two variables, including matrices and linear spaces, and familiarity with basic probability theory, computer programming, and elementary data structures. Numerical and statistical methods and advanced algorithms are described as needed as well as material on geometry in two and three dimensions. For some sections in the book, more mathematical background is needed. Such sections can be omitted by readers not interested in the rigorous formalization. We have made efforts to provide intuitive concepts, even for mathematical sections, that will help a reader understand the basic elements without the details.

An introductory text is based on material from several sources. This book also contains material from research papers, books, and other places. We have made no attempt to exhaustively list all original sources. We do provide some pointers to readers who are interested in exploring topics more deeply in each chapter. The references at the end of the book provide a list of sources that were directly used in the preparation of the book.

We strongly encourage readers to send any comments and corrections by mail to one of the authors or electronically to jain@ece.ucsd.edu.

Ramesh Jain
Rangachar Kasturi
Brian G. Schunck

Acknowledgments

This book has benefited from the comments and suggestions of many people. Professors Kim Boyer, Glenn Healey, William Thompson, and Mohan Trivedi provided extensive comments and suggestions on the book. We would also like to thank Kevin Bowyer, Rattikorn Hewett, Ting-Chuen Pong, and Mubarak Shah for their reviews. We are also thankful to Anil Jain, Lawrence O'Gorman, Dorothea Blostein, Jason Daida, Robert Bolles, and John Barron for their help in various ways. Eric Munson played a key role in this book by maintaining its schedule; his persistent but very kind and gentle reminders were most important in our finishing this project in a reasonable time. Thanks also to the rest of the staff at McGraw-Hill, New York, and at Publication Services, Inc., of Champaign, Illinois.

Numerous students directly contributed to the preparation of this book. The most important help and contributions came from David Kosiba. David helped us in all aspects of the book, including many of the illustrations and the concept for the cover art. This book has been influenced significantly by his energy, interest in machine vision, and perseverance. Sue Lott and Dino Terzides also helped with the illustrations. Other students who went out of their way to provide help included Sandy Bartlett, James Han, Patrick Kelly, Dan Sebald, Nilesh Patel, Francis Quek, Todd Elvins, Arun Katkere, Saied Moezzi, and Jennifer Schlenzig. We sincerely appreciate the secretarial support provided by Suzie Mostoller at Penn State, Kathy Dewitt at

University of Michigan, and Edna Nerona at UC, San Diego. Brian Schunck would also like to thank his colleagues at Perceptron, Charles Wu at the Ford Scientific Research Laboratories, and Chuck Meyer at the University of Michigan for supporting his research and providing facilities while this book was being written. On a more personal side, the authors are very thankful to their families for their patience and understanding.

MACHINE VISION

Chapter 1

Introduction

1.1 Machine Vision

The goal of a machine vision system is to create a model of the real world from images. *A machine vision system recovers useful information about a scene from its two-dimensional projections.* Since images are two-dimensional projections of the three-dimensional world, the information is not directly available and must be recovered. This recovery requires the inversion of a many-to-one mapping. To recover the information, knowledge about the objects in the scene and projection geometry is required.

To consider applications of a machine vision system and type of processing required, let us consider the three images shown in Figures 1.1 to 1.3. These figures show three different applications of a machine vision system. The information recovered by a vision system in the three cases is different. In the first figure, we are interested in diagnosis of a disease using computed tomography images. Machine vision systems help a physician to recover information by enhancing the images. Quantitative measurements on regions of interest can also be made easily available. Such systems are being developed for all imaging modes useful in different aspects of health care. Similar applications are being developed for inspection of industrial, agricultural, and other products. Machine vision systems have been used for quality control of products ranging from pizza to turbine blades, from submicron structures on wafers to auto-body panels, and from apples to oranges.

Figure 1.2 shows two pairs of images acquired by a mobile robot. Each pair represents a stereo pair at a particular time instant. These images are

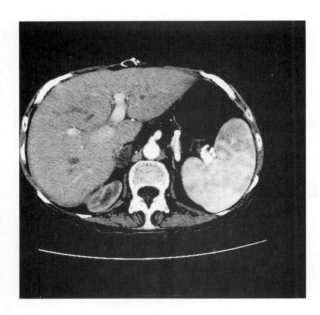

Figure 1.1: Medical images may be processed by a computer vision system to assist in diagnosis. This image is a contrast-enhanced CT (transverse X-ray computed tomography) image of a human liver displayed at the soft tissue window settings. (*Courtesy of Prof. Charles R. Meyer and Dr. Peyton Bland, Department of Radiology, University of Michigan, Ann Arbor.*)

used to recover three-dimensional structure of the environment by the robot for autonomously navigating in its environment. The information obtained from stereo and motion is combined to get a robust map of the environment at a resolution that is sufficient for the task. Such techniques are useful in autonomous navigation of automobiles, airplanes, tanks, and robots. We will see that there are many different methods to recover depth to points in images. Usually the results of several of these must be combined to recover reliable depth values.

Figure 1.3 shows an image of an arctic region taken from a satellite. Different regions in this image correspond to ice floes of different age. Such images are routinely taken by satellites. These images are extremely important in weather forecasting, global change analysis, agriculture and forestry, and other applications. Machine vision systems are playing an increasingly important role in analysis and information management of the exceedingly large volume of data collected by satellites.

Figure 1.2: Two pairs of stereo images acquired by a mobile robot. These will be used in recovering the layout of the environment by the robot. (*Images provided by Robert Bolles at SRI.*)

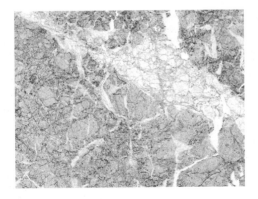

Figure 1.3: An image of the arctic region. This image should be analyzed to find the age and size of ice floes and other objects. (*Arctic image provided by Jason Daida at the University of Michigan.*)

1.2   Relationships to Other Fields

Many fields are related to machine vision. As we will see, techniques developed from many areas are used for recovering information from images. In this section, we briefly describe some very closely related fields. No effort is made to relate machine vision (also called computer vision) to other fields exhaustively.

Image processing is a well-developed field. Image processing techniques usually transform images into other images; the task of information recovery is left to a human user. This field includes topics such as image enhancement, image compression, and correcting blurred or out-of-focus images. On the other hand, machine vision algorithms take images as inputs but produce other types of outputs, such as representations for the object contours in an image. Thus, emphasis in machine vision is on recovering information automatically, with minimal interaction with a human. Image processing algorithms are useful in early stages of a machine vision system. They are usually used to enhance particular information and suppress noise.

Computer graphics generates images from geometric primitives such as lines, circles, and free-form surfaces. Computer graphics techniques play a significant role in visualization and virtual reality. Machine vision is the inverse problem: estimating the geometric primitives and other features from the image. Thus, computer graphics is the synthesis of images, and machine vision is the analysis of images. In the early days of these two fields, there was not much relationship between them, but in the last few years these two fields have been growing closer. Machine vision is using curve and surface representations and several other techniques from computer graphics, and computer graphics is using many techniques from machine vision to enter models into the computer for creating realistic images. Visualization and virtual reality are bringing these two fields closer.

Pattern recognition classifies numerical and symbolic data. Many statistical and syntactical techniques have been developed for classification of patterns. Techniques from pattern recognition play an important role in machine vision for recognizing objects. In fact, many industrial applications rely heavily on pattern recognition. Object recognition in machine vision usually requires many other techniques. We will discuss some aspects of statistical pattern recognition briefly in the discussion of object recognition.

Artificial intelligence is concerned with designing systems that are intelligent and with studying computational aspects of intelligence. Artificial intelligence is used to analyze scenes by computing a symbolic representation of the scene contents after the images have been processed to obtain features. Artificial intelligence may be viewed as having three stages: perception, cognition, and action. Perception translates signals from the world into symbols, cognition manipulates symbols, and action translates symbols into signals that effect changes in the world. Many techniques from artificial intelligence play important roles in all aspects of computer vision. In fact, computer vision is often considered a subfield of artificial intelligence.

Design and analysis of *neural networks* has become a very active field in the last decade [147, 148]. Neural networks are being increasingly applied to solve some machine vision problems. Since this field is in its infancy, there are no established techniques for machine vision yet. We will not discuss any neural network–based techniques in this book.

Psychophysics, along with cognitive science, has studied human vision for a long time. Many techniques in machine vision are related to what is known about human vision. Many researchers in computer vision are more interested in preparing computational models of human vision than in designing machine vision systems. Many of the techniques discussed in this book have a strong similarity to those in psychophysics. Our emphasis in this book will be on designing machine vision systems; we will not make any effort to relate the techniques discussed here to those in psychophysics.

Machine vision produces measurements or abstractions from geometrical properties. It may be useful to remember the equation

$$\text{Vision} = \text{Geometry} + \text{Measurement} + \text{Interpretation}. \qquad (1.1)$$

As you will see in this book, machine vision comprises techniques for estimating features in images, relating feature measurements to the geometry of objects in space, and interpreting this geometric information.

1.3 Role of Knowledge

Decision making always requires knowledge of the application or goal. As we will see, at every stage in machine vision decisions must be made by the system. Emphasis in machine vision systems is on maximizing automatic operation at each stage, and these systems should use knowledge to accomplish this. The knowledge used by the system includes models of features, image formation, models of objects, and relationships among objects. Without explicit use of knowledge, machine vision systems can be designed to work only in a very constrained environment for very limited applications. To provide more flexibility and robustness, knowledge is represented explicitly and used by the system. Our goal in this book is to point out the types of knowledge used by machine vision systems at different stages in order to make the reader aware of the issues that should be considered to make the system more adaptive and robust. We will see that knowledge is used by designers of systems in many implicit as well as explicit forms. The efficacy and efficiency of a system is usually governed by the quality of the knowledge used by the system. Difficult problems are often solvable only by identifying the proper source of knowledge and appropriate mechanisms to use it in the system.

1.4   Image Geometry

There are two parts to the image formation process:

1. The geometry of image formation, which determines where in the image plane the projection of a point in the scene will be located.

2. The physics of light, which determines the brightness of a point in the image plane as a function of scene illumination and surface properties.

This section introduces the first of these two parts, the geometry of image formation. Although an understanding of the physics of light is not necessary for understanding the fundamentals of most vision algorithms, such knowledge is often useful in building vision systems. For this reason, optics will be covered briefly in Chapter 8, and the physics of image formation, called radiometry, will be discussed in Chapter 9.

The basic model for the projection of points in the scene onto the image plane is diagrammed in Figure 1.4. In this model, the imaging system's center of projection coincides with the origin of the three-dimensional coordinate system. The coordinate system for points in the scene is the three-dimensional space spanned by the unit vectors $\mathbf{x}$, $\mathbf{y}$, and $\mathbf{z}$ that form the axes

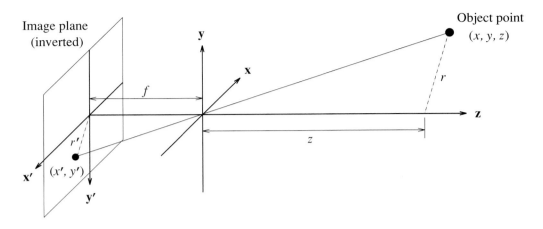

Figure 1.4: The point on the image plane that corresponds to a particular point in the scene is found by following the line that passes through the scene point and the center of projection.

of the coordinate system. A point in the scene has coordinates $(\mathbf{x}, \mathbf{y}, \mathbf{z})$. The x coordinate is the horizontal position of the point in space as seen from the camera, the y coordinate is the vertical position of the point in space as seen from the camera, and the z coordinate is the distance from the camera to the point in space along a line parallel to the $\mathbf{z}$ axis. The *line of sight* of a point in the scene is the line that passes through the point of interest and the center of projection. The line drawn in Figure 1.4 is a line of sight.

The image plane is parallel to the $\mathbf{x}$ and $\mathbf{y}$ axes of the coordinate system at a distance f from the center of projection, as shown in Figure 1.4. Note that the image plane in an actual camera is at a distance of f behind the center of projection (as shown in Figure 1.4) and the projected image is inverted. It is customary to avoid this inversion by assuming that the image plane is in front of the center of projection as shown in Figure 1.5. The image plane is spanned by the vectors $\mathbf{x}'$ and $\mathbf{y}'$ to form a two-dimensional coordinate system for specifying the position of points in the image plane. The position of a point in the image plane is specified by the two coordinates x' and y'. The point $(0,0)$ in the image plane is the origin of the image plane. The position in the image plane of a point in the scene is found by intersecting the line of sight with the image plane according to the projection scheme as described in the following sections.

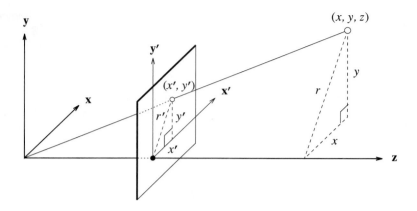

Figure 1.5: An illustration showing the line of sight that is used to calculate the projected point (x', y') from the object point (x, y, z).

1.4.1   Perspective Projection

The position (x', y') in the image plane of a point at position (x, y, z) in the scene is found by computing the coordinates (x', y') of the intersection of the line of sight passing through the scene point (x, y, z) with the image plane as shown in Figure 1.5.

The distance of the point (x, y, z) from the **z** axis is $r = \sqrt{x^2 + y^2}$, and the distance of the projected point (x', y') from the origin of the image plane is $r' = \sqrt{x'^2 + y'^2}$. The **z** axis, the line of sight to point (x, y, z), and the line segment of length r from point (x, y, z) to the **z** axis (perpendicular to the **z** axis) form a triangle. The **z** axis, the line of sight to point (x', y') in the image plane, and the line segment of length r' from point (x', y') to the **z** axis (perpendicular to the **z** axis) form another triangle. The two triangles are similar, so the ratios of the corresponding sides of the triangles must be the same:

$$\frac{f}{z} = \frac{r'}{r}. \tag{1.2}$$

The triangle formed from the x and y coordinates and the perpendicular distance r and the triangle formed from the image plane coordinates x', y' and the perpendicular distance r' are also similar triangles:

$$\frac{x'}{x} = \frac{y'}{y} = \frac{r'}{r}. \tag{1.3}$$

Combining Equations 1.2 and 1.3 yields the equations for perspective projection:

$$\frac{x'}{x} = \frac{f}{z} \quad \text{and} \quad \frac{y'}{y} = \frac{f}{z}. \tag{1.4}$$

The position of a point (x, y, z) in the image plane is given by the equations

$$x' = \frac{f}{z} x \tag{1.5}$$

$$y' = \frac{f}{z} y. \tag{1.6}$$

1.4.2 Coordinate Systems

In this presentation, we have assumed that the center of projection coincides with the origin of the three-dimensional space and that the camera axes are aligned with the coordinate system used to specify the location of a point in the scene. In general, the camera is displaced and rotated with respect to the three-dimensional coordinate system used for specifying the coordinates of points in the scene. The coordinates (x_a, y_a, z_a) in the absolute coordinate system must be transformed into the coordinates (x_c, y_c, z_c) of the point in the camera coordinate system before projecting the points onto the image plane. The general case is presented in Chapter 12 on calibration, which covers the mathematics of transforming coordinates between coordinate systems. Absolute coordinates are also called world coordinates.

Individual objects may have their own coordinate system, called model coordinates. The scene consists of object models that have been placed (rotated and translated) into the scene, yielding object coordinates in the coordinate system of the scene (absolute coordinates). The scene coordinates in the absolute coordinate system are transformed to camera coordinates before projection onto the image plane.

Throughout the book, subscripts are used to denote the coordinate system, and primes are used to denote the image plane coordinates of a point after projection onto the image plane. Subscripts are omitted when the coordinate system is clear, and the primes are omitted when it is clear that coordinates are in the image plane. These conventions allow us to be precise about the coordinate system when necessary and to distinguish between a point in camera coordinates and its projection onto the image plane, but they allow the notation to be simplified to the common usage for point coordinates in analytic geometry when we are presenting the equations for curves and surfaces.

In keeping with the convention, we will omit primes from image plane coordinates for the rest of this chapter and for Chapters 2 through 6, which cover machine vision algorithms that are restricted to the image plane.

1.5   Sampling and Quantization

A continuous function cannot be represented exactly in a digital computer. The interface between the optical system that projects a scene onto the image plane and the computer must sample the image at a finite number of points and represent each sample within the finite word size of the computer. This is sampling and quantization. Each image sample is called a pixel.

We will initially assume that images are sampled on a regular grid of squares so that the horizontal and vertical distances between pixels are the same throughout the image. We will generalize this assumption in Chapter 12. Each pixel is represented in the computer as a small integer. Frequently, the pixel is represented as an unsigned 8-bit integer in the range $[0, 255]$, with 0 corresponding to black, 255 corresponding to white, and shades of gray distributed over the middle values.

Many cameras acquire an analog image, which is then sampled and quantized to convert it to a digital image. The sampling rate determines how many pixels the digital image will have (the image resolution), and quantization determines how many intensity levels will be used to represent the intensity value at each sample point. As shown in Figures 1.6 and 1.7, an image looks very different at different sampling rates and quantization levels. In most machine vision applications, the sampling and quantizing rates are predetermined due to the limited choice of available cameras and image acquisition hardware; but in many applications it may be important to know the effects of sampling and quantizing. The image processing book [121] discusses the factors that should be considered in selecting appropriate sampling and quantization rates to retain the important information in images.

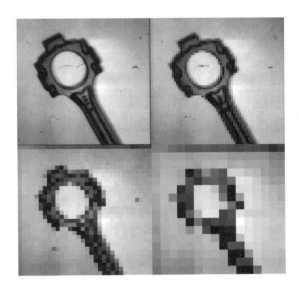

Figure 1.6: An image shown at many different spatial resolutions. *Top left:* Original image sampled at 256×256 and 128 gray levels. *Top right:* 64×64. *Bottom left:* 32×32. *Bottom right:* 16×16.

Figure 1.7: An image shown at many different gray level resolutions. *Top left:* Image sampled at 32 gray levels. *Top right:* 16 gray levels. *Bottom left:* 8 gray levels. *Bottom right:* 4 gray levels.

1.6   Image Definitions

It is important to understand the relationship between the geometry of image formation, described in previous sections, and the representation for images in the computer. There must be a bridge from the mathematical notation used to develop machine vision algorithms to the algorithmic notation used in programs.

A pixel is a sample of the image intensity quantized to an integer value. An image is a two-dimensional array of pixels. The row and column indices $[i, j]$ of a pixel are integer values that specify the row and column in the array of pixel values. Pixel $[0, 0]$ is located at the top left corner of the image. The index i points down, and j points to the right. This index notation corresponds closely to the array syntax used in computer programs. The positions of points in the image plane have x and y coordinates. The y coordinate corresponds to the vertical direction, and the x coordinate corresponds to the horizontal direction. The y axis points up, and the x axis points to the right. Note that the directions corresponding to the two indices i and j in the pixel index $[i, j]$ are the reverse of the directions corresponding to the respective coordinates in the position (x, y).

The x and y coordinates are real numbers, stored as floating-point numbers in the computer. Image plane coordinates (x, y) can be computed from pixel coordinates $[i, j]$ of an n by m pixel array using the formulas

$$x = j - \frac{m-1}{2} \qquad (1.7)$$

$$y = -\left(i - \frac{n-1}{2}\right), \qquad (1.8)$$

which assume that the origin of the image plane coordinate system corresponds to the center of the image array.

In an imaging system, each pixel occupies some finite area on the image plane. Machine vision algorithms that depend on the exact shape of the pixel footprint will not be covered in this book, so we may assume for concreteness that the pixels partition the image plane into equal-sized squares. Positions in the image plane can be represented to fractions of a pixel. The coordinates (x_{ij}, y_{ij}) of the pixel with indices $[i, j]$ are the location of the center of the pixel in the coordinate system of the image plane. Since we are concerned only with the location of the center of the pixel in the image plane (the location at which the image sample was taken), the pixel may be further abstracted to a point in the image plane. The array of pixels in the computer program corresponds to the grid of image plane locations at which the samples were obtained, as illustrated in Figure 1.8.

In diagramming images, we may show the image as a square tessellation of a rectangular region, with each square shaded to indicate the image intensity of that pixel. This is purely a technique for visualization and does not imply that the shape of the pixel matters to the algorithm being discussed. For the purposes of nearly all of the algorithms discussed in this book, the image can be modeled as a square grid of samples of the image intensity, represented in the computer as an array of pixel values. The camera and digitizing electronics are designed to ensure that this assumption is satisfied. Some variations, such as different spacing between the rows and columns in the grid, distortions due to lens imperfections, and errors in the construction of the camera, can be removed through calibration without changing the algorithms that process the image (see Section 12.9).

To summarize, a pixel is both a gray value, which is a quantized sample of the continuous image intensity, and an image location, specified as the row and column indices in the image array. The image array is obtained by sampling the image intensity at points on a rectangular grid. Points in the image plane, specified with coordinates x and y, may lie between the grid locations at which pixels were sampled.

1.7 Levels of Computation

An image usually contains several objects. A vision application usually involves computing certain properties of an object, not the image as a whole. To compute properties of an object, individual objects must first be identified as separate objects; then object properties can be computed by applying calculations to the separate objects.

Definitions and algorithms for connectivity and segmentation that will allow different objects to be represented as distinct subimages will be presented in Chapters 2 and 3. For now, consider computer vision algorithms from the viewpoint of locality of computation. Consider each algorithm in terms of its input-output characteristics. Here our aim is to characterize operations so that we can discuss the nature of input and output and how best

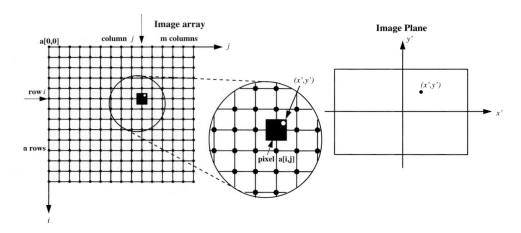

Figure 1.8: Relationship between image plane coordinates and image array indices. Note that the location of the origin of the x–y plane is arbitrary with respect to the image array.

to implement these operations. Note that the input to a computer vision system is an image, and the output, unlike that of image processing systems, is some symbolic quantity denoting identity or location of an object, for instance. The amount of data processed by a vision system is very large, and that makes the computational requirements of a computer vision system very demanding. The last few years have witnessed many special architectures designed for computer vision. Since we want to discuss characteristics of operations to predict their computational requirements, we classify the levels of operations and study their general characteristics.

1.7.1   Point Level

Some operations produce an output based on only a point in an image. Thresholding is an example. A thresholding algorithm produces output values that depend only on the input value, for a preset threshold. Thus,

$$f_B[i, j] = O_{\text{point}}\{f_A[i, j]\}, \qquad (1.9)$$

where f_A and f_B are the input and output images, respectively. This operation can be efficiently implemented using a lookup table (see Figure 1.9).

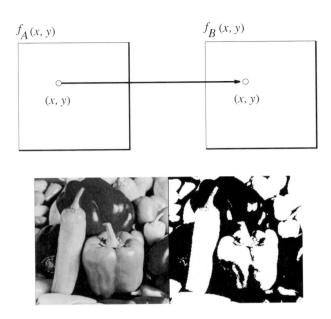

$$f_B[i, j] = O_{\text{point}}\{f_A[i, j]\}$$

Figure 1.9: *Top:* Point operations are applied to individual image pixels and produce an output image as the result. *Bottom left:* Original image. *Bottom right:* Thresholded image where pixels from the original image with a gray level value greater than 128 are set to white, while the remaining pixels are set to black.

1.7.2 Local Level

A local operation produces an output image in which the intensity at a point depends on the neighborhood of the corresponding point in the input image. Thus,

$$f_B[i, j] = O_{\text{local}}\{f_A[i_k, j_l]; [i_k, j_l] \in N[i, j]\}. \tag{1.10}$$

An example of such an operation is shown in Figure 1.10. Smoothing and edge detection are local operations. Since these operations require values from a neighborhood in the input image, array processors or Single Instruction, Multiple Data (SIMD) machines may be suitable for implementing these operations. In general, these operations can be easily implemented on parallel machines and can often be performed in real time.

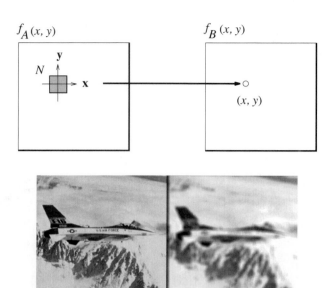

$$f_B[i,j] = O_{\text{local}}\{f_A[i_k, j_l];\quad [i_k, j_l] \in N[i,j]\}$$

Figure 1.10: *Top:* Local operations are applied to pixel neighborhoods and produce an output image as the result. *Bottom left:* Original image. *Bottom right:* Smoothed image where each pixel value is the average gray value calculated from its 5×5 local neighborhood in the original image.

1.7.3   Global Level

The output of certain operators depends on the whole picture. Such operations are called global operations:

$$P = O_{\text{global}}\{f[i,j]\}. \tag{1.11}$$

This operation is shown in Figure 1.11.

The output of these operators may be an image or it may be symbolic output. A histogram of intensity values and the Fourier transform are global operations. Global operations are responsible for the slowness of vision systems. We will see that most operations at higher levels are global in nature and pose the biggest challenge to designers of algorithms and architectures.

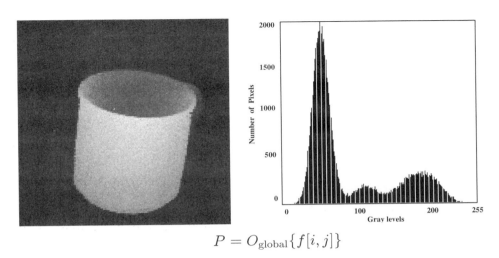

$$P = O_{\text{global}}\{f[i,j]\}$$

Figure 1.11: An example of a global operation: an image (left) and its histogram (right). A histogram is a plot of the number of pixels at each gray value contained in the image.

1.7.4 Object Level

Most applications of computer vision require properties to be computed at the object level. Size, average intensity, shape, and other characteristics of an object must be computed for the system to recognize it. Many other characteristics of an object must be determined for defect detection. Operators restricted to those pixels belonging to an object are occasionally applied to determine these properties. This leads to very difficult questions: What is an object? How do we find objects?

We will see that an object is defined in a particular context. In fact, many operations in machine vision are performed to find where a particular object is located in an image. Objects in images pose a *catch-22* situation. We must use all points that belong to an object to compute some of its characteristics, but we must use those characteristics to identify those points. We will see that significant efforts are spent to solve the *figure-ground* problem (separation of foreground pixels from background pixels) to group points into objects.

At this point, we should just remember that to understand the contents of an image, a machine vision system must perform several operations at the object level.

1.8   Road Map

This book provides a practical introduction to machine vision. The chapters are presented in a sequence that builds the application of machine vision from the simple binary world to the spatiotemporal world that we want our systems to understand. Roughly, the first half of the book works primarily in two dimensions, and the second half expands the coverage to 3-D. Chapter 2, on binary images, introduces basic terms and concepts used in machine vision. The techniques discussed here are used in all aspects of a vision system. We introduce the morphological approach here as well. Chapter 3 presents techniques to find regions in images. This chapter also covers techniques to represent regions. All vision systems have to use regions for their tasks, and hence the techniques covered in this chapter are essential in implementing systems. Chapter 4, on filtering, introduces techniques to enhance images and discusses fundamental aspects of filtering techniques. Edge detection techniques are very important and are an essential step in many machine vision systems. Chapter 5, on edge detection, introduces several edge detection techniques to show how such a simple operation involves complex processes. The next chapter presents techniques to represent contours. Edges are local; to use them one must group them into meaningful objects and represent them. Texture also plays an important role in many machine vision tasks such as surface inspection, scene classification, surface orientation, and shape determination. These topics are discussed in Chapter 7. The essentials of geometrical optics and radiometry are covered in Chapter 8 and 9 respectively. Color is an essential component of images in many applications and is discussed in Chapter 10.

Chapter 11 presents techniques to recover depth information from images using passive and active methods. The techniques discussed here take us to the three-dimensional from *two-dimensional* images. In the next chapter, we present techniques for camera calibration. To recover three-dimensional information from images, it is essential that we know the location and orientation of the camera and its parameters. Various techniques for camera calibration are discussed in this chapter. Representation and characteristics of curves and surfaces in space are discussed in Chapter 13. Techniques for interpolation and approximation are discussed in this chapter. Chapter 14 discusses dynamic vision. Techniques to detect changes in images, segment images based on motion characteristics, and track objects are presented. Structure from motion is briefly introduced here. Object recognition is one of the most common tasks that a vision system performs. We present fundamental aspects of object recognition in Chapter 15.

Most of the material included here can be covered in a semester. In a quarter course, some chapters may have to be omitted. In fact, the book may form a very good two-quarter sequence with inclusion of some advanced material. The sequence of chapters in a particular course depends on the application emphasis of the course. The first six chapters and Chapter 15, on object recognition, cover basic techniques, and we believe that they should be covered independent of the emphasis. In courses emphasizing image applications, Chapters 7 and 10 must also be covered. Courses emphasizing robotics and other three-dimensional information should cover depth, Chapter 11, and camera calibration, Chapter 12. People oriented toward mobile robotics and video will find Chapter 14 essential. Those interested in visualization, reverse engineering, and inspection will benefit from the material covered in Chapter 13, on curves and surfaces.

Our recommendation is to use the first six chapters in the sequence in which they are presented and then select later chapters based on your interest and emphasis. The chapters on camera calibration and curves and surfaces are more mathematical in nature, due to the nature of the problems addressed, than other chapters.

We consider the exercises a very important part of the course. We have presented conceptual problems, regular problems to test the understanding of techniques presented here, and programming exercises. We recommend that this course be made a laboratory-oriented course. Students should implement and experiment with many of the techniques that are presented here. Projects to implement simple vision systems are very important to a real understanding of machine vision techniques.

Further Reading

Machine vision is still a very active research area. Several conference proceedings (*Conf. on Computer Vision and Pattern Recognition, Int. Conf. on Computer Vision, Int. Conf. on Pattern Recognition, Int. Conf. on Robotics and Automation, Int. Conf. on Document Analysis and Recognition, Workshop on Computer Vision*, and numerous conferences of *SPIE*) describe the state of the art in this area. Several archival journals cover this research area (*IEEE Trans. Pattern Analysis and Machine Intelligence; Computer Vision, Graphics, and Image Processing; IEEE Trans. Image Processing; IEEE Trans. Systems, Man, and Cybernetics; Machine Vision and Applications; Int. J. Computer Vision; J. Opt. Soc. Am. A; Image and Vision Computing*; and *Pattern Recognition*) and are very good sources for studying the state of the art. Many special issues of journals, proceedings of conferences and workshops on special topics, and research books are also published every year, and these are usually good sources of information on particular aspects of machine vision. A list of such sources is provided in [140, pp. 702–706]. The volumes [139, 140] provide a collection of the best papers from the last decade.

Several introductory books have been written on this topic. Rosenfeld and Kak [206], Gonzalez and Woods [90], Pratt [196], Ballard and Brown [19], Horn [109], Jain [121], Levine [155], Wechsler [248], Nalwa [177], Haralick and Shapiro [103], and Schalkoff [211] are some of the books that have been used in introductory courses and by people interested in self-learning. The topics covered in this chapter are discussed in many books. Aspects of digitization are usually covered at length in books on image processing [121]. There are several excellent books on image processing [91, 121, 196] and computer graphics [81, 204] that provide information useful in machine vision. Readers interested in knowing more about pattern recognition should refer to [70]. There are also many good books on artificial intelligence [201, 251].

Readers interested in reading about human vision may find that some books in the above list provide information on this topic also. A few particularly good sources for human vision from a computational perspective are Marr [163], Mayhew and Frisby [166], Rock [203], and Gibson [86, 87]. Many books cover special application areas. A list of such books is provided in [139, 140]. Readers interested in computational vision with strong emphasis on human vision should see [155, 163].

Exercises

1.1 We interpret a variety of images effortlessly. In each of these cases, many types of knowledge are used. Consider the following applications and try to list all possible sources of knowledge that you use in interpretation of these images.

  a. Reading English text

  b. Image of an indoor scene

  c. Image of a road scene

  d. Image of an airport

  e. Remotely sensed image of an area

  f. Microscopic image of an organism

  g. X-ray image

  h. Other

1.2 List the levels of computations on images. Give two examples of each level of computation.

1.3 Determination of the distance to different objects in a scene is an important operation in machine vision. Many techniques will be discussed in this book. Here let us consider the simple projection model. In stereo vision, two cameras are used to determine the distance to a point using triangulation. Assume that you have two cameras and they are placed such that their optical axes are parallel to each other. There is a prominent point P in the scene such that its projections in two images are P_1 and P_2, respectively. Assuming that you know the location of the cameras and that the point appears in both images, how can you determine the distance to this point using perspective projections? Derive the relationship you can use to determine the distance. If you have freedom to select the location of the cameras, will you place them close to each other or far from each other? Justify your answer.

1.4 For a given picture, assume that you sample it using several sampling rates to represent it as an $N \times N$ array of pixels. At each pixel, you use the same number of bits to represent intensity values. Prepare a table showing the memory requirements for the image at different resolutions. Start at 16×16 and go to 4096×4096. Plot the memory requirements against N.

1.5 In the above exercise, now assume that pixels can be represented using different numbers of bits. One may use only 1 bit to represent a binary image and 24 bits to represent a color image. Though in most applications, single-channel intensity values are represented using only 8 bits, one may use 16 or more bits. Plot memory requirements for different numbers of bits; assume a fixed array size of 512×512 pixels.

1.6 Scan a picture using a tabletop scanner. You can select different sampling rates, expressed usually as dots per inch (dpi), and different intensity or color levels. Scan the image at many different resolutions and intensity ranges. See the memory requirements. Considering the memory requirements, at what resolution and number of intensity level will you save the image?

Now take at least two other images. These images should contain different amounts of detail. Repeat the above experiment. Would you store all these images using the same parameters or would you use different parameters? Why?

1.7 A surveillence camera is embedded in one of the walls of a room as shown in Figure 1.12. The optical axis of the camera is perpendicular to the wall, and the lens center is in the plane of the wall. The focal length of the lens is 0.05 meters. The $\mathbf{x}$–$\mathbf{z}$ plane of the camera is parallel to the $\mathbf{x}$–$\mathbf{y}$ plane of the world coordinate system. The image plane is behind the wall. Find the image plane coordinates of **(a)** the room corner A, and **(b)** the head of a person 2 meters tall standing at a distance of 3 m $\times$ 2 m from the corner B as shown in Figure 1.12.

Computer Projects

1.1 Imaging geometry:

a. Consider a camera with the origin of its image plane located at (x_a^0, y_a^0, z_a^0) with reference to the world coordinates as shown in Figure 1.13. The camera axis passes through a point (x_a^c, y_a^c, z_a^c) in world coordinates. Assume that the $\mathbf{x}'$ axis of the camera coordinate system is parallel to the $\mathbf{x}_a$–$\mathbf{y}_a$ plane. The lens center is

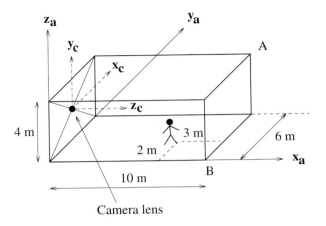

Figure 1.12: Diagram of surveillence camera and room setup for Exercise 1.7.

located at a distance of f from the origin of the image coordinates along the $\mathbf{z}_c$ direction.

b. A polyhedral object with vertices P_1, P_2, ... is located in the world coordinate system at (x_a^1, y_a^1, z_a^1), (x_a^2, y_a^2, z_a^2), ..., respectively. The number of edges in the object is N. Assume that the pairs of vertices which are connected are known.

c. Write a program which will generate the image formed on the image plane. Assume that the object is a "stick figure" (i.e., you need not be concerned about hidden line removal). The image plane is limited in extent to $+1$ and -1 along both the x' and y' directions. Rescale this image to a 256×256 image.

d. In summary, your program should do the following:

- Ask for data file name.
- Read (x_a^0, y_a^0, z_a^0), (x_a^C, y_a^C, z_a^C), and f.
- Read the number of edges and vertices in the object.
- Read the coordinates of each vertex.
- Read the list of vertices which are connected to each other.

For example, consider a unit cube with one of its corners located at the origin and three of its edges along the positive $\mathbf{x}_a$, $\mathbf{y}_a$, $\mathbf{z}_a$ directions. Let the image plane origin be located at $(10, 10, 10)$

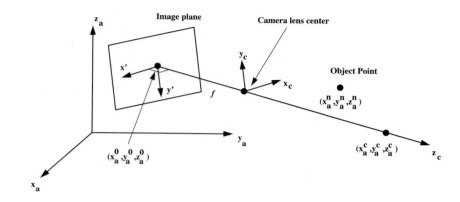

Figure 1.13: The camera coordinate system for Computer Project 1.1.

and the camera pointing towards $(0, 0, 0)$. Let the camera's f be 3. The corresponding data file will be as follows:

(x_a^0, y_a^0, z_a^0), (x_a^C, y_a^C, z_a^C), and f:	10	10	10	0	0	0	3
Number of edges and vertices:	12	8					
Coordinates of each vertex:	0	0	0				
	0	0	1				
	0	1	0				
	1	0	0				
	1	1	1				
	1	1	0				
	1	0	1				
	0	1	1				
Connected vertices:	1	2					
	1	3					
	1	4					
	5	6					
	5	7					
	5	8					
	2	7					
	2	8					
	3	6					
	3	8					
	4	6					
	4	7					

Chapter 2

Binary Image Processing

An image contains a continuum of intensity values before it is quantized to obtain a digital image. The information in an image is in these gray values. To interpret an image, the variations in the intensity values must be analyzed. The most commonly used number of quantization levels for representing image intensities is 256 different gray levels. It is not uncommon, however, to see digital images quantized to 32, 64, 128, or 512 intensity levels for certain applications, and even up to 4096 (12 bits) are used in medicine. Clearly, more intensity levels allow better representation of the scene at the cost of more storage.

In the early days of machine vision, the memory and computing power available was very limited and expensive. These limitations encouraged designers of vision applications to focus their efforts on binary vision systems. A binary image contains only two gray levels. The difference this makes in the representation of a scene is shown in Figure 2.1.

In addition, designers noted that people have no difficulty in understanding line drawings, silhouettes, and other images formed using only two gray levels. Encouraged by this human capability, they used binary images in many applications.

Even though computers have become much more powerful, binary vision systems are still useful. First of all, the algorithms for computing properties of binary images are well understood. They also tend to be less expensive and faster than vision systems that operate on gray level or color images. This is due to the significantly smaller memory and processing requirements of binary vision. The memory requirements of a gray level system working

25

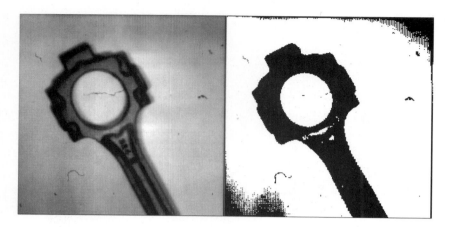

Figure 2.1: A gray level image and its corresponding binary image.

with 256 gray levels will be eight times that of a system working with a binary image of the same size. The storage size may be further reduced by using techniques such as run-length encoding, covered in Section 2.4. The processing time requirements are lower because many operations on binary images may be performed as logical operations instead of integer arithmetic operations.

Smaller memory requirements and faster execution times are not the only reasons for studying binary vision systems. Many techniques developed for these systems are also applicable to vision systems which use gray scale images. A convenient way to represent an object in a gray level or color image is to use its mask. The mask of an object is a binary picture in which the object points are 1 and other points are 0. After an object has been separated from the background, its geometric and topological properties may be required in decision making. These properties can be computed from its binary image. All the techniques discussed in this chapter can be applied to a region in a gray image. Thus, though we will discuss these techniques in the context of binary images, their application is not limited to binary images.

In general, binary vision systems are useful in cases where a silhouette contains enough information to allow recognition of an object and where the environment can be adequately controlled. To obtain a good silhouette, the objects must be easily separated from the background. This can be achieved by using special illumination techniques and by having only a few objects in the scene. There are many industrial situations that fulfill these requirements. For example, binary vision systems have found application in optical character recognition, chromosome analysis, and recognition of industrial parts. In these cases, the binary vision system usually uses a threshold to separate objects from the background. The proper value of this threshold depends on illumination and on reflectance characteristics of objects. The resulting binary picture allows computation of geometric and topological properties (*features*) of objects for the given task. In many applications, these characteristics are enough for recognition of objects.

It should be mentioned here, however, that with the increase in the complexity of applications, more and more vision systems are using gray scale images. This is due to the fact that in many material handling and assembly tasks, the illumination cannot be controlled to obtain good contrast between objects and background. Care has to be exercised to make the system insensitive to small changes in illumination and reflectance characteristics of other objects in a scene. In many applications, this becomes a formidable task. Similarly, in inspection tasks, it may not be possible to recover subtle information using only two intensity levels. Internal details of an object may be lost in thresholding and may make the task of detecting surface defects very difficult.

Certain generally used conventions concerning binary images will be followed in this chapter. Object pixels will have the value 1 and background pixels will have 0. In displaying pictures, 0 is white and 1 is black; thus, in binary images, the background is white and objects are black. We will also assume that pictures are of size $n \times m$ pixels and are represented in a computer as a two-dimensional array. This representation allows us to visualize images with the spatial relationships between points maintained in the form familiar to people.

The techniques discussed in this chapter, though simple, have played a very important role in industrial vision. We will study the following aspects of binary vision systems in this chapter:

- Formation of binary images

- Geometric properties

- Topological properties

- Object recognition in binary images

Many concepts discussed here are used in all aspects of machine vision. Many definitions are related to digital geometry and are useful in discussions related to sampled images. In general, after an image has been segmented into several objects, each object is represented as a region. Discussions related to these object regions use the terminology and concepts discussed in this chapter.

2.1   Thresholding

One of the most important problems in a vision system is to identify the subimages that represent objects. This operation, which is so natural and so easy for people, is surprisingly difficult for computers. The partitioning of an image into regions is called segmentation. Ideally, a partition represents an object or part of an object. Formally, segmentation can be defined as a method to partition an image, $F[i, j]$, into subimages, called regions, $P_1, \ldots, P_k$, such that each subimage is an object candidate.

Definition 2.1 *A region is a subset of an image.*

Definition 2.2 *Segmentation is grouping pixels into regions, such that*

- $\cup_{i=1}^{k} P_i = $ Entire image ($\{P_i\}$ is an *exhaustive* partitioning.)

- $P_i \cap P_j = \emptyset$, $i \neq j$ ($\{P_i\}$ is an *exclusive* partitioning.)

- Each region P_i satisfies a predicate; that is, all points of the partition have *some* common property.

- Pixels belonging to adjacent regions, when taken jointly, do not satisfy the predicate.

As shown above, a partition satisfies a predicate. This predicate may be as simple as *has uniform intensity* but is more complex in most applications. Segmentation is a very important step in understanding images.

A binary image is obtained using an appropriate segmentation of a gray scale image. If the intensity values of an object are in an interval and the intensity values of the background pixels are outside this interval, a binary image can be obtained using a thresholding operation that sets the points in that interval to 1 and points outside that range to 0. Thus, for binary vision, segmentation and thresholding are synonymous. Many cameras have been designed to perform this thresholding operation in hardware. The output of such a camera is a binary image. In most applications, however, cameras give a gray scale image and the binary image is obtained using thresholding.

Thresholding is a method to convert a gray scale image into a binary image so that objects of interest are separated from the background. For thresholding to be effective in *object-background separation*, it is necessary that the objects and background have sufficient contrast and that we know the intensity levels of either the objects or the background. In a fixed thresholding scheme, these intensity characteristics determine the value of the threshold.

Let us assume that a binary image $B[i,j]$ is the same as a thresholded gray image $F_T[i,j]$ which is obtained using a threshold T for the original gray image $F[i,j]$. Thus,

$$B[i,j] = F_T[i,j] \qquad (2.1)$$

where for a darker object on a lighter background

$$F_T[i,j] = \begin{cases} 1 & \text{if } F[i,j] \le T \\ 0 & \text{otherwise.} \end{cases} \qquad (2.2)$$

If it is known that the object intensity values are in a range $[T_1, T_2]$, then we may use

$$F_T[i,j] = \begin{cases} 1 & \text{if } T_1 \le F[i,j] \le T_2 \\ 0 & \text{otherwise.} \end{cases} \qquad (2.3)$$

A general thresholding scheme in which the intensity levels for an object may come from several disjoint intervals may be represented as

$$F_T[i,j] = \begin{cases} 1 & \text{if } F[i,j] \in Z \\ 0 & \text{otherwise} \end{cases} \qquad (2.4)$$

where Z is a set of intensity values for object components. The results of producing an image using different thresholds are shown in Figure 2.2.

Note how knowledge about the application domain is incorporated into the thresholding algorithm. It has, in fact, been tailored for the domain; therefore, the same threshold values may not work in a new domain. The

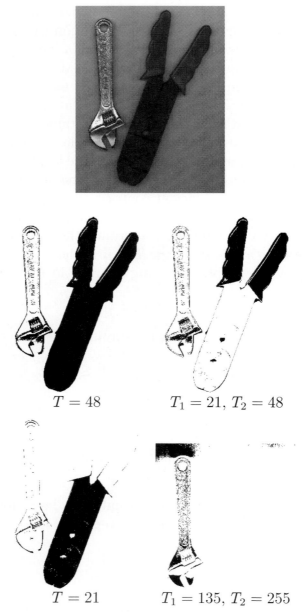

Figure 2.2: A gray level image and its resulting binary images using different thresholds. *Top:* Original gray-level image. *Middle left:* Original image thresholded with $T = 48$. *Middle right:* $T_1 = 21$ and $T_2 = 48$. *Bottom left:* $T = 21$. *Bottom right:* $T = 135$ and $T_2 = 255$.

threshold is usually selected on the basis of experience with the application domain. In some cases, the first few runs of the system may be used for interactively analyzing a scene and determining an appropriate value for the threshold.

Automatic thresholding of images is often the first step in the analysis of images in machine vision systems. Many techniques have been developed for utilizing the intensity distribution in an image and the knowledge about the objects of interest for selecting a proper threshold value automatically. This was briefly introduced in Figure 1.11, where an image and its histogram are given. Many such automatic methods for thresholding an image are presented in Section 3.2.

2.2 Geometric Properties

Suppose that a thresholding scheme has given us objects in an image. The next step is to recognize and locate objects. In most industrial applications, the camera location and the environment are known. Using simple geometry, one may find the three-dimensional locations of objects from their two-dimensional positions in images. Moreover, in most applications the number of different objects is not large. If the objects are different in size and shape, the size and shape features of objects may be determined from their images to help the system recognize them. Many applications in industry have utilized some simple features of regions for determining the locations of objects and for recognizing them (e.g., size, position, orientation).

If there are several objects, one can compute these features for each object. An object is usually represented by a connected component or a region. The concept of connectedness and the algorithms for finding connected components in an image will be discussed later in this chapter. For the present discussion, let us assume that an image has only one object.

2.2.1 Size

In general, for a binary image it is well known that the area A is given by

$$A = \sum_{i=1}^{n} \sum_{j=1}^{m} B[i,j]. \tag{2.5}$$

This is the zeroth-order moment.

2.2.2   Position

The position of an object in an image plays an important role in many applications. There are different ways to specify the position of an object, such as using its enclosing rectangle or centroid. In industrial applications, objects usually appear on a known surface, such as a table, and the position of the camera is known with respect to the table. In such cases, an object's position in the image determines its spatial location. The position of an object in an image may be defined using the center of area of the object image. Though other methods such as the enclosing rectangle of the object image may be used, the center of area is a point and is relatively insensitive to noise in the image.

The center of area in binary images is the same as the center of mass if we consider the intensity at a point as the mass at that point. To calculate the position of the object, we use

$$\bar{x} \sum_{i=0}^{n-1} \sum_{j=0}^{m-1} B[i,j] = \sum_{i=0}^{n-1} \sum_{j=0}^{m-1} j B[i,j] \qquad (2.6)$$

$$\bar{y} \sum_{i=0}^{n-1} \sum_{j=0}^{m-1} B[i,j] = -\sum_{i=0}^{n-1} \sum_{j=0}^{m-1} i B[i,j] \qquad (2.7)$$

where $\bar{x}$ and $\bar{y}$ are the coordinates of the center of the region measured with respect to the top left pixel. Thus, the position of an object is

$$\bar{x} = \frac{\sum_{i=0}^{n-1} \sum_{j=0}^{m-1} j B[i,j]}{A} \qquad (2.8)$$

$$\bar{y} = \frac{-\sum_{i=0}^{n-1} \sum_{j=0}^{m-1} i B[i,j]}{A}. \qquad (2.9)$$

These are the first-order moments. The position calculated using first moments is not necessarily an integer and usually lies between the integer values of the image array indices. We emphasize that this does *not* imply that the calculated position is better than the resolution of pixel coordinates.

2.2.3 Orientation

Calculating the orientation of an object is a little more complex than calculating its position. For some shapes (such as circles), orientation is not unique. To define unique orientation, an object must be elongated. If so, the orientation of the axis of elongation can be used to define the orientation of the object. Usually, the axis of least second moment, which in 2-D is equivalent to the axis of least inertia, is used as the axis of elongation.

The axis of second moment for an object image is that line for which the sum of the squared distances between object points and the line is minimum. Given a binary image, $B[i,j]$, compute the least-squares fit of a line to the object points in the binary image. Minimize the sum of the squared perpendicular distances of all object points from the line,

$$\chi^2 = \sum_{i=0}^{n-1} \sum_{j=0}^{m-1} r_{ij}^2 B[i,j], \qquad (2.10)$$

where r_{ij} is the perpendicular distance from an object point $[i,j]$ to the line. To avoid numerical problems when the line is nearly vertical, represent the line in polar coordinates:

$$\rho = x \cos\theta + y \sin\theta. \qquad (2.11)$$

As shown in Figure 2.3, θ is the orientation of the normal to the line with the x axis, and ρ is the distance of the line from the origin. The distance r of a point (x,y) is obtained by plugging the coordinates of the point into the equation for the line:

$$r^2 = (x \cos\theta + y \sin\theta - \rho)^2. \qquad (2.12)$$

Plugging the representation of the line into the minimization criterion yields the regression problem for fitting a straight line to the object points. Determine the model parameters ρ and θ by minimizing

$$\chi^2 = \sum_{i=0}^{n-1} \sum_{j=0}^{m-1} (x_{ij} \cos\theta + y_{ij} \sin\theta - \rho)^2 B[i,j]. \qquad (2.13)$$

Setting the derivative of χ^2 with respect to ρ to zero and solving for ρ yields

$$\rho = (\bar{x} \cos\theta + \bar{y} \sin\theta), \qquad (2.14)$$

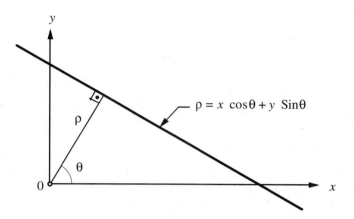

Figure 2.3: Polar representation of a straight line.

which shows that the regression line passes through the center of object points $(\bar{x}, \bar{y})$. After substituting this value of ρ in the above equation for χ^2 and replacing

$$x' = x - \bar{x} \qquad y' = y - \bar{y}, \tag{2.15}$$

the minimization problem becomes

$$\chi^2 = a \cos^2 \theta + b \sin \theta \cos \theta + c \sin^2 \theta. \tag{2.16}$$

The parameters

$$a = \sum_{i=0}^{n-1} \sum_{j=0}^{m-1} (x'_{ij})^2 B[i, j] \tag{2.17}$$

$$b = 2 \sum_{i=0}^{n-1} \sum_{j=0}^{m-1} x'_{ij} y'_{ij} B[i, j] \tag{2.18}$$

$$c = \sum_{i=0}^{n-1} \sum_{j=0}^{m-1} (y'_{ij})^2 B[i, j] \tag{2.19}$$

are the second-order moments. The expression for χ^2 can be rewritten as

$$\chi^2 = \frac{1}{2}(a + c) + \frac{1}{2}(a - c) \cos 2\theta + \frac{1}{2} b \sin 2\theta. \tag{2.20}$$

Differentiating χ^2, setting the result to zero, and solving for θ yields

$$\tan 2\theta = \frac{b}{a - c}. \tag{2.21}$$

The orientation of the axis is given by

$$\sin 2\theta = \pm \frac{b}{\sqrt{b^2 + (a - c)^2}} \tag{2.22}$$

$$\cos 2\theta = \pm \frac{a - c}{\sqrt{b^2 + (a - c)^2}}. \tag{2.23}$$

The axis of orientation is obtained for the minimal value of χ^2. Note that if $b = 0$ and $a = c$, the object does not have a unique axis of orientation. The elongation E of the object is the ratio of the largest to smallest values for χ:

$$E = \frac{\chi_{\max}}{\chi_{\min}}. \tag{2.24}$$

When the expressions for $\sin 2\theta$ and $\cos 2\theta$ are substituted into Equation 2.20, their signs determine whether χ^2 is a maximum or minimum. Note that the elongation is 1 for a circle and that this is the lower bound on E.

2.3 Projections

The projection of a binary image onto a line may be obtained by partitioning the line into bins and finding the number of 1 pixels that are on lines perpendicular to each bin. A simple example of this is shown in Figure 2.4. Projections are compact representations of images, since much useful information is retained in the projection. However, projections are not unique in the sense that more than one image may have the same projection. Horizontal and vertical projections can be easily obtained by finding the number of 1 pixels for each bin in the vertical and horizontal directions, respectively, as shown in Figure 2.5.

The projection $H[i]$ along the rows and the projection $V[j]$ along the columns of a binary image are given by

$$H[i] = \sum_{j=0}^{m-1} B[i, j] \tag{2.25}$$

$$V[j] = \sum_{i=0}^{n-1} B[i, j]. \tag{2.26}$$

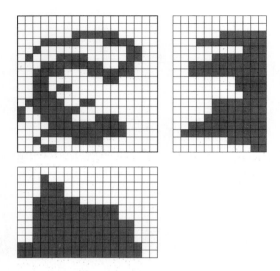

Figure 2.4: The binary image of a lizard with its horizontal and vertical projections.

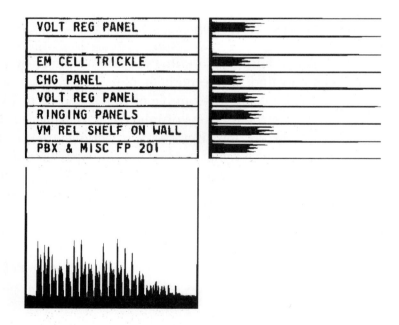

Figure 2.5: Horizontal and vertical projections of an image.

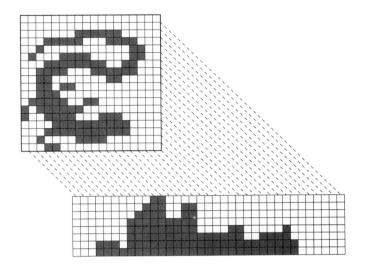

Figure 2.6: The binary lizard image with its diagonal projection.

A general projection onto any line may be defined. An example of a diagonal projection is given in Figure 2.6.

It can be shown that the first moments of an image equal the first moments of its projections. Since calculating the position of an object requires only the first moments, the position can be computed from the horizontal and vertical projections. Thus,

$$A = \sum_{j=0}^{m-1} V[j] = \sum_{i=0}^{n-1} H[i] \qquad (2.27)$$

$$\bar{y} = \frac{\sum_{i=0}^{n-1} i H[i]}{A} \qquad (2.28)$$

$$\bar{x} = \frac{\sum_{j=0}^{m-1} j V[j]}{A}. \qquad (2.29)$$

As we saw in the preceding section, the orientation of an object requires knowledge of the second moment. The second moment can be computed from the diagonal projection of an image. Thus, orientation may be computed using horizontal, vertical, and diagonal projections.

In some applications, projections can be used as features for recognition of objects. Projections offer a compact representation and allow application of fast algorithms.

The trick to updating the diagonal projection is to compute the index for the histogram bucket for the current row and column. Let the row and column be denoted by i and j, respectively. Suppose that the dimensions of the image are n rows and m columns, so i and j range from 0 to $n-1$ and 0 to $m-1$, respectively, and assume that the index d for the diagonal can be computed by an affine transformation (linear combination plus constant) of the row and column:

$$d = ai + bj + c. \tag{2.30}$$

The diagonal projection will require $n+m-1$ buckets. The affine transformation should map the upper right pixel into the first position of the diagonal projection, and the lower left pixel into the last position. Solve the equations

$$a \cdot 0 + b\,(m-1) + c \; = \; 0 \tag{2.31}$$
$$a\,(n-1) + b \cdot 0 + c \; = \; n+m-2 \tag{2.32}$$
$$a \; = \; -b \tag{2.33}$$

to obtain the formula

$$d = i - j + m - 1. \tag{2.34}$$

2.4 Run-Length Encoding

Another compact representation of a binary image is its run-length encoding. In this representation, numbers indicating the lengths of the runs of 1 pixels in the image are used to represent the image. This coding has been used for image transmission. Additionally, some properties, such as the area of objects, may be directly computed from their run-length codes.

Two approaches are commonly used in run-length encoding. In the first, the start position and lengths of runs of 1s for each row are used. The other approach uses only lengths of runs, starting with the length of the 1 run. We will use the second convention and represent run lengths for each row of an image. Thus, $r_{i,k}$ denotes the length of the kth run in the ith row of an image. The run-length codes for an image are shown in Figure 2.7.

Binary image:

1	1	1	0	0	0	1	1	0	0	0	1	1	1	1	0	1	1	0	1	1	1
0	0	0	0	1	1	1	1	1	1	1	1	1	1	1	1	1	0	1	1	1	1
1	1	1	0	0	0	0	0	0	0	0	0	0	0	0	0	1	1	1	1	1	1

Start and length of 1 runs: $(1,3)\ (7,2)\ (12,4)\ (17,2)\ (20,3)$

$(5,13)\ (19,4)$

$(1,3)\ (17,6)$

Length of 1 and 0 runs: $3,3,2,3,4,1,2,1,3$

$0,4,13,1,4$

$3,13,6$

Figure 2.7: Run-length codes for a simple binary image.

The area of all objects can be obtained by summing the lengths of all 1 runs.

$$A = \sum_{i=0}^{n-1} \sum_{k=0}^{\left(\frac{m_i-1}{2}\right)} r_{i,2k+1} \tag{2.35}$$

where m_i is the number of runs in the ith row.

The horizontal projection can be easily computed from the run-length code, without generating the image (see Figure 2.8). The vertical and diagonal projections can also be computed from the run-length code without generating the image by using clever code. Area and first and second moments can be calculated from the projections as explained in Section 2.3. Calculating other properties of images may require partial or complete regeneration of the image.

2.5 Binary Algorithms

Segmenting object pixels from background pixels is a difficult problem. We will not address this problem here. Let us assume here that somehow an object can be defined and, using a predicate, the points of an image belonging

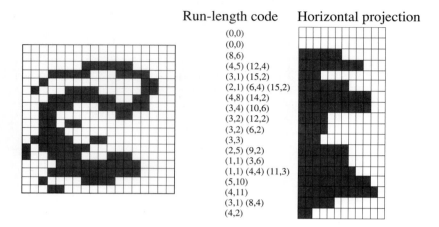

Figure 2.8: Horizontal projection calculated from the run-length code.

to an object may be labeled. The problem then is to group together all points of an image that are labeled as object points into an object image. In this chapter we will assume that all such points are spatially close. This notion of *spatial proximity* requires a more precise definition so that an algorithm may be devised to group spatially close points into a component. For this purpose, let us introduce some definitions.

2.5.1   Definitions

Neighbors

A pixel in a digital image is spatially close to several other pixels. In a digital image represented on a square grid, a pixel has a common boundary with four pixels and shares a corner with four additional pixels. We say that two pixels are *4-neighbors* if they share a common boundary. Similarly, two pixels are *8-neighbors* if they share at least one corner. For example, the pixel at location $[i, j]$ has 4-neighbors $[i + 1, j]$, $[i - 1, j]$, $[i, j + 1]$, and $[i, j - 1]$. The 8-neighbors of the pixel include the 4-neighbors plus $[i+1, j+1]$, $[i+1, j-1]$, $[i-1, j+1]$ and $[i-1, j-1]$. A pixel is said to be *4-connected* to its 4-neighbors and *8-connected* to its 8-neighbors (see Figure 2.9).

4-neighbors $[i+1, j]$, $[i-1, j]$, $[i, j+1]$, $[i, j-1]$

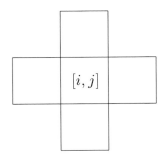

8-neighbors $[i+1, j+1]$, $[i+1, j-1]$, $[i-1, j+1]$, $[i-1, j-1]$ plus all of the 4-neighbors

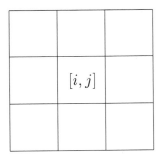

Figure 2.9: The 4- and 8-neighborhoods for a rectangular image tessellation. Pixel $[i, j]$ is located in the center of each figure.

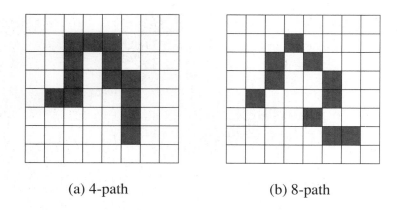

(a) 4-path   (b) 8-path

Figure 2.10: Examples of a 4-path and an 8-path.

Path

A *path* from the pixel at $[i_0, j_0]$ to the pixel at $[i_n, j_n]$ is a sequence of pixel indices $[i_0, j_0], [i_1, j_1], [i_2, j_2], \ldots, [i_n, j_n]$ such that the pixel at $[i_k, j_k]$ is a neighbor of the pixel at $[i_{k+1}, j_{k+1}]$ for all k with $0 \le k \le n - 1$. If the neighbor relation uses 4-connection, then the path is a 4-path; for 8-connection, the path is an 8-path. Simple examples of these are shown in Figure 2.10.

Foreground

The set of all 1 pixels in an image is called the *foreground* and is denoted by S.

Connectivity

A pixel $p \in S$ is said to be *connected* to $q \in S$ if there is a path from p to q consisting entirely of pixels of S.

Note that connectivity is an equivalence relation. For any three pixels p, q, and r in S, we have the following properties:

1. Pixel p is connected to p (reflexivity).

2. If p is connected to q, then q is connected to p (commutativity).

3. If p is connected to q and q is connected to r, then p is connected to r (transitivity).

Connected Components

A set of pixels in which each pixel is connected to all other pixels is called a *connected component.*

Background

The set of all connected components of $\bar{S}$ (the complement of S) that have points on the border of an image is called the *background.* All other components of $\bar{S}$ are called *holes.*

Let us consider the simple picture shown below:

	1	
1		1
	1	

How many objects and how many holes are in this figure? If we consider 4-connectedness for both foreground and background, there are four objects that are 1 pixel in size and there is one hole. If we use 8-connectedness, then there is one object and no hole. Intuitively, in both cases we have an ambiguous situation. A similar ambiguous situation arises in a simple case like:

1	0
0	1

where if the 1s are connected, then the 0s should not be.

To avoid this awkward situation, different connectedness should be used for objects and backgrounds. If we use 8-connectedness for S, then 4-connectedness should be used for $\bar{S}$.

Boundary

The *boundary* of S is the set of pixels of S that have 4-neighbors in $\bar{S}$. The boundary is usually denoted by S'.

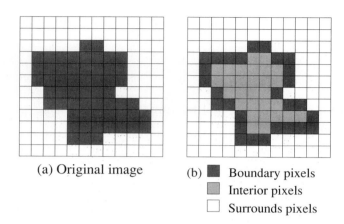

(a) Original image   (b) ■ Boundary pixels
                              ▨ Interior pixels
                              ☐ Surrounds pixels

Figure 2.11: A binary image with its boundary, interior, and surrounds.

Interior

The *interior* is the set of pixels of S that are not in its boundary. The interior of S is $(S - S')$.

Surrounds

Region T *surrounds* region S (or S is inside T), if any 4-path from any point of S to the border of the picture must intersect T. Figure 2.11 shows an example of a simple binary image with its boundary, interior, and surrounds.

2.5.2   Component Labeling

One of the most common operations in machine vision is finding the connected components in an image. The points in a connected component form a candidate region for representing an object. As mentioned earlier, in computer vision most objects have surfaces. Points belonging to a surface project to spatially close points. The notion of "spatially close" is captured by connected components in digital images. It should be mentioned here that connected component algorithms usually form a bottleneck in a binary vision system. The algorithm is sequential in nature, because the operation of finding connected components is a global operation. If there is only one

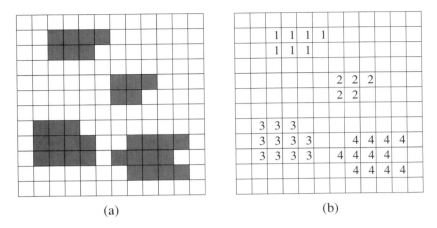

Figure 2.12: An image (a) and its connected component image (b).

object in an image, then there may not be a need for finding the connected components; however, if there are many objects in an image and the object properties and locations need to be found, then the connected components must be determined.

A component labeling algorithm finds all connected components in an image and assigns a unique label to all points in the same component. Figure 2.12 shows an image and its labeled connected components. In many applications, it is desirable to compute characteristics (such as size, position, orientation, and bounding rectangle) of the components while labeling these components. There are two algorithms for component labeling: recursive and sequential.

Recursive Algorithm

A recursive algorithm is given as Algorithm 2.1. On a sequential processor, this algorithm is very inefficient. Due to its inefficiency, this algorithm is rarely used on general-purpose computers; but is commonly used on parallel machines.

Algorithm 2.1   Recursive Connected Components Algorithm

 1. *Scan the image to find an unlabeled 1 pixel and assign it a new label L.*

 2. *Recursively assign a label L to all its 1 neighbors.*

 3. *Stop if there are no more unlabeled 1 pixels.*

 4. *Go to step 1.*

Sequential Algorithm

The sequential algorithm usually requires two passes over the image. Since this algorithm works with only two rows of an image at a time, it can be used even when images are stored as a file and space limitations do not allow the full image to be brought into memory. This algorithm, given as Algorithm 2.2, looks at the neighborhood of a pixel and tries to assign already used labels to a 1 pixel. In case of two different labels in the neighborhood of a pixel, an *equivalence table* is prepared to keep track of all labels that are equivalent. This table is used in the second pass to assign a unique label to all pixels of a component.

In the algorithm, there are three cases of interest when scanning an image left to right and top to bottom. The algorithm looks at only two of a pixel's 4-neighbors: the ones above and to the left of it. Note that these two pixels have already been seen by the algorithm. If none of these pixels is 1, then the pixel requires a new label. If only one of the two pixels is 1 and has been assigned a label L, then the pixel will be assigned L. If both pixels are 1s and have been assigned the same label L, then the new pixel will be assigned L; however, in the case where the neighbors have been assigned different labels M and N, then the two labels have been used for the same component and they must be merged. In this case the pixel is assigned one of the labels, usually the smaller label, and both labels are recorded as equivalent labels in the equivalence table.

The equivalence table contains the information to assign unique labels to each connected component. In the first scan, all those labels that belong to one component are declared equivalent. In the second pass, one label from an equivalent set is selected to be assigned to all pixels of a component. Usually

Algorithm 2.2 Sequential Connected Components Algorithm using 4-connectivity

1. *Scan the image left to right, top to bottom.*

2. *If the pixel is 1, then*

   (a) *If only one of its upper and left neighbors has a label, then copy the label.*

   (b) *If both have the same label, then copy the label.*

   (c) *If both have different labels, then copy the upper's label and enter the labels in the equivalence table as equivalent labels.*

   (d) *Otherwise assign a new label to this pixel and enter this label in the equivalence table.*

3. *If there are more pixels to consider, then go to step 2.*

4. *Find the lowest label for each equivalent set in the equivalence table.*

5. *Scan the picture. Replace each label by the lowest label in its equivalent set.*

the smallest label is assigned to a component. The second pass assigns an unique label to each component.

After all of the connected components have been found, the equivalence table should be renumbered so that gaps in the labels are eliminated; then the image is rescaned using the equivalence table as a lookup table for renumbering the labels in the image.

Area, first moments, and second moments can be calculated for each component as part of the sequential connected components algorithm. Of course, separate variables must be used to accumulate the moment information for each region. When regions are merged, the moment accumulations for each region are just added together.

2.5.3 Size Filter

It is very common to use thresholding for finding a binary image. In most cases, there are some regions in an image that are due to noise. Usually, such

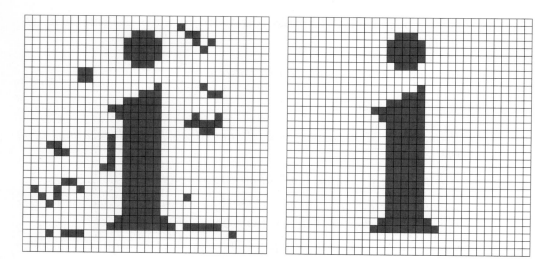

Figure 2.13: A noisy binary image of the letter "i" (left) and the resulting image after the application of a size filter (right). $T = 10$.

regions are small. In many applications, it is known that objects of interest are of size greater than T_0 pixels. In such cases one may use a size filter to remove noise after component labeling. All components below T_0 in size are removed by changing the corresponding pixels to 0. This simple filtering mechanism is very effective in removing noise. Figures 2.13 and 2.14 show two examples of the application of a size filter to a noisy character image.

2.5.4   Euler Number

In many applications, the genus or *Euler number* is used as a feature of an object. *Genus* is defined as the number of components minus the number of holes. Thus,

$$E = C - H \tag{2.36}$$

where E, C, and H are the Euler number, the number of connected components, and the number of holes, respectively. This provides a simple topological feature that is invariant to translation, rotation, and scaling. Several algorithms have been developed that allow computation of the genus using local structures such as runs. Figure 2.15 gives examples of objects with their corresponding Euler numbers.

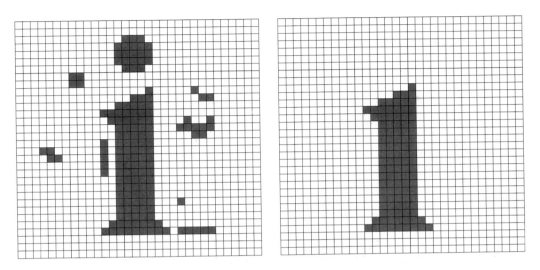

Figure 2.14: The possible error resulting from a poor choice of threshold for the size filter ($T = 25$). Note that the "dot" is missing from the letter "i."

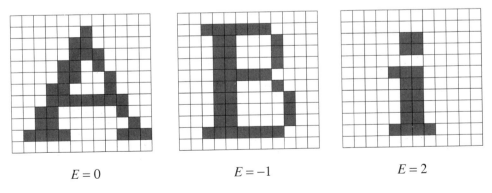

$E = 0$ $E = -1$ $E = 2$

Figure 2.15: The letters "A," "B," and "i" and their Euler numbers. Note that 8-connectivity is used for the foreground and 4-connectivity for the background.

2.5.5   Region Boundary

The boundary of a connected component S is the set of pixels of S that are adjacent to $\bar{S}$. A simple local operation may be used to find pixels on the boundary. In most applications, one wants to track pixels on the boundary in a particular order. One common approach is to track all pixels of a region in a clockwise sequence. Here we discuss a simple boundary-following algorithm.

The boundary-following algorithm selects a starting pixel $s \in S$ and tracks the boundary until it comes back to the starting pixel, assuming that the boundary is not on the edge of an image. The algorithm is given as Algorithm 2.3. This algorithm will work for all regions whose size is greater than 1. The boundary found by this algorithm for an 8-connected region is given in Figure 2.16.

Algorithm 2.3   Boundary-Following Algorithm

1. *Find the starting pixel $s \in S$ for the region using a systematic scan, say from left to right and from top to bottom of the image.*

2. *Let the current pixel in boundary tracking be denoted by c. Set $c = s$ and let the 4-neighbor to the west of s be $b \in \bar{S}$.*

3. *Let the eight 8-neighbors of c starting with b in clockwise order be $n_1, n_2, \ldots, n_8$. Find n_i, for the first i that is in S.*

4. *Set $c = n_i$ and $b = n_{i-1}$.*

5. *Repeat steps 3 and 4 until $c = s$.*

2.5.6   Area and Perimeter

As we discussed earlier, the area is the number of pixels in S. If there are several components $S_1, S_2, \ldots, S_n$, then the area of each component is the number of pixels in that component. The number of pixels in each component may be obtained along with the labeling of the components. In a general case, the area of each of the n components may be obtained in one scan of an image.

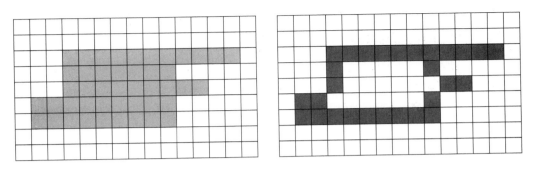

Figure 2.16: Results of a boundary-following algorithm. *Left:* Original binary object. *Right:* Calculated boundary.

The perimeter of a component may be defined in many different ways. Some common definitions are:

1. The sum of lengths of the "cracks" separating pixels of S from pixels of $\bar{S}$. A crack is a line that separates a pair of pixels p and q such that $p \in S$ and $q \in \bar{S}$.

2. The number of steps taken by a boundary-following algorithm.

3. The number of boundary pixels of S.

The measured perimeter will be very different according to different definitions. In general, the perimeter obtained using definition 1 is much longer than the perimeter measurements obtained using the other two definitions.

2.5.7 Compactness

It is well known that the compactness of a continuous geometric figure is measured by the isoperimetric inequality

$$\frac{P^2}{A} \geq 4\pi \qquad (2.37)$$

where P and A are the figure's perimeter and area, respectively. A circle is the most compact figure (i.e., has the smallest compactness value) according to this measure. In the case of a circle, the ratio P^2/A achieves its minimal value of 4π; for other figures the ratio is larger. Consider a circle at an angle relative to the viewer. As the circle is tilted and assumes the shape of an ellipse, the area decreases but the perimeter does not decrease as rapidly, so the compactness increases. At the extreme angle of tilt, the ellipse is squeezed into a line and the compactness goes to infinity. For digital pictures, P^2/A is computed by dividing the square of the length of the boundary by the size (number of pixels). This provides a good measure of dispersedness or compactness. This ratio has been used in many applications as a feature of a region.

Another way of viewing compactness is that a more compact region encloses a larger amount of area for a given perimeter. Note that a square is more compact than a rectangle with the same perimeter.

2.5.8   Distance Measures

In many applications, it is necessary to find the distance between two pixels or two components of an image. Unfortunately, there is no unique method of defining distance in digital images. One can define distance in many different ways. For all pixels p, q, and r, any distance metric must satisfy all of the following three properties:

1. $d(p, q) \geq 0$ and $d(p, q) = 0$ iff $p = q$

2. $d(q, p) = d(p, q)$

3. $d(p, r) \leq d(p, q) + d(q, r)$

Several distance functions have been used in digital geometry. Some of the more common distance functions are:

Euclidean

$$d_{\text{Euclidean}}([i_1, j_1], [i_2, j_2]) = \sqrt{(i_1 - i_2)^2 + (j_1 - j_2)^2} \qquad (2.38)$$

City-block

$$d_{\text{city}} = |i_1 - i_2| + |j_1 - j_2| \qquad (2.39)$$

Chessboard

$$d_{\text{chess}} = \max(|i_1 - i_2|, |j_1 - j_2|) \qquad (2.40)$$

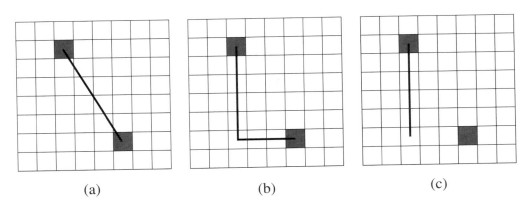

Figure 2.17: Examples of (a) Euclidean, (b) city-block, and (c) chessboard distance measures.

The city-block and chessboard distance measures are preferred over Euclidean due to their simplicity of calculation. These three functions are illustrated in Figure 2.17.

A set of pixels at distance $\leq k$ according to a distance metric is called a *disc* of radius k under that metric. Figure 2.18 shows discs of size 3 according to the above three measures. Note that the shapes of the neighborhood are significantly different. An important consideration in using a distance measure is the fact that though the Euclidean distance is closest to the continuous case, it is computationally the most expensive and results in real-valued distances. The integer-valued square of the Euclidean distance can also be used as a distance measure.

2.5.9 Distance Transforms

In certain applications, such as character recognition, the minimum distance between a pixel of an object component and the background is used. Thus, given an object region, S, we must compute the distance to the background region, $\bar{S}$, for all pixels in S. The transform for obtaining an image representing such distances is called a *distance transform*. A parallel iterative algorithm to compute the distance transform is obtained using the equations

$$f^0[i,j] = f[i,j] \tag{2.41}$$
$$f^m[i,j] = f^0[i,j] + \min\left(f^{m-1}[u,v]\right) \tag{2.42}$$

Euclidean distance:

$$
\begin{array}{ccccccc}
 & & & 3 & & & \\
 & \sqrt{8} & \sqrt{5} & 2 & \sqrt{5} & \sqrt{8} & \\
 & \sqrt{5} & \sqrt{2} & 1 & \sqrt{2} & \sqrt{5} & \\
3 & 2 & 1 & 0 & 1 & 2 & 3 \\
 & \sqrt{5} & \sqrt{2} & 1 & \sqrt{2} & \sqrt{5} & \\
 & \sqrt{8} & \sqrt{5} & 2 & \sqrt{5} & \sqrt{8} & \\
 & & & 3 & & &
\end{array}
$$

City-block distance:

$$
\begin{array}{ccccccc}
 & & & 3 & & & \\
 & & 3 & 2 & 3 & & \\
 & 3 & 2 & 1 & 2 & 3 & \\
3 & 2 & 1 & 0 & 1 & 2 & 3 \\
 & 3 & 2 & 1 & 2 & 3 & \\
 & & 3 & 2 & 3 & & \\
 & & & 3 & & &
\end{array}
$$

Chessboard distance:

$$
\begin{array}{ccccccc}
3 & 3 & 3 & 3 & 3 & 3 & 3 \\
3 & 2 & 2 & 2 & 2 & 2 & 3 \\
3 & 2 & 1 & 1 & 1 & 2 & 3 \\
3 & 2 & 1 & 0 & 1 & 2 & 3 \\
3 & 2 & 1 & 1 & 1 & 2 & 3 \\
3 & 2 & 2 & 2 & 2 & 2 & 3 \\
3 & 3 & 3 & 3 & 3 & 3 & 3
\end{array}
$$

Figure 2.18: Different distance measures.

$$
\begin{array}{cccccc}
1 & 1 & 1 & 1 & 1 & 1 \\
1 & 1 & 1 & 1 & 1 & 1 \\
1 & 1 & 1 & 1 & 1 & 1 \\
1 & 1 & 1 & 1 & 1 & 1 \\
1 & 1 & 1 & 1 & 1 & 1 \\
\end{array}
\quad \rightarrow \quad
\begin{array}{cccccc}
1 & 1 & 1 & 1 & 1 & 1 \\
1 & 2 & 2 & 2 & 2 & 1 \\
1 & 2 & 2 & 2 & 2 & 1 \\
1 & 2 & 2 & 2 & 2 & 1 \\
1 & 1 & 1 & 1 & 1 & 1 \\
\end{array}
\quad \rightarrow \quad
\begin{array}{cccccc}
1 & 1 & 1 & 1 & 1 & 1 \\
1 & 2 & 2 & 2 & 2 & 1 \\
1 & 2 & 3 & 3 & 2 & 1 \\
1 & 2 & 2 & 2 & 2 & 1 \\
1 & 1 & 1 & 1 & 1 & 1 \\
\end{array}
$$

Figure 2.19: Distance transform of an image after the first and second iterations.

where m is the iteration number for all pixels $[u, v]$, such that $d([u, v], [i, j]) = 1$. Note that this only uses the 4-neighbors of $[i, j]$.

This algorithm does not change pixels of $\bar{S}$. In the first iteration, all of the pixels that are not adjacent to $\bar{S}$ are changed to 2. In the succeeding iterations, pixels farther away from $\bar{S}$ change. No pixel changes after the distances to all pixels are obtained. The operation of this algorithm is shown in Figure 2.19.

2.5.10 Medial Axis

We say that the distance $d([i, j], \bar{S})$ from the pixel $[i, j]$ in S to $\bar{S}$ is locally maximum if

$$d([i, j], \bar{S}) \geq d([u, v], \bar{S}) \tag{2.43}$$

for all pixels $[u, v]$ in the neighborhood of $[i, j]$. The set of pixels in S with distances from $\bar{S}$ that are locally maximum is called the skeleton, symmetric axis, or medial axis of S and is usually denoted by S^*. Some examples of the medial axis transform when using 4-neighbors are given in Figures 2.20 and 2.21. Figure 2.22 shows that a small amount of noise in the original image can cause a significant difference in the resulting medial axis transform.

The original set S can be reconstructed from S^* and the distances of each pixel of S^* from $\bar{S}$. S^* is a compact representation of S. S^* is used to represent the shape of a region. By deleting pixels of S^* whose distances from $\bar{S}$ are small, we can create a simplified version of S^*.

The medial axis has been used for compact representation of objects. However, a region in a binary image may also be represented using its boundary. A boundary-following algorithm may be used to obtain a sequence of

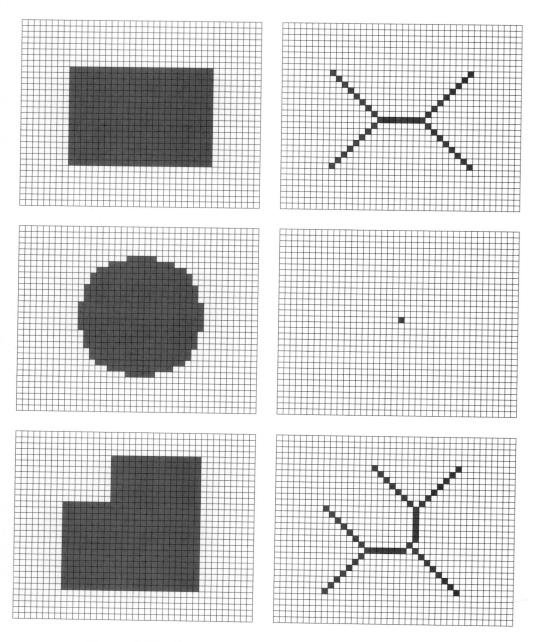

Figure 2.20: Examples of the medial axis transform.

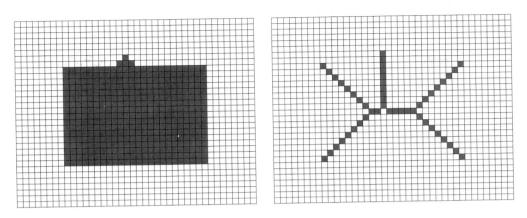

Figure 2.21: The result of the medial axis transform on a noisy image. Note that this result is very different from the first example in Figure 2.20.

pixels representing the boundary. The boundary may be very compactly represented using chain codes, to be discussed in a later section. For arbitrary objects, a boundary is a more compact representation of a region. However, to find whether a given pixel is in the region or not, the medial axis is a better representation. This is due to the fact that pixels that are within a region represented by the medial axis can be easily detected using pixels on the axis and the maximal disc at each pixel, as given by the distance transform.

2.5.11 Thinning

Thinning is an image-processing operation in which binary-valued image regions are reduced to lines that approximate their center lines, also called "skeletons" or core lines. The purpose of thinning is to reduce the image components to their essential information so that further analysis and recognition are facilitated. Although the thinning operation can be applied to binary images containing regions of any shape, it is useful primarily for "elongated" shapes as opposed to convex, or "bloblike" shapes. Thinning is commonly used in the preprocessing stage of document analysis applications for representing lines in a drawing or character strokes in a text image.

The thinning requirements are formally stated as follows:

1. Connected image regions must thin to connected line structures.

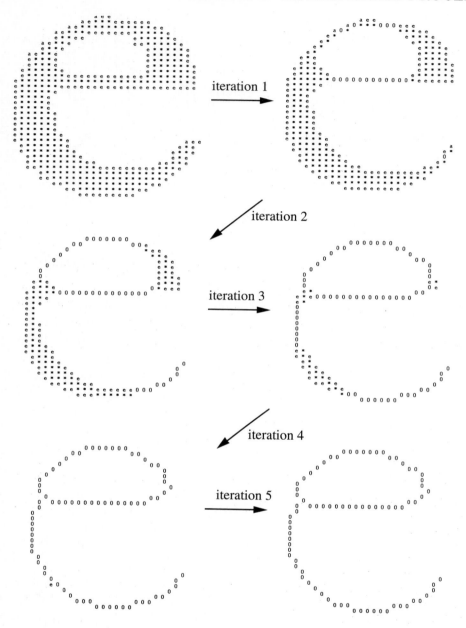

Figure 2.22: Sequence of five iterations of thinning the letter "e." On each iteration a layer of the boundary is peeled off. On iteration 5, the end is detected because no pixels change. (*Courtesy of Lawrence O'Gorman, AT&T Bell Laboratories.*)

2. The thinned result should be minimally 8-connected (explained below).

3. Approximate endline locations should be maintained.

4. The thinning results should approximate the medial lines.

5. Extraneous spurs (short branches) caused by thinning should be minimized.

That the results of thinning must maintain connectivity as specified by requirement 1 is essential. This guarantees a number of connected line structures equal to the number of connected regions in the original image. Requirement 2 stipulates that the resulting lines should always contain a minimal number of pixels that maintain 8-connectedness. Requirement 3 states that the locations of endlines should be maintained. Since thinning can be achieved by iteratively removing outer boundary pixels, it is important not to also iteratively remove the last pixels of a line. This would shorten the line and fail to preserve its location. Requirement 4 states that the resultant lines should best approximate the medial lines of the original regions. Unfortunately, in digital space, the true medial lines can only be approximated. For example, for a 2-pixel-wide vertical or horizontal line, the true medial line should run at the half-pixel spacing along the middle of the original line. Since it is impossible to represent this in digital images, the result will be a single line running at one side of the original. With respect to requirement 5, it is obvious that noise should be minimized, but it is often difficult to say what is noise and what isn't. We don't want spurs to result from every small bump on the original region, but we do want to recognize when a somewhat larger bump is a feature. Though some thinning algorithms have parameters to remove spurs, we believe that thinning and noise removal should be performed separately. Since one person's undesired spur may be another's desired short line, it is best to perform thinning first and then, in a separate process, remove any spurs whose length is less than a specified minimum.

A common thinning approach is to examine each pixel in the image within the context of its neighborhood region of 3×3 pixels and to "peel" the region boundaries, one pixel layer at a time, until the regions have been reduced to thin lines. This process is performed iteratively: on each iteration, every image pixel is inspected within 3×3 windows, and single-pixel-thick boundaries that are not required to maintain connectivity or the position of a line end are erased. In Figure 2.22 you can see how, on each iteration, the outside layer of a 1-valued region is peeled off in this manner. When no changes are made on an iteration, the process is complete and the image is thinned.

2.5.12   Expanding and Shrinking

A component of an image can be systematically expanded or contracted. When a component is allowed to change such that some background pixels are converted to 1, the operation is called expanding. If object pixels are systematically deleted or converted to 0, then it is called shrinking. A simple implementation of expanding and shrinking may be:

Expanding: Change a pixel from 0 to 1 if any neighbors of the pixel are 1.

Shrinking: Change a pixel from 1 to 0 if any neighbors of the pixel are 0.

Thus, shrinking may be considered as expanding the background. Example of these operations are given in Figure 2.23.

It is interesting that simple operations like expanding and shrinking can be used to do some very useful and seemingly complex operations on images. Let us denote:

$S^{(k)}$: S expanded k times
$S^{(-k)}$: S shrunk k times

It can be shown that the following properties hold:

$$(S^m)^{-n} \neq (S^{-n})^m$$
$$\neq S^{(m-n)}$$

$$S \subset (S^k)^{-k}$$
$$S \supset (S^{-k})^k$$

Expanding followed by shrinking can be used for filling undesirable holes, and shrinking followed by expanding can be used for removing isolated noise pixels (see Figure 2.24). Expanding and shrinking can be used to determine isolated components and clusters. In morphological image processing and dilation and erosion operations, generalized forms of expanding and shrinking are used extensively to do many tasks.

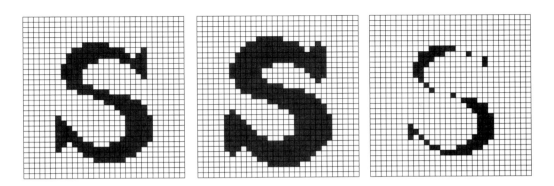

Figure 2.23: An example of expanding and shrinking operations on the letter "s." *Left*: The original image. *Middle*: Expanded image. *Right*: Shrunken image.

2.6 Morphological Operators

Mathematical morphology gets its name from the study of shape. This approach exploits the fact that in many machine vision applications, it is natural and easy to think in terms of shapes when designing algorithms. A morphological approach facilitates shape-based, or iconic, thinking. The fundamental unit of pictorial information in the morphological approach is the binary image.

The *intersection* of any two binary images A and B, written $A \cap B$, is the binary image which is 1 at all pixels p which are 1 in both A and B. Thus,

$$A \cap B = \{p | p \in A \text{ and } p \in B\}. \tag{2.44}$$

The *union* of A and B, written $A \cup B$, is the binary image which is 1 at all pixels p which are 1 in A or 1 in B (or 1 in both). Symbolically,

$$A \cup B = \{p | p \in A \text{ or } p \in B\}. \tag{2.45}$$

Let Ω be a universal binary image (all 1) and A a binary image. The *complement* of A is the binary image which interchanges the 1s and 0s in A. Thus,

$$\overline{A} = \{p | p \in \Omega \text{ and } p \notin A\}. \tag{2.46}$$

The *vector sum* of two pixels p and q with indices $[i, j]$ and $[k, l]$ is the pixel $p + q$ with indices $[i + k, j + l]$. The *vector difference* $p - q$ is the pixel with indices $[i - k, j - l]$.

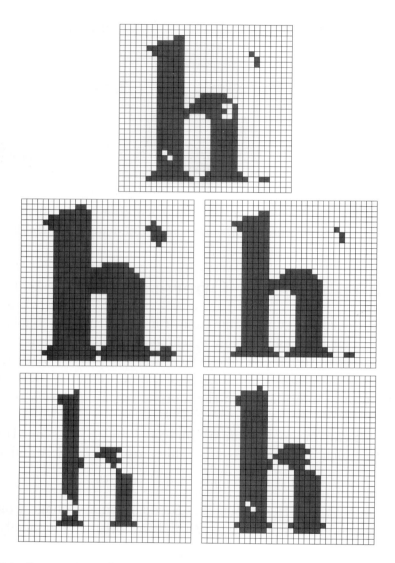

Figure 2.24: Sequences of expanding and shrinking the letter "h." *Top:* The original noisy image. *Middle:* Expanding followed by shrinking. *Bottom:* Shrinking followed by expanding. Note that expanding followed by shrinking effectively filled the holes but did not eliminate the noise. Conversely, shrinking followed by expanding eliminated the noise but did not fill the holes.

If A is a binary image and p is a pixel, then the *translation* of A by p is an image given by

$$A_p = \{a + p | a \in A\}. \tag{2.47}$$

Dilation

Translation of a binary image A by a pixel p shifts the origin of A to p. If $A_{b_1}, A_{b_2}, \ldots, A_{b_n}$ are translations of the binary image A by the 1 pixels of the binary image $B = \{b_1, b_2, \ldots, b_n\}$, then the union of the translations of A by the 1 pixels of B is called the *dilation* of A by B and is given by

$$A \oplus B = \bigcup_{b_i \in B} A_{b_i} = \bigcup_{a_i \in A} B_{a_i} \tag{2.48}$$

Dilation has both associative and commutative properties. Thus, in a sequence of dilation steps the order of performing operations is not important. This fact allows breaking a complex shape into several simpler shapes which can be recombined as a sequence of dilations.

Erosion

The opposite of dilation is *erosion*. The erosion of a binary image A by a binary image B is 1 at a pixel p if and only if every 1 pixel in the translation of B to p is also 1 in A. Erosion is given by

$$A \ominus B = \{p | B_p \subseteq A\}. \tag{2.49}$$

Often the binary image B is a regular shape which is used as a probe on image A and is referred to as a *structuring element*. Erosion plays a very important role in many applications. Erosion of an image by a structuring element results in an image that gives all locations where the structuring element is contained in the image.

Figures 2.25 through 2.28 illustrate the dilation and erosion operations with a simple binary object and an upside-down "T-shaped" structuring element. Figure 2.26 shows examples of translations of the structuring element to 1 pixels of the original figure where the entire structuring element does *not* fit entirely inside the original object. In this case, during a dilation, every pixel in the structuring element will be present in the final dilated image, including the pixel not contained in the original object (shown as a lightly

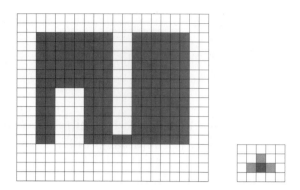

Figure 2.25: The original test image A (left) and structuring element B (right). Note that the origin of the structuring element is darker than the other pixels in B.

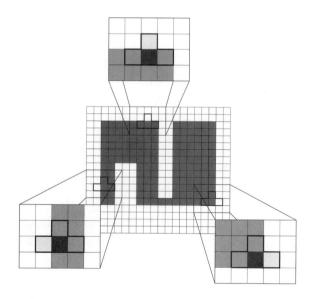

Figure 2.26: Translations of the structuring element B to 1 pixels in A where the entire structuring element is *not* contained within A. During a dilation operation, every pixel in the structuring element will be present in the final image. During an erosion operation, the pixel at the origin of the structuring element will be deleted.

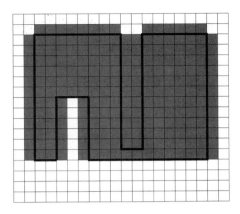

$$A \oplus B = \bigcup_{b_i \in B} A_{b_i}$$

Figure 2.27: The dilation of A by B. The boundary of the original figure A is shown as a bold line.

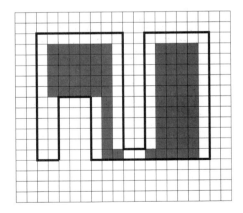

$$A \ominus B = \{p | B_p \subseteq A\}$$

Figure 2.28: The erosion of A by B. The boundary of the original figure A is shown as a bold line.

shaded pixel). But during an erosion operation the pixel at the origin of the structuring element will be removed because the entire structuring element is not within the object. Conversely, in the case where the entire structuring element *does* fit within the original object, there will be no change to the final dilated or eroded image (i.e., no pixels will be added or deleted at that point).

Dilation and erosion exhibit a dual nature that is geometric rather than logical and involves a geometric complement as well as a logical complement. The geometric complement of a binary image is called its *reflection*. The reflection of a binary image B is that binary image B' which is symmetric with B about the origin, that is

$$B' = \{-p|p \in B\}. \tag{2.50}$$

The geometric duality of dilation and erosion is expressed by the relationships

$$\overline{A \oplus B} = \overline{A} \ominus B' \tag{2.51}$$

and

$$\overline{A \ominus B} = \overline{A} \oplus B'. \tag{2.52}$$

Geometric duality contrasts with logical duality:

$$\overline{A \cup B} = \overline{A} \cap \overline{B} \tag{2.53}$$

and

$$\overline{A \cap B} = \overline{A} \cup \overline{B}, \tag{2.54}$$

also called deMorgan's law. The duality of dilation and erosion are illustrated in Figures 2.29 through 2.31.

Erosion and dilation are often used in filtering images. If the nature of noise is known, then a suitable structuring element can be used and a sequence of erosion and dilation operations can be applied for removing the noise. Such filters affect the shape of the objects in the image.

The basic operations of mathematical morphology can be combined into complex sequences. For example, an erosion followed by a dilation with the same structuring element (probe) will remove all of the pixels in regions which are too small to contain the probe, and it will leave the rest. This sequence

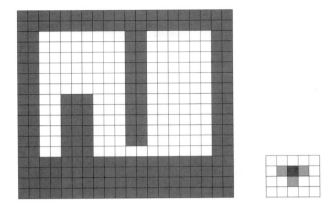

Figure 2.29: The complement of A and the reflection of B. Note that the origin of the reflected structuring element is still the darker pixel.

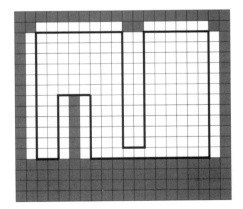

$$\overline{A} \ominus B'$$

Figure 2.30: The dual of dilation: the result of eroding the background of A with the reflection of B. The original boundary is shown as a bold line.

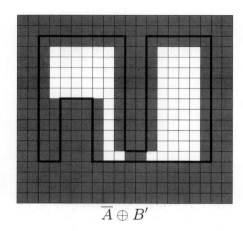

$$\overline{A} \oplus B'$$

Figure 2.31: The dual of erosion: the result of dilating the background of A with the reflection of B. The original boundary is shown as a bold line.

is called *opening*. As an example, if a disc-shaped probe image is used, then all of the convex or isolated regions of pixels smaller than the disc will be eliminated. This forms a filter which suppresses positive spatial details. The remaining pixels show where the structuring element is contained in the foreground. The difference of this result and the original image would show those regions which were too small for the probe, and these could be the features of interest, depending on the application.

The opposite sequence, a dilation followed by an erosion, will fill in holes and concavities smaller than the probe. This is referred to as *closing*. These operations are illustrated in Figures 2.32 and 2.33 with the same T-shaped structuring element. Again, what is removed may be just as important as what remains. Such filters can be used to suppress spatial features or discriminate against objects based upon their size. The structuring element used does not have to be compact or regular, and can be any pattern of pixels. In this way, features made up of distributed pixels can be detected.

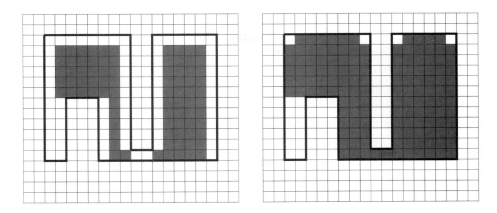

Figure 2.32: Opening operation. *Left:* Initial erosion. *Right:* Succeeding dilation. The boundary of the original figure A is shown as a bold line.

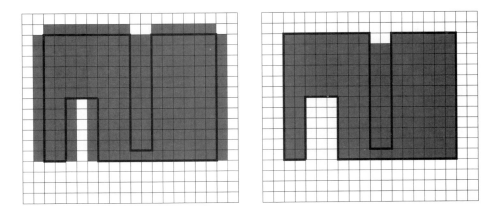

Figure 2.33: Closing operation. *Left:* Initial dilation. *Right:* Succeeding erosion. The boundary of the original figure A is shown as a bold line.

2.7   Optical Character Recognition

Morphological operations can be used for optical character recognition when the characters are relatively noise-free and are of the same font and size. First, extract the character to be recognized from one of the images in which the character appears. Next, use expanding or closing to fill holes and cavities. Then shrink the character image to remove unwanted regions and to reduce the size of the character so that it will fit comfortably inside an instance of the character. This processed image is the model for the character.

To recognize instances of the character in other images, use that character model as a probe and perform erosion. The images may have to be cleaned (holes filled and unwanted clutter removed) before performing erosion. After erosion, compute connected components, apply the size filter to discard regions that are too small, and compute the position of each region that passes through the size filter. This provides the position of each recognized instance of the character model in the image. Good character models obtained after cleaning, filling, and shrinking will match most instances of the character, including instances in a slightly different font and size. However, omnifont recognition has been a very challenging problem for researchers. Optical character recognition can be implemented in real time with special-purpose hardware.

Further Reading

Several aspects of binary vision are given in the books by Rosenfeld and Kak [206] and Horn [109]. Morphological image processing is discussed in the books [68, 103]. An example of the use of morphological operations in a practical application is provided in the paper by Mitchell and Gillies on reading hand-printed zip codes [168]. For a tutorial on optical character recognition and document analysis see O'Gorman and Kasturi [188].

Exercises

2.1 We saw in this chapter that the zeroth-, first-, and second-order moments for a binary region provide important information about the image's size, location, and orientation. Define higher moments of a region. Do you think that these moments will provide any useful characteristics for the region? If so, what will be the nature of information provided by these higher moments?

2.2 In many applications an early step is to determine where in an image are interesting objects. After these objects are located, most processing is focused only on these areas. How can you use projections for such focus-of-attention computations?

2.3 A component labeling algorithm is a computation bottleneck in many applications. This can be considered a bridge between lower levels and higher (semantic) levels in a vision system. How can you develop a fast algorithm to compute connected components? Can you develop a parallel algorithm?

Computer Projects

2.1 Use a thresholding program (or write your own) and see an image at many thresholding levels. Find a suitable threshold to get the best representation of an object in the image. Now select another object and find the best threshold for it. Repeat this for each object. Are these thresholds the same? Why? Repeat this experiment with several images.

2.2 Develop algorithms to compute area and first and second moments of a region from its run-length code.

2.3 Develop a medial axis algorithm. Apply it to several binary images of irregularly shaped objects to study the strengths and weaknesses of this technique to represent shapes of objects.

2.4 In many robotic applications, for a given domain a medial axis can be used for path planning. Consider the map of a building floor. Find the medial axis for the hallways and see which hallways can be navigated by a robot of specific size.

2.5 Construct algorithms to implement expanding and shrinking operations. Use these algorithms to implement different types of noise removal in binary images.

2.6 Modify your shrink algorithm to make it intelligent so that it does not completely eliminate a region. This feature can then be used to compute the number of connected components in a region. Implement this algorithm.

2.7 Design a machine vision system to identify objects from their binary images. Consider common objects such as coins, pens, notebooks, and other desk accessories. Develop a recognition strategy based on the features that you studied in this chapter. Implement all operations and test your system.

Chapter 3

Regions

A region in an image is a group of connected pixels with similar properties. Regions are important for the interpretation of an image because they may correspond to objects in a scene. An image may contain several objects and, in turn, each object may contain several regions corresponding to different parts of the object. For an image to be interpreted accurately, it must be partitioned into regions that correspond to objects or parts of an object. However, due to segmentation errors, the correspondence between regions and objects will not be perfect, and object-specific knowledge must be used in later stages for image interpretation.

3.1 Regions and Edges

Consider the simple image shown in Figure 3.1. This figure contains several objects. The first step in the analysis and understanding of this image is to partition the image so that regions representing different objects are explicitly marked. Such partitions may be obtained from the characteristics of the gray values of the pixels in the image. Recall that an image is a two-dimensional array and the values of the array elements are the gray values. Pixels, gray values at specified indices in the image array, are the observations, and all other attributes, such as region membership, must be derived from the gray values. There are two approaches to partitioning an image into regions: region-based segmentation and boundary estimation using edge detection.

Figure 3.1: This figure shows an image with several regions. Note that regions and boundaries contain the same information because one representation can be derived from the other.

In the region-based approach, all pixels that correspond to an object are grouped together and are marked to indicate that they belong to one region. This process is called *segmentation*. Pixels are assigned to regions using some criterion that distinguishes them from the rest of the image. Two very important principles in segmentation are *value similarity* and *spatial proximity*. Two pixels may be assigned to the same region if they have similar intensity characteristics or if they are close to one another. For example, a specific measure of value similarity between two pixels is the difference between the gray values, and a specific measure of spatial proximity is Euclidean distance. The variance of gray values in a region and the compactness of a region can also be used as measures of value similarity and spatial proximity of pixels within a region, respectively.

The principles of similarity and proximity come from the assumption that points on the same object will project to pixels in the image that are spatially close and have similar gray values. Clearly, this assumption is not satisfied in many situations. We can, however, group pixels in the image using these simple assumptions and then use domain-dependent knowledge to match regions to object models. In simple situations, segmentation can be done with thresholding and component labeling, as discussed in Chapter 2. Complex images may require more sophisticated techniques than thresholding to assign pixels to regions that correspond to parts of objects.

Segmentation can also be done by finding the pixels that lie on a region boundary. These pixels, called edges, can be found by looking at neighboring pixels. Since edge pixels are on the boundary, and regions on either side of the boundary may have different gray values, a region boundary may be found by measuring the difference between neighboring pixels. Most edge detectors use only intensity characteristics as the basis for edge detection, although derived characteristics, such as texture and motion, may also be used.

In ideal images, a region will be bounded by a closed contour. In principle, region segmentation and edge detection should yield identical results. Edges (closed contours) may be obtained from regions using a boundary-following algorithm. Likewise, regions may be obtained from edges using a region-filling algorithm. Unfortunately, in real images it is rare to obtain correct edges from regions and vice versa. Due to noise and other factors, neither region segmentation nor edge detection provides perfect information.

In this chapter, we will discuss the basic concepts of regions, concentrating on two issues:

- Segmenting an image into regions

- Representing the regions

This chapter begins with a discussion of automatic thresholding and histogram methods for segmentation, followed by a discussion of techniques for representing regions. Then more sophisticated techniques for region segmentation will be presented. Edge detection techniques will be discussed in Chapter 5.

In the following section, we will discuss techniques for region formation. Thresholding is the simplest region segmentation technique. After discussing thresholding, we will present methods to judge the similarity of regions using their intensity characteristics. These methods may be applied after an initial region segmentation using thresholding. It is expected that these algorithms will produce a region segmentation that corresponds to an object or its part. Motion characteristics of points can also be used to form and refine regions. Knowledge-based approaches may be used to match regions to object models. The use of motion will be discussed in Chapter 14.

3.2   Region Segmentation

The segmentation problem, first defined in Section 2.1, is now repeated for ease of reference: Given a set of image pixels $\mathcal{I}$ and a homogeneity predicate $P(\cdot)$, find a partition S of the image $\mathcal{I}$ into a set of n regions R_i,

$$\bigcup_{i=1}^{n} R_i = \mathcal{I}.$$

The homogeneity predicate and partitioning of the image have the properties that any region satisfies the predicate

$$P(R_i) = \text{True}$$

for all i, and any two adjacent regions cannot be merged into a single region that satisfies the predicate

$$P(R_i \cup R_j) = \text{False}.$$

The homogeneity predicate $P(\cdot)$ defines the conformity of all points in the region R_i to the region model.

The process of converting a gray value image into a binary image is a simple form of segmentation where the image is partitioned into two sets. The algorithms for thresholding to obtain binary images can be generalized to more than two levels. The thresholds in the algorithm discussed in Chapter 2 were chosen by the designer of the system. To make segmentation robust to variations in the scene, the algorithm should be able to select an appropriate threshold automatically using the samples of image intensity present in the image. The knowledge about the gray values of objects should not be hard-wired into an algorithm; the algorithm should use knowledge about the relative characteristics of gray values to select the appropriate threshold. This simple idea is useful in many computer vision algorithms.

3.2.1   Automatic Thresholding

To make segmentation more robust, the threshold should be automatically selected by the system. Knowledge about the objects in the scene, the application, and the environment should be used in the segmentation algorithm in a form more general than a fixed threshold value. Such knowledge may include

- Intensity characteristics of objects

- Sizes of the objects

- Fractions of an image occupied by the objects

- Number of different types of objects appearing in an image

A thresholding scheme that uses such knowledge and selects a proper threshold value for each image without human intervention is called an automatic thresholding scheme. Automatic thresholding analyzes the gray value distribution in an image, usually by using a histogram of the gray values, and uses the knowledge about the application to select the most appropriate threshold. Since the knowledge employed in these schemes is more general, the domain of applicability of the algorithm is increased.

Suppose that an image contains n objects $O_1, O_2, \ldots, O_n$, including the background, and gray values from different populations $\pi_1, \ldots, \pi_n$ with probability distributions $p_1(z), \ldots, p_n(z)$. In many applications, the probabilities $P_1, \ldots, P_n$ of the objects appearing in an image may also be known. Using this knowledge, it is possible to rigorously formulate the threshold selection problem. Since the illumination geometry of a scene controls the probability distribution of intensity values $p_i(z)$ in an image, one cannot usually precompute the threshold values. As we will see, most methods for automatic threshold selection use the size and probability of occurrence and estimate intensity distributions by computing histograms of the image intensities.

Many automatic thresholding schemes have been used in different applications. Some of the common approaches are discussed in the following sections. To simplify the presentation, we will follow the convention that objects are dark against a light background. In discussing thresholds, this allows us to say that gray values below a certain threshold belong to the object and gray values above the threshold are from the background, without resorting to more cumbersome language. The algorithms that we present in the following sections can easily be modified to handle other cases such as light objects against a dark background, medium gray objects with background values that are light and dark, or objects with both light and dark gray values against a medium gray background. Some algorithms can be generalized to handle object gray values from an arbitrary set of pixel values.

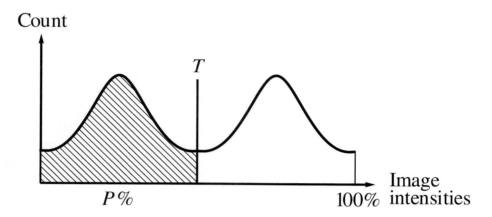

Figure 3.2: The shaded areas in the histogram represent p percent of the image area. The threshold is selected so that p percent of the histogram is assigned to the object.

P-Tile Method

The p-tile method uses knowledge about the area or size of the desired object to threshold an image. Suppose that in a given application objects occupy about p percent of the image area. By using this knowledge to partition the gray value histogram of the input image, one or more thresholds can be chosen that assign p percent of the pixels to the object. Figure 3.2 gives an example of a binary image formed using this technique.

Clearly, this method is of very limited use. Only a few applications, such as page readers, allow such an estimate of the area in a general case.

Mode Method

If the objects in an image have the same gray value, the background has a different gray value, and the image pixels are affected by zero-mean Gaussian noise, then we may assume that the gray values are drawn from two normal distributions with parameters (μ_1, σ_1) and (μ_2, σ_2). The histogram for an image will then show two separate peaks, as shown in Figure 3.3. In the ideal case of constant intensity values, $\sigma_1 = \sigma_2 = 0$, there will be two spikes in the histogram and the threshold can be placed anywhere between the spikes. In practice, the two peaks are not so well separated. In this case, we may detect

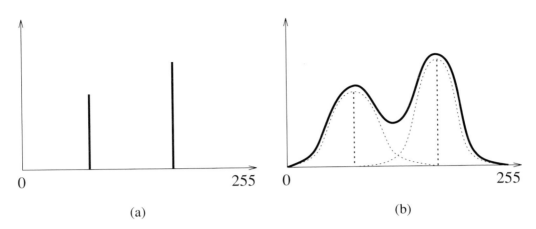

Figure 3.3: Ideally, the intensities of the background and objects will be widely separated. In the ideal case, the threshold T can be anywhere between the two peaks, as shown in (a). In most images, the intensities will overlap, resulting in histograms as shown in (b).

peaks and valleys in the histogram, and the threshold may be set to the pixel value corresponding to the valley. It can be shown that the probability of misclassification is minimized by this choice of the threshold when the size of the object is equal to that of the background (see Exercise 3.2). In most applications, since the histogram is sparsely populated near the valley, the segmentation is not sensitive to the threshold value.

The determination of peaks and valleys is a nontrivial problem, and many methods have been proposed to solve it. For an automatic thresholding scheme, we should have a measure of the *peakiness* and *valleyness* of a point in a histogram. A computationally efficient method is given in Algorithm 3.1. This method ignores local peaks by considering peaks that are at some minimum distance apart. The peakiness is based on the height of the peaks and the depth of the valleys; the distance between the peaks and valleys is ignored.

This approach can be generalized to images containing many objects with different mean gray values. Suppose there are n objects with normally distributed gray values with parameters $(\mu_1, \sigma_1), (\mu_2, \sigma_2), \ldots, (\mu_n, \sigma_n)$, and the background is also normally distributed with parameters (μ_0, σ_0). If the means are significantly different, the variances are small, and none of the ob-

Algorithm 3.1 Peakiness Detection for Appropriate Threshold Selection

1. *Find the two highest local maxima in the histogram that are at some minimum distance apart. Suppose these occur at gray values g_i and g_j.*

2. *Find the lowest point g_k in the histogram H between g_i and g_j.*

3. *Find the peakiness, defined as $\min(H(g_i), H(g_j))/H(g_k)$.[1]*

4. *Use the combination (g_i, g_j, g_k) with highest peakiness to threshold the image. The value g_k is a good threshold to separate objects corresponding to g_i and g_j.*

jects is very small in size, then the histogram for the image will contain $n+1$ peaks. The valley locations $T_1, T_2, \ldots, T_n$ can be determined, and pixels with gray values in each interval $(T_i, T_{i+1}]$ can be assigned to the corresponding object (see Figure 3.4).

Iterative Threshold Selection

An iterative threshold selection method starts with an approximate threshold and then successively refines this estimate. It is expected that some property of subimages resulting from the threshold can be used to select a new threshold value that will partition the image better than the first threshold. The threshold modification scheme is critical to the success of this approach. The method is given in Algorithm 3.2.

Adaptive Thresholding

If the illumination in a scene is uneven, then the above automatic thresholding schemes may not be suitable. The uneven illumination may be due to shadows or due to the direction of illumination. In all such cases, the same threshold value may not be usable throughout the complete image (see Figure 3.5). *Non-adaptive* methods analyze the histogram of the entire image.

---

[1]If the valley region shows a great deal of "spikiness" with many empty bins, then a certain amount of smoothing may be required to remove the possibility of division by zero.

Count

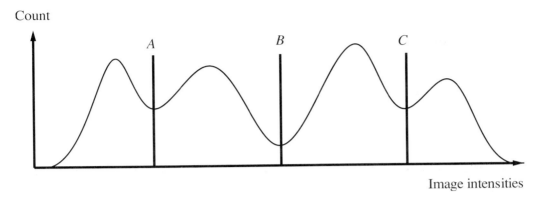

Image intensities

Figure 3.4: Histogram for an image containing several objects with differing intensity values.

Methods to deal with uneven illumination or uneven distribution of gray values in the background should look at a small region of an image and then analyze this subimage to obtain a threshold for only that subimage. Some techniques have been developed to deal with this kind of situation.

A straightforward approach to segment such images is to partition the image into $m \times m$ subimages and select a threshold T_{ij} for each subimage based on the histogram of the ijth subimage ($1 \leq i, j \leq m$). The final segmentation of the image is the union of the regions of its subimages. The results of this method are shown in Figure 3.6.

Variable Thresholding

Another useful technique in the case of uneven illumination is to approximate the intensity values of the image by a simple function such as a plane or biquadratic. The function fit is determined in large part by the gray value of the background. Histogramming and thresholding can be done relative to the base level determined by the fitted function. This technique is also called background normalization. For example, Figure 3.7 shows a three-dimensional plot of the box image with uneven illumination. If a planar function were fitted to this function, it would lay somewhere between the two rough surfaces representing the box and the background. Now, by using this fitted plane as the basis for thresholding, the box can be easily segmented. Any points in the image that are above the plane will be part of the box, and anything below will be part of the background.

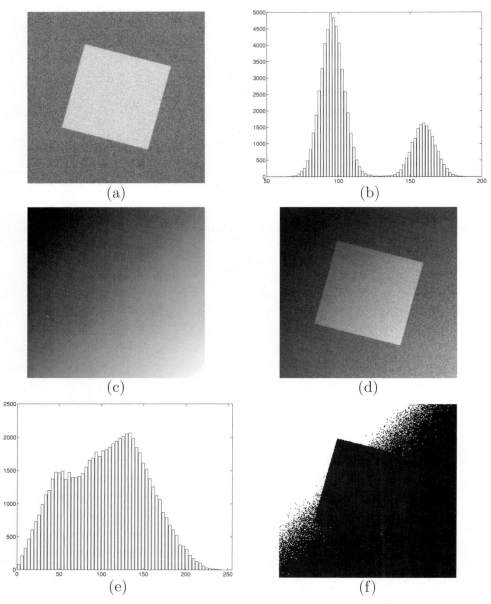

(a)

(b)

(c)

(d)

(e)

(f)

Figure 3.5: An example of an image with uneven illumination which is not amenable to regular thresholding. (a) Original image with uniform illumination. (b) Histogram of original. (c) Simulated uneven illumination. (d) Box with uneven illumination. (e) Histogram of box with uneven illumination. (f) Box thresholded at approximate valley of histogram, $T = 72$. Note that regular thresholding is not effective in segmenting the object from the background.

Algorithm 3.2 Iterative Threshold Selection

1. *Select an initial estimate of the threshold, T. A good initial value is the average intensity of the image.*

2. *Partition the image into two groups, R_1 and R_2, using the threshold T.*

3. *Calculate the mean gray values μ_1 and μ_2 of the partitions R_1 and R_2.*

4. *Select a new threshold:*

$$T = \frac{1}{2}(\mu_1 + \mu_2).$$

5. *Repeat steps 2–4 until the mean values μ_1 and μ_2 in successive iterations do not change.*

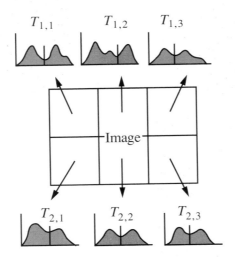

Figure 3.6: The process of adaptive thresholding on an image which is not amenable to regular thresholding.

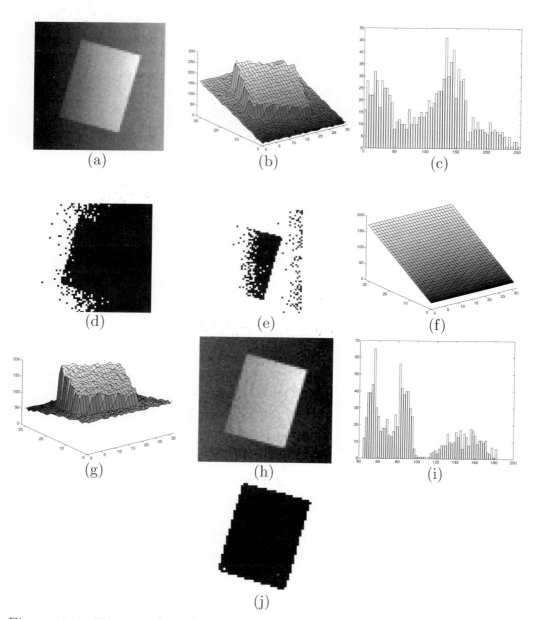

Figure 3.7: The results of variable thresholding on an image that is not amenable to regular thresholding. (a) Original image with uneven illumination. (b) 3-D plot of original image. (c) Histogram of original image. (d) Thresholded original, $T = 85$. (e) Thresholded original, $T = 165$. (f) Approximated planar function. (g) Plot of difference between original image and fitted plane. This is the *normalized* image. (h) The normalized image. (i) Histogram of normalized image. (j) Thresholded result, $T = 110$.

Algorithm 3.3 Double Thresholding for Region Growing

1. *Select two thresholds T_1 and T_2.*

2. *Partition the image into three regions: R_1, containing all pixels with gray values below T_1; R_2, containing all pixels with gray values between T_1 and T_2, inclusive; and R_3, containing all pixels with gray values above T_2.*

3. *Visit each pixel assigned to region R_2. If the pixel has a neighbor in region R_1, then reassign the pixel to region R_1.*

4. *Repeat step 3 until no pixels are reassigned.*

5. *Reassign any pixels left in region R_2 to region R_3.*

Double Thresholding

In many applications, it is known that certain gray values belong to objects. However, there may be additional gray values that belong to either objects or the background. In such a case, one may use a conservative threshold T_1 to obtain the *core* of the object and then use some method to *grow* the object regions. The methods used for growing these regions will depend on the specific application. Common approaches include using another threshold to accept pixels if they have a neighbor that is a core pixel or by using intensity characteristics, such as a histogram, to determine points to be included in the object region. A simple approach is to accept all points that are below a second threshold T_2 and are connected to the original set of points. This approach is outlined in Algorithm 3.3, and the results are shown in Figure 3.8.

In Algorithm 3.3, region R_1 is the core region, region R_2 is the fringe region (also called the intermediate or transition region), and region R_3 is the background. The core region is grown by adding pixels from the fringe that are neighbors of core pixels. After region growing, any pixels that are not in the core region are background pixels. The double thresholding algorithm for region growing implements the principles of value similarity and spatial proximity. Fringe pixels have values that are close to the values of core pixels since the two sets of pixels are adjacent in the histogram, and fringe pixels are spatially close to core pixels since they are neighbors.

3.2.2   Limitations of Histogram Methods

As discussed above, one may use information in the histogram of an image to select an appropriate threshold for segmentation. This approach is useful in those applications where objects have constant gray values. If the illumination is different in different parts of a scene, then a single threshold may not be sufficient to segment the image, even if the image contains only one object. In such cases one must use techniques that effectively partition an image, arbitrarily, and select thresholds for each subimage independently. We also saw some other heuristics for using histogram-based segmentation. If the images are complex, these approaches will also perform poorly.

The most basic limitation of the histogram-based approaches is due to the fact that a histogram throws away spatial information about the intensity values in an image. The histogram describes the global intensity distribution. Several images with very different spatial distributions of gray values may have similar histograms. For example, one cannot distinguish between a random distribution of black and white points, a binary checkerboard, and an image that is half black and half white just on the basis of their histograms. The global nature of a histogram limits its applicability to complex scenes. It does not exploit the important fact that points from the same object are usually spatially close due to *surface coherence*.

3.3   Region Representation

Regions are used in many contexts and can be represented in many alternative forms. Different representations are suitable in different applications. Some applications require computations only for a single region, while others require relationships among different regions of an image. In this section, we will discuss a few commonly used representations of regions and study their features. It must be mentioned here that regions can also be represented as closed contours. We will discuss those representations separately in Chapter 6.

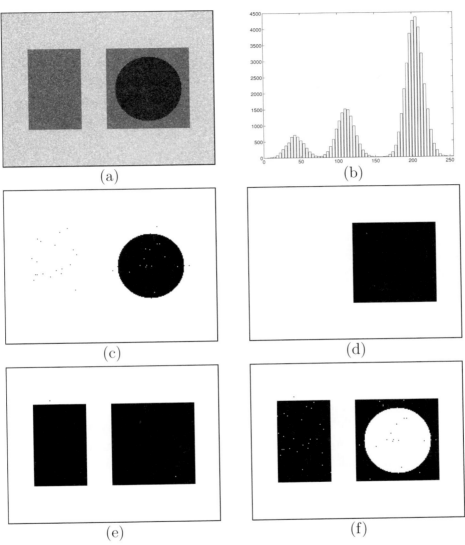

Figure 3.8: An image that is difficult to segment using a single threshold and its segmentation using double thresholding. (a) Original image. (b) Histogram of original image. (c) Core region, $T_1 = 70$. (d) Results after growing the core region using a second threshold, $T_2 = 155$. Compare this with the results using regular thresholding shown in (e) and (f). (e) Threshold $= T_2$. (f) $T_1 \leq$ Threshold $\leq T_2$. Note that it is impossible to segment the entire right square without using region growing threshold.

Most region representations can be classified into one of the following three classes:

1. Array representations

2. Hierarchical representations

3. Symbolic representations

3.3.1   Array Representation

The basic representation for regions is to use an array of the same size as the original image with entries that indicate the region to which a pixel belongs. Referring back to Chapter 2, Figure 2.12(b) is a representation of regions in the image shown in Figure 2.12(a). Thus, if element i, j of the array has the value α, then the corresponding pixel in the image belongs to region α. The simplest example of this representation is a binary image where each pixel belongs to either region 0, the background, or region 1, the foreground.

Another scheme uses membership arrays (images), commonly called masks or bitmaps. Each region is associated with a binary image, its mask, that indicates which pixels belong to that region. By overlaying masks on the original image, intensity characteristics of regions can be found. One advantage to this scheme is that ambiguous situations, where the region membership of a pixel cannot be definitely decided, can be handled by allowing the pixel to be a member of more than one region. The corresponding pixel will be 1 in more than one mask. Array representations contain region information in iconic or image form. Symbolic information is not explicitly represented.

The last few years have seen increased use of gray value images in computer applications. This popularity is due to widespread use of raster graphics terminals and decreased memory costs. Array representation preserves all details of regions required in most applications. This has made binary masks a very popular representation in computers and much hardware support is available to manipulate them.

3.3.2   Hierarchical Representations

Images can be represented at many different resolutions. Clearly, by reducing an image's resolution, thus reducing the size of the array, some data is lost, making it more difficult to recover information. However, reduction in resolution results in reduced memory and computing requirements. Hierarchical representation of images allows representation at multiple resolutions. In many applications, one can compute properties of images first at a low resolution and then perform additional computations over a selected area of the image at a higher resolution. Hierarchical representations are also used for browsing in images. We present two commonly used forms of hierarchical representations: pyramids and quad trees.

Pyramids

A pyramid representation of an $n \times n$ image contains the image and k reduced versions of the image. Usually n is a power of 2 and the other images are $n/2 \times n/2$, $n/4 \times n/4$, ..., 1×1. In a pyramid representation of an image, the pixel at level l is obtained by combining information from several pixels in the image at level $l + 1$. The whole image is represented as a single pixel at the top level, level 0, and the bottom level is the original (unreduced) image. A pixel at a level represents aggregate information represented by several pixels at the next level. Figure 3.9 shows an image and its reduced versions in a pyramid. Here the pyramid is obtained by simply averaging the gray values in 2×2 neighborhoods. It is possible, however, to devise other strategies to form reduced-resolution versions. Similarly, it is possible to taper the pyramid in nonlinear ways.

An implementational point is that the entire pyramid fits into a linear array of size $2(2^{2 \times \text{level}})$.

Quad Trees

A quad tree may be considered an extension of pyramids for binary images. A quad tree contains three types of nodes: white, black, and gray. A quad tree is obtained by recursive splitting of an image. A region in an image is split into four subregions of identical size, as shown in Figure 3.10. For each subregion, if all points in the region are either white or black, then this region is no longer considered as a candidate for splitting; if it contains pixels of both kinds, it is considered to be a "gray region" and is further split into four subregions. An image obtained using this recursive splitting is represented in a tree structure. The splitting process is repeated until there are no gray

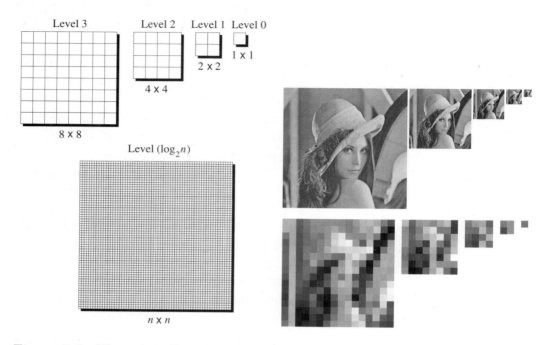

Figure 3.9: The original image is a 512×512 image; its reduced-resolution versions are successively obtained by averaging four points. All successive versions are shown here. Note that the low resolution images have been enlarged for display.

regions in the tree. Each node in this structure is either a leaf node or has four children—thus the name *quad tree*.

Quad trees are finding increasing application in spatial databases. Several algorithms have been developed for converting a raster array to a quad tree and a quad tree to a raster array. Algorithms for computing several pictorial properties have also been developed. The last few years have seen efforts to represent a quad tree using codes to reduce the memory required by pointers.

3.3.3 Symbolic Representations

A region can be represented using symbolic characteristics. Some commonly used symbolic characteristics are:

- Enclosing rectangle

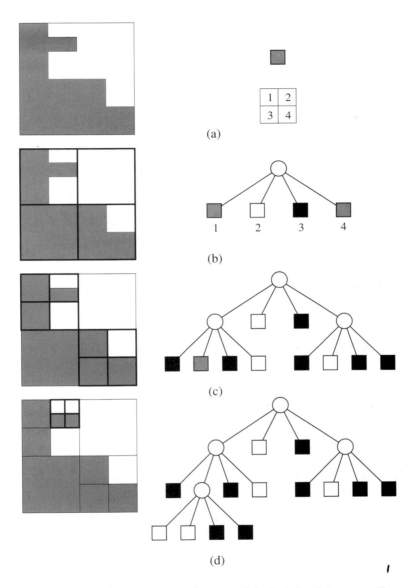

Figure 3.10: The building of a quad tree. (a) Original image, "gray region." (b) Original split into four subregions (the left node in the tree corresponds to the top left region in the image). Note that two of these regions are also gray regions. (c) Splitting the gray regions from (b) into four subregions. One of these regions is still a gray region. (d) Splitting of the last gray region and the final quad tree.

- Centroid

- Moments

- Euler number

Other characteristics such as the mean and variance of intensity values, although not symbolic representations, are also commonly used to represent regions in an image. These measures are straightforward generalizations of the calculations introduced in Chapter 2. In addition, other application-dependent features of a region may be represented. When representing an image for interpretation, we may be required to represent relationships among neighboring regions also.

3.3.4   Data Structures for Segmentation

When implementing region merging and splitting algorithms for segmenting an image (discussed in Section 3.4), information about the regions which have been formed must be maintained in some data structure. Since merge and split operations use information about boundaries between regions, as well as general characteristics of the regions, several data structures have been proposed to allow easy manipulation of region characteristics. In this section, we will discuss a few data structures that facilitate region merging and splitting.

Region Adjacency Graphs

A region adjacency graph (RAG) is used to represent regions and relationships among them in an image. The emphasis is on the partitions of an image in the form of regions and the characteristics of each partition. As shown in Figure 3.11, the following conventions are used. Nodes are used to represent regions, and arcs between nodes represent a common boundary between regions. Different properties of regions may be stored in the node data structure. The RAG emphasizes the adjacency of regions and plays a vital role in segmentation. After an initial segmentation based on primitive characteristics such as intensity values, the results may be represented in a RAG, and then regions may be combined to obtain a better segmentation. A method to generate the RAG is given in Algorithm 3.4.

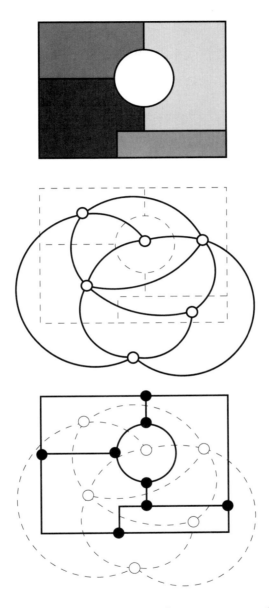

Figure 3.11: The region adjacency graph for a segmented image. *Top:* Segmented image. *Middle:* Region adjacency graph. *Bottom:* The dual of the region agacency graph.

Algorithm 3.4   Region Adjacency Graph

1. *Scan the membership array a and perform the following steps at each pixel index $[i, j]$.*

2. *Let $r_1 = a[i, j]$.*

3. *Visit the neighbors $[k, l]$ of the pixel at $[i, j]$. For each neighbor, perform the following step.*

4. *Let $r_2 = a[k, l]$. If $r_1 \neq r_2$, add an arc between nodes r_1 and r_2 in the region adjacency graph.*

The dual of a region adjacency graph may also be used in some situations. In the dual representation, nodes represent boundaries and arcs represent the regions that are separated by the boundaries.

Picture Trees

The picture tree emphasizes the inclusion of a region within another region. This representation, shown in Figure 3.12, is usually recursive. A version of this representation, the quad tree discussed in an earlier section, has received significant attention. In a picture tree, the emphasis is on nesting regions, while in a quad tree, rectangular regions are split into four rectangles of equal size, independent of the location of regions. A picture tree is usually produced by recursively splitting an image into component parts. Splitting stops when a region with constant characteristics has been reached. This representation will be discussed in detail in a later section on shape representations.

Super Grid

In some applications, it is desirable to store segmentation information in an image array. The representation of boundaries in this situation creates some problems. Intuitively, boundaries should be located *between* the pixels of two adjacent regions. However, often boundaries become actual pixels in image representations. This dilemma is solved by introducing a super grid on the image grid (see Figure 3.13). If the original image is $N \times N$, then the

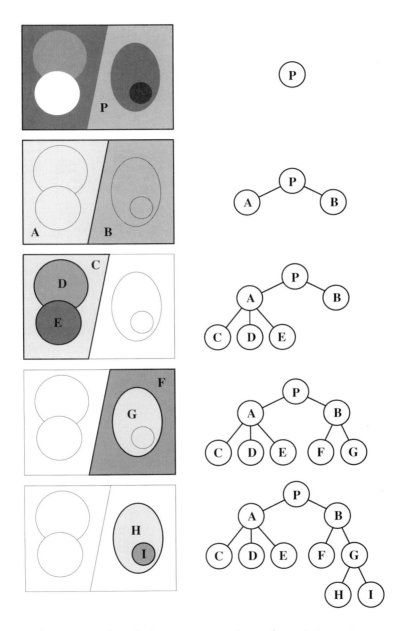

Figure 3.12: An example of the construction of a picture tree representing the inclusion relationships among regions.

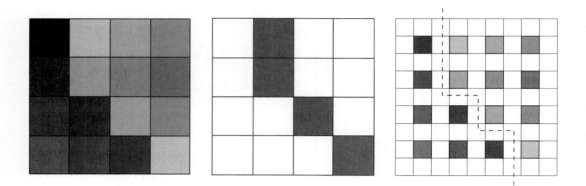

Figure 3.13: A super grid region representation. *Left:* Picture grid. *Middle:* Traditional boundary representation. *Right:* Super grid representation. Note that the boundary is now *between* the two regions, unlike the traditional representation.

supergrid is $(2N + 1) \times (2N + 1)$. Each pixel is surrounded by eight nonpixel points on the super grid. Nonpixel points are used to indicate whether or not there is a boundary between two pixels, and in what direction the boundary runs. This representation simplifies merge and split operations by explicitly representing the boundary between any two pixels of an image.

3.4   Split and Merge

A simple intensity-based segmentation usually results in too many regions. Even in images where most humans see very clear regions with constant gray value, the output of a thresholding algorithm may contain many extra regions. The main reasons for this problem are high-frequency noise and a gradual transition between gray values in different regions.

After the initial intensity-based region segmentation, the regions may need to be refined or reformed. Several approaches have been proposed for postprocessing such regions obtained from a simple segmentation approach. Some of these approaches use domain-dependent knowledge, while other approaches use knowledge about the imaging process. The refinement may be done interactively by a person or automatically by a computer. In an automatic system, the segmentation will have to be refined based on object characteristics and general knowledge about the images.

Automatic refinement is done using a combination of split and merge operations. Split and merge operations eliminate false boundaries and spurious regions by merging adjacent regions that belong to the same object, and they add missing boundaries by splitting regions that contain parts of different objects. Some possible approaches for refinement include:

- Merge adjacent regions with similar characteristics.

- Remove questionable edges.

- Use topological properties of the regions.

- Use shape information about objects in the scene.

- Use semantic information about the scene.

The first three approaches use only information about image intensity combined with other domain-independent characteristics of regions. A discussion of approaches for region refinement follows.

3.4.1   Region Merging

The merge operation combines regions that are considered similar. A high-level merge algorithm is given in Algorithm 3.5. The algorithm can be adapted to any of the measures of region similarity discussed in the following sections.

However, when applying a simple algorithm such as this, one may still get into trouble. Consider the following example. We have an image with three adjacent regions A, B, and C. The similarity predicate determines separately that A and B are similar and that B and C are similar. However, A and C are *not*. When merging similar regions, local decisions to merge A and B, and separately B and C, will collapse the three regions into a single region even when A and C are not similar. In this case, one must take additional region characteristics into consideration before merging similar regions.

The most important operation in the merge algorithm is to determine the similarity between two regions. Many approaches have been proposed to judge the similarity of regions. Broadly, the approaches to judge the similarity are based either on the gray value of regions or on the weakness

Algorithm 3.5   Region Merging

1. *Form initial regions in the image using thresholding (or a similar approach) followed by component labeling.*

2. *Prepare a region adjacency graph (RAG) for the image.*

3. *For each region in an image, perform the following steps:*

   (a) *Consider its adjacent region and test to see if they are similar.*

   (b) *For regions that are similar, merge them and modify the RAG.*

4. *Repeat step 3 until no regions are merged.*

of boundaries between the regions, and may include the spatial proximity of regions.

Two approaches to judging the similarity of adjacent regions are:

1. Compare their mean intensities. If the mean intensities do not differ by more than some predetermined value, the regions are considered similar and should be candidates for merging. A modified form of this approach uses surface fitting to determine whether the regions may be approximated by one surface.

2. Assume that the intensity values are drawn from a probability distribution. Consider whether or not to merge adjacent regions based on the probability that they will have the same statistical distribution of intensity values. This approach uses hypothesis testing to judge the similarity of adjacent regions and is discussed in more detail below.

Merging Statistically Similar Regions

This approach considers statistical characteristics of two adjacent regions to decide whether or not they should be merged. Assume that the regions in an image have constant gray value corrupted by statistically independent, additive, zero-mean Gaussian noise, so that the gray values are drawn from normal distributions. Suppose that two adjacent regions R_1 and R_2 contain m_1 and m_2 points, respectively. There are two possible hypotheses:

H_0: Both regions belong to the same object. In this case, the intensities are all drawn from a single Gaussian distribution with parameters (μ_0, σ_0^2).

H_1: The regions belong to different objects. In this case, the intensities of each region are drawn from separate Gaussian distributions with parameters (μ_1, σ_1^2) and (μ_2, σ_2^2).

In general, these parameters are not known but are estimated using the samples. For example, when a region contains n pixels having gray levels, g_i, $i = 1, 2, \ldots, n$, drawn from a normal distribution given by

$$p(g_i) = \frac{1}{\sqrt{2\pi}\sigma} e^{-\frac{(g_i - \mu)^2}{2\sigma^2}} \tag{3.1}$$

the Maximum Likelihood estimation equations for the parameters are given by

$$\hat{\mu} = \frac{1}{n} \sum_{i=1}^{n} g_i \tag{3.2}$$

$$\hat{\sigma}^2 = \frac{1}{n} \sum_{i=1}^{n} (g_i - \hat{\mu})^2 \tag{3.3}$$

Under the hypothesis H_0, all pixels are independently drawn from a single distribution, $N(\mu_0, \sigma_0^2)$. The joint probability density under H_0 is given by

$$p(g_1, g_2, \ldots, g_{m_1+m_2} \mid H_0) = \prod_{i=1}^{m_1+m_2} p(g_i \mid H_0) \tag{3.4}$$

$$= \prod_{i=1}^{m_1+m_2} \frac{1}{\sqrt{2\pi}\sigma_0} e^{-\frac{(g_i - \mu_0)^2}{2\sigma_0^2}} \tag{3.5}$$

$$= \frac{1}{(\sqrt{2\pi}\sigma_0)^{m_1+m_2}} e^{-\frac{\sum_{i=1}^{m_1+m_2}(g_i - \mu_0)^2}{2\sigma_0^2}} \tag{3.6}$$

$$= \frac{1}{(\sqrt{2\pi}\sigma_0)^{m_1+m_2}} e^{-\frac{(m_1+m_2)}{2}} \tag{3.7}$$

Under the hypothesis H_1, m_1 pixels belong to region 1 with a distribution $N(\mu_1, \sigma_1)$ and m_2 pixels belong to region 2 with a distribution $N(\mu_2, \sigma_2^2)$. Under this hypothesis, the joint density function is given by

$$p(g_1, g_2, \ldots, g_{m_1}, g_{m_1+1}, \ldots, g_{m_1+m_2} \mid H_1) = \frac{1}{(\sqrt{2\pi}\sigma_1)^{m_1}} e^{-\frac{m_1}{2}} \cdot \frac{1}{(\sqrt{2\pi}\sigma_2)^{m_2}} e^{-\frac{m_2}{2}}$$

$$(3.8)$$

The likelihood ratio, L, is then defined as the ratio of the probability densities under the two hypotheses

$$L = \frac{p(g_1, g_2, \ldots \mid H_1)}{p(g_1, g_2, \ldots \mid H_0)} \tag{3.9}$$

$$= \frac{\sigma_0^{m_1+m_2}}{\sigma_1^{m_1} \cdot \sigma_2^{m_2}} \tag{3.10}$$

The values of the parameters σ_0, σ_1, and σ_2 in the above equation are estimated from Eqs. 3.2 and 3.3 using all the $(m_1 + m_2)$ pixels, m_1 pixels from region 1 and m_2 pixels from region 2, respectively. If the likelihood ratio L is below a threshold value, there is strong evidence for the likelihood that there is only one region and the two regions may be merged.

This approach may also be used for edge detection. Since the likelihood ratio indicates when two regions should be considered to be separate, it indicates when there should be a boundary between the two regions. For edge detection, the likelihood ratio between neighborhoods on either side of a point may be used to detect the presence of edges.

There are other possible modifications to this ratio which can play an important role in many applications. The likelihood ratio was derived under the assumption that a region contains constant gray values which, due to noise, have a normal distribution. It is possible to assume that the underlying intensity distribution is not constant but rather planar or quadratic. The likelihood ratio in these cases may be derived and used in a similar way.

3.4.2   Removing Weak Edges

Another approach to merging is to combine two regions if the boundary between them is weak. This approach attempts to remove weak edges between adjacent regions by considering not only the intensity characteristics, but also the length of the common boundary. The common boundary is dissolved if

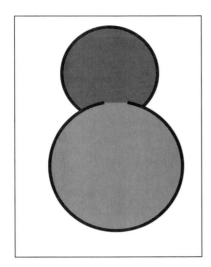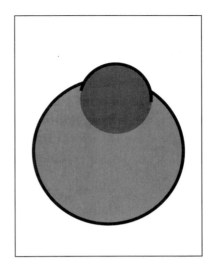

Figure 3.14: Approach 1: Merging by removing weak boundaries if the ratio of the weak boundary to the minimum region perimeter is above some threshold. *Left:* The two regions should *not* be merged because the weak boundary is very short as compared to the perimeter of the smaller region. *Right:* The two regions should be merged into a single region because the weak part of the common boundary is a significant fraction of the perimeter of the smaller region. Note that the strong boundary is marked by a bold line whereas the weak boundary is marked simply by a change in color of the two regions.

the boundary is weak and the resulting boundary (of the merged region) does not change gray value too quickly. The notation used in the following discussion is illustrated in Figures 3.14 and 3.15.

A weak boundary is one for which the intensities on either side differ by less than an amount T. Other criteria, such as edgeness values, may be used to determine the strength of an edge point that is on the boundary separating two regions. In an algorithm to merge regions by dissolving weak boundaries between regions, one must consider the relative lengths of the weak and complete boundaries between the regions. Two approaches which consider the relative lengths are:

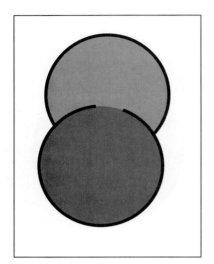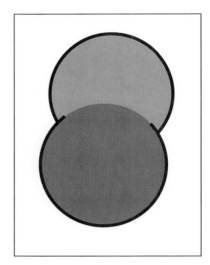

Figure 3.15: Approach 2: Merging by removing weak boundaries if the ratio of the weak boundary to the total common boundary is above some threshold. *Left:* The two regions should *not* be merged because the weak boundary is small when compared to the total common boundary. *Right:* The two regions should be merged into a single region because the weak part of the boundary is almost the same as the total common boundary. Note that the strong boundary is marked by a bold line whereas the weak boundary is marked simply by a change in color of the two regions.

1. Merge adjacent regions R_1 and R_2 if

$$\frac{W}{S} > \tau,$$

where W is the length of the weak part of the boundary, τ is a threshold, and $S = \min(S_1, S_2)$ is the minimum of the perimeters of the two regions as shown in Figure 3.14. In this algorithm, the performance depends on τ. For a small value of τ, there will be too many region merges. For a high value, the algorithm becomes too conservative. A good heuristic value for τ is 0.5.

2. Merge adjacent regions R_1 and R_2 if

$$\frac{W}{S} > \tau,$$

where S is now the common boundary as shown in Figure 3.15. The parameter value $\tau = 0.75$ usually yields satisfactory results.

3.4.3 Region Splitting

If some property of a region is not constant, the region should be split. The segmentation based on the split approach starts with large regions. In many cases, one may start with the whole image as the starting region. A split algorithm is given in Algorithm 3.6.

Several decisions must be made before a region is split. The problem is usually in deciding when a property is not constant over a region and how to split a region so that the property for each of the resulting components is constant. These questions are usually application-dependent and require knowledge of the characteristics of regions in that application. In some applications, the variance of the intensity values is used as a measure for how close the gray values are to being constant. In other applications, a function is fitted to approximate the underlying intensity values. The error between this function and the actual intensity values is used as the measure of region similarity.

More difficult than deciding if the gray values are constant across a region is deciding where to split a region. One approach used to determine the best boundary for dividing a region is to consider the measures of edge strength within the region. The easiest methods for splitting regions are those that divide the region into a fixed number of equal-sized regions; these are called regular decomposition methods. The quad tree representation for images, discussed in Section 3.3.2, is an example of regular decomposition.

Algorithm 3.6 Region Splitting

1. *Form initial regions in the image.*

2. *For each region in an image, recursively perform the following steps:*

   (a) *Compute the variance in the gray value for the region.*

   (b) *If the variance is above a threshold, split the region along the appropriate boundary.*

Algorithm 3.7   Split and Merge Region Segmentation

1. *Start with the entire image as a single region.*

2. *Pick a region R. If $P(R)$ is false, then split the region into four subregions.*

3. *Consider any two or more neighboring subregions, $R_1, R_2, \ldots, R_n$, in the image. If $P(R_1 \cup R_2 \cup \cdots \cup R_n)$ is true, merge the n regions into a single region.*

4. *Repeat these steps until no further splits or merges take place.*

    The quad tree approach, however, must be modified for use in segmenting nonbinary (gray value) images. Instead of considering black and white regions, we must use the variance of the image intensity to decide whether a region should be split. Due to the numerous ways in which a region may be split, splitting regions is generally more difficult than merging them.

3.4.4   Split and Merge

Split and merge operations may be used together. After a presegmentation based on thresholding, a succession of splits and merges may be applied to refine the segmentation. Combined split and merge algorithms are useful for segmenting complex scenes. Domain knowledge may be introduced to guide the split and merge operations.

    Suppose that an image is partitioned into a set of regions, $\{R_k\}$, for $k = 1, 2, \ldots, m$. All of the pixels in a region will be homogeneous according to some property defined by a predicate P applied to the region. The predicate represents the similarity between the pixels in a region. For example, the predicate could be defined using the variance in gray values within a region:

$$P(R) = \begin{cases} 1 & \text{if the variance is small} \\ 0 & \text{otherwise.} \end{cases} \tag{3.11}$$

    The split and merge algorithm for region segmentation is outlined in Algorithm 3.7.

3.5 Region Growing

In many images, the gray values of individual regions are not nearly constant and more sophisticated techniques must be used for segmentation. The best techniques are those based on the assumption that the image can be partitioned into regions that can be modeled by simple functions. This idea can be applied naturally for region segmentation.

The segmentation problem set forth in Section 3.2 leads to an algorithm that starts with seed regions and then grows the regions to form larger regions satisfying these constraints. For the example in this text, the homogeneity predicate is based on fitting planar and biquadratic functions to the gray values in a region. However, in general, the homogeneity predicate can be based on any characteristic of the regions in the image such as average intensity, variance, texture, or color. The algorithm is given here as Algorithm 3.8.

The algorithm begins by partitioning the image into $n \times n$ regions where n is typically between 5 and 9. Regions are merged if a single planar or biquadratic function can be fit to the gray values in both regions. The planar and biquadratic models are a linear combination of basis functions. The basis functions span the variable-order bivariate polynomials, so the model is

$$f(x, y, a, m) = \sum_{i+j \leq m} a_{ij} x^i y^j, \tag{3.12}$$

where the order m of the model is restricted to $0 \leq m \leq 2$. This means that the region models are restricted to planar and biquadratic functions.

The homogeneity predicate is based on the distance of points in a region from the function that models the region:

$$\chi^2(R, a, m) = \sum_{(x,y) \in R} d^2(x, y, a, m) \tag{3.13}$$

where the distance is ordinary Euclidean distance:

$$d^2(x, y, a, m) = [g(x, y) - f(x, y, a, m)]^2 . \tag{3.14}$$

The gray value $g(x, y)$ at point (x, y) in the image plane is the gray value of the pixel at that image location. Given a set of points R, the problem is to find the order m of the model and the model parameters a that minimize the error function $\chi^2(R, a, m)$. This is a least-squares problem that can be

Algorithm 3.8   Region Growing Using Planar and Biquadratic Models

1. *Partition the image into initial seed regions $R_i^{(0)}$.*

2. *Fit a planar model to each seed region. If the chi-squared error is small enough, accept the seed region and its model; otherwise, reject the seed region.*

3. *For each region, find all points that are compatible with the region by extrapolating the region model to the neighbors of the region. Compatible points are defined as*

$$C_i^{(k)} = \{(x,y)\,|\,d^2(x,y,a,m) \leq \epsilon \text{ and } (x,y) \text{ is a 4-neighbor of } R_i^{(k)} \cup C_i^{(k)}\}$$

where ϵ is the compatibility threshold.

4. *If there were no compatible points, increase the order of the model: $m \leftarrow m+1$. If the model order is larger than the maximum model order, then do not grow the region further; otherwise, continue region growing by returning to step 3.*

5. *Form the new region $R_i^{(k+1)} = R_i^{(k)} \cup C_i^{(k)}$, refit the model at the same order to the new region, and compute the goodness of fit $\chi^2(R_i^{(k+1)}, a, m)$.*

6. *Compute the difference between the old and new goodness of fit for the region model:*

$$\rho^{(k+1)} = \chi^2(m, a^{(k+1)}, R_i^{(k+1)}) - \chi^2(m, a^{(k)}, R_i^{(k)})$$

7. *If $\rho^{(k+1)} < T_1$, continue region growing by returning to step 3.*

8. *Increase the order of the model: $m \leftarrow m+1$. If the model order is larger than the maximum model order, then do not grow the region further.*

9. *Refit the region model at the new model order m. If the error of fit decreases, accept the new model order and continue region growing by returning to step 3; otherwise, do not grow region R_i further.*

solved using singular value decomposition. Refer to [196] and Appendix B for a complete discussion on singular value decomposition.

The reason that the compatibility of points is checked against the surface patch before the surface is refit to the combined set of points is that least-squares fitting is very sensitive to outliers. If an outlier is added to the region before fitting, then the surface patch can be so severely distorted by the outlier that it no longer fits even the points that actually belong to the region.

It is not possible to know in advance how the image should be partitioned into regions that are modeled by distinct surface patches, since this is the segmentation problem itself. One can use a rigorous approach to find a conservative seed region. It is possible to use a conservative thresholding for finding the seed. One may also use a sophisticated approach to find initial seeds using the domain knowledge or the nature of images. For example, in range images differential geometric characteristics may be used to find initial seed regions. A computationally efficient approach to do this is to use some general image characteristics to identify such seed regions. One may partition the image into 7×7 seed regions and then fit surface patches to these seed regions. The seed regions are accepted based on the chi-squared error in the fit of the surface patch. If a 7×7 patch is rejected, then it is replaced by overlapping 5×5 surface patches to get higher resolution. Regions are grown by acquiring compatible points. Points are compatible with a region if the surface patch for the region can be extended to include the point such that the value of the point is not far from the surface. The surface patch is refit over the new domain which includes the original points plus the compatible points.

One may allow points to be associated with more than one region. The ambiguity is resolved by a postprocessing step that performs model selection. In classic region growing, each point is associated with at most one region. This constraint can be enforced by modifying the algorithm so that one region takes a point away from another region only if reassigning the point improves the fit of the surface patch model for both regions. More precisely, reassign a point only if the combined error for surface fits for both regions is lower. Figure 3.16 shows the results of this approach.

The classical region segmentation techniques described earlier can be viewed as region growing techniques where the surface patches are restricted to constants. In other words, the assumption is that regions are nearly constant

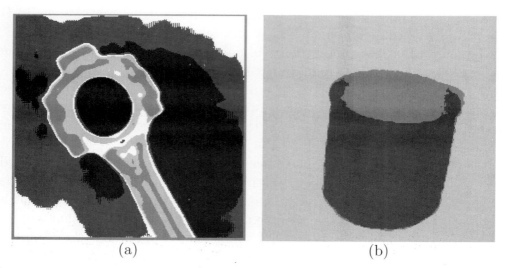

Figure 3.16: The segmentation of an image using the region growing algorithm is given here. Part (a) shows the results for an intensity image and (b) shows results for a range image.

in image intensity and the image can be partitioned into regions of piecewise constant image intensity. More sophisticated segmentation techniques generalize the assumptions on the variation in image intensity to allow more realistic intensity variations due to shading.

Further Reading

After more than two decades of intensive research efforts, segmentation still remains a difficult problem in machine vision. Several approaches do not use any domain-specific information in the early stages of vision [163]. On the other hand, some psychophysicists believe that every aspect of perception uses domain information extensively [203]. Computer vision research has been influenced by both these views. Haralick and Shapiro [101] present a review of early segmentation techniques.

A discussion of thresholding methods for region formation is given in the books by Rosenfeld and Kak [206], and Haralick and Shapiro [103]. For an introduction to quad trees, see the book by Samet [210]. Pyramid representations are discussed in several sources; for an introduction see [230].

Region growing has been a popular topic. A statistical approach for merging regions was proposed by Yakimovsky [258]. An algorithm for region growing was developed by Besl and Jain [28].

Color can be used for segmentation. A representative treatment of segmentation techniques using color information is given in [105]. For an early work based on histograms of colors, see [189]. Geman and Geman's [85] stochastic relaxation approach for segmentation and restoration of images has attracted significant attention. The relaxation method, also known as simulated annealing, has many applications in image analysis.

Explicit application of knowledge for segmentation has been addressed in many systems [30, 119, 165, 183]. Knowledge-based approaches definitely have strong appeal and will potentially be very useful for segmentation of complex images. So far their success has been limited due to poor performance of operators in domain-independent early processing. Model-based reasoning can help [49]. The model-based reasoning will emerge as a powerful approach in many applications of machine vision.

Considerable work on region growing for land use classification and related applications has been done in the field of remote sensing [208].

Exercises

3.1 Regions and boundaries may be considered duals of each other. They also contain equivalent information. Do you see advantages of one over the other in machine vision applications? Why?

3.2 In the automatic thresholding approach based on the mode method, the bottom of the valley is usually used as the threshold value. Assuming that the object and the background are of equal size in the image and their intensity values are drawn from two different Gaussian probability distributions $N(\mu_1, \sigma)$ and $N(\mu_2, \sigma)$, show that the recommended threshold indeed results in the optimal threshold because it minimizes the misclassification error. Obtain an expression for the optimal threshold for the general case when the object pixels cover $p\%$ of the image area and the object and background distributions are given by $N(\mu_1, \sigma_1)$ and $N(\mu_2, \sigma_2)$, respectively.

3.3 What knowledge about the objects in images is used in the p-tile and mode methods for automatically selecting the threshold value?

3.4 What is the limitation of histogram-based methods for finding regions in an image? How can this be overcome in most applications? Can you suggest a postprocessing method for correcting the results obtained using a histogram-based method for images that contain objects that do not have Gaussian distributions for their intensity values?

3.5 Pyramid representations are appearing in many different forms. In many common multimedia applications thumbnail sketches are used. What are thumbnail sketches? Can you consider these schemes a subset of pyramid schemes?

3.6 Find the quad tree representation for the following object.

0	0	1	1	1	1	0	0
0	0	1	1	1	1	0	0
0	1	1	1	1	1	1	0
0	1	1	1	1	1	1	0
0	1	1	1	1	1	1	0
1	1	1	1	1	1	1	1
1	1	0	0	0	0	1	1
1	1	0	0	0	0	1	1

3.7 Kodak's photo-CD form of image representation has become popular in several applications. What is this representation? What are its useful features?

3.8 Quad tree representations have a serious limitation for applications in machine vision. A connected component in an image may be distributed at distant nodes in this representation. Can you suggest a scheme to modify quad trees to remove this limitation?

3.9 Another serious problem with quad trees is the rotation and scaling of regions. What will happen to the quad tree of an image if a region in it is translated, rotated, or scaled? Can you suggest some solutions to overcome these problems?

3.10 Show that the pyramid representation of an $n \times n$ image actually does fit within a linear array of size $2(2^{2 \times \text{level}})$.

3.11 Assume that regions in an image have an intensity distribution that is planar combined with Gaussian noise. Thus, intensity value at a point of the region can be represented as

$$I(x, y) = Ax + By + C + N(0, \sigma^2). \qquad (3.15)$$

Derive the likelihood ratio-based merge criterion under this assumption.

Computer Projects

3.1 Consider three different types of images: an office scene, an outdoor scene, and maybe a scene containing just one simple object. Implement an automatic thresholding scheme and study the results for these images.

3.2 Put two simple objects on a desk. Select one light and one dark object. Illuminate the scene such that light comes from the direction of the light object. Develop a suitable automatic thresholding scheme to segment the image as best as you can. Now change the direction of illumination by moving your light source so that the illumination is from the direction of the dark object. See whether your algorithm still works. If not, experiment and make it work in these and similar other cases.

3.3 Implement the likelihood ratio-based approach for merging regions. Use it as a postprocessing step in your earlier experiments.

3.4 Implement the region growing approach for finding regions in an image. Use it in your earlier experiments. Now compare the results of all these approaches and find the strengths and weaknesses of all these approaches for different situations in images.

Chapter 4

Image Filtering

When an image is acquired by a camera or other imaging system, often the vision system for which it is intended is unable to use it directly. The image may be corrupted by random variations in intensity, variations in illumination, or poor contrast that must be dealt with in the early stages of vision processing.

This chapter discusses methods for image enhancement aimed at eliminating these undesirable characteristics. The chapter begins with histogram modification, followed by a brief review of discrete linear systems and frequency analysis, and then coverage of various filtering techniques. The Gaussian smoothing filter is covered in depth.

4.1 Histogram Modification

Many images contain unevenly distributed gray values. It is common to find images in which all intensity values lie within a small range, such as the image with poor contrast shown in Figure 4.1. Histogram equalization is a method for stretching the contrast of such images by uniformly redistributing the gray values. This step may make threshold selection approaches more effective. In general, histogram modification enhances the subjective quality of an image and is useful when the image is intended for viewing by a human observer.

A simple example of histogram modification is image scaling: the pixels in the range $[a, b]$ are expanded to fill the range $[z_1, z_k]$. The formula for

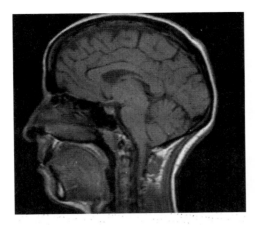

Figure 4.1: An image with poor contrast.

mapping a pixel value z in the original range into a pixel value z' in the new range is

$$z' = \frac{z_k - z_1}{b - a}(z - a) + z_1$$
$$= \frac{z_k - z_1}{b - a}z + \frac{z_1 b - z_k a}{b - a}. \tag{4.1}$$

The problem with this scheme is that when the histogram is stretched according to this formula, the resulting histogram has gaps between bins (see Figure 4.2). Better methods stretch the histogram while filling all bins in the output histogram continuously.

If the desired gray value distribution is known a priori, the following method may be used. Suppose that p_i is the number of pixels at level z_i in the original histogram and q_i is the number of pixels at level z_i in the desired histogram. Begin at the left end of the original histogram and find the value k_1 such that

$$\sum_{i=1}^{k_1-1} p_i \leq q_1 < \sum_{i=1}^{k_1} p_i. \tag{4.2}$$

The pixels at levels $z_1, z_2, \ldots, z_{k_1-1}$ map to level z_1 in the new image. Next,

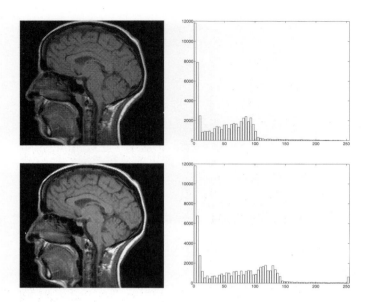

Figure 4.2: The original image has very poor contrast since the gray values are in a very small range. Histogram scaling improves the contrast but leaves gaps in the final histogram. *Top:* Original image and histogram. *Bottom:* Image and resulting histogram after histogram scaling.

find the value k_2 such that

$$\sum_{i=1}^{k_2-1} p_i \leq q_1 + q_2 < \sum_{i=1}^{k_2} p_i. \tag{4.3}$$

The next range of pixel values, $z_{k_1}, \ldots, z_{k_2-1}$, maps to level z_2. This procedure is repeated until all gray values in the original histogram have been included. The results of this approach are shown in Figure 4.3.

If the histogram is being expanded, then pixels having the same gray value in the original image may have to be spread to different gray values in the new image. The simplest procedure is to make random choices for which output value to assign to each input pixel. Suppose that a uniform random number generator that produces numbers in the range $[0, 1)$ is used to spread pixels evenly across an interval of n output values $q_k, q_{k+1}, \ldots, q_{k+n-1}$. The output pixel number can be computed from the random number r using the formula

$$k + \lfloor n * r \rfloor. \tag{4.4}$$

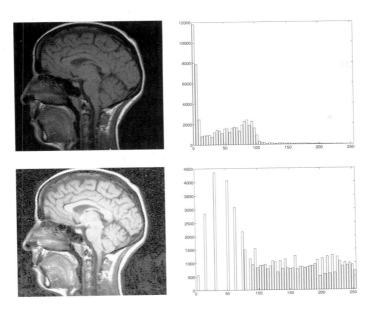

Figure 4.3: The original image has very poor contrast since the gray values are in a very small range. Histogram equalization improves the contrast by mapping the gray values to an approximation to a uniform distribution. However, this method still leaves gaps in the final histogram unless pixels having the same gray levels in the input image are spread across several gray levels in the output image. *Top:* Original image and histogram. *Bottom:* Image and resulting histogram after histogram equalization.

In other words, for each decision, draw a random number, multiply by the number of output values in the interval, round down to the nearest integer, and add this offset to the lowest index.

4.2 Linear Systems

Many image processing operations can be modeled as a *linear system*:

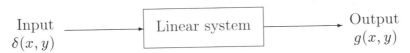

For a linear system, when the input to the system is an impulse $\delta(x, y)$ centered at the origin, the output $g(x, y)$ is the system's *impulse response*. Furthermore, a system whose response remains the same irrespective of the position of the input pulse is called a *space invariant* system:

Input
$\delta(x-x_0, y-y_0)$ $\longrightarrow$ Linear space invariant system $\longrightarrow$ Output
$g(x-x_0, y-y_0)$

A linear space invariant (LSI) system can be completely described by its impulse response $g(x, y)$ as follows:

Input
$f(x, y)$ $\longrightarrow$ LSI system
$g(x, y)$ $\longrightarrow$ Output
$h(x, y)$

where $f(x, y)$ and $h(x, y)$ are the input and output images, respectively. The above system must satisfy the following relationship:

$$a \cdot f_1(x, y) + b \cdot f_2(x, y) \implies a \cdot h_1(x, y) + b \cdot h_2(x, y)$$

where $f_1(x, y)$ and $f_2(x, y)$ are the input images, $h_1(x, y)$ and $h_2(x, y)$ are the output images corresponding to f_1 and f_2, and a and b are constant scaling factors.

For such a system, the output $h(x, y)$ is the *convolution* of $f(x, y)$ with the impulse response $g(x, y)$ denoted by the operator $\star$ and is defined as:

$$\begin{aligned}
h(x, y) &= f(x, y) \star g(x, y) \\
&= \int_{-\infty}^{\infty}\int_{-\infty}^{\infty} f(x', y')\, g(x - x', y - y')\, dx'\, dy'. \quad (4.5)
\end{aligned}$$

For discrete functions, this becomes:

$$\begin{aligned}
h[i, j] &= f[i, j] \star g[i, j] \\
&= \sum_{k=1}^{n}\sum_{l=1}^{m} f[k, l]\, g[i - k, j - l]. \quad (4.6)
\end{aligned}$$

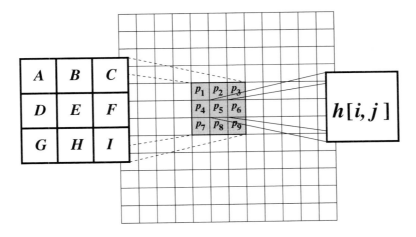

$$h[i,j] = A\,p_1 + B\,p_2 + C\,p_3 + D\,p_4 + E\,p_5 + F\,p_6 + G\,p_7 + H\,p_8 + I\,p_9$$

Figure 4.4: An example of a 3×3 convolution mask. The origin of the convolution mask corresponds to location E and the weights $A, B, \ldots I$ are the values of $g[-k, -l]$, $k, l = -1, 0, +1$.

If f and h are images, convolution becomes the computation of weighted sums of the image pixels. The impulse response, $g[i,j]$, is referred to as a *convolution mask*. For each pixel $[i,j]$ in the image, the value $h[i,j]$ is calculated by translating the convolution mask to pixel $[i,j]$ in the image, and then taking the weighted sum of the pixels in the neighborhood about $[i,j]$ where the individual weights are the corresponding values in the convolution mask. This process is illustrated in Figure 4.4 using a 3×3 mask.

Convolution is a linear operation, since

$$g[i,j] \star \{a_1 f_1[i,j] + a_2 f_2[i,j]\} = a_1\{g[i,j] \star f_1[i,j]\} + a_2\{g[i,j] \star f_2[i,j]\}$$

for any constants a_1 and a_2. In other words, the convolution of a sum is the sum of the convolutions, and the convolution of a scaled image is the scaled convolution. Convolution is a spatially invariant operation, since the same filter weights are used throughout the image. A spatially varying filter requires different filter weights in different parts of the image and hence cannot be represented by convolution.

Fourier Transform

An $n \times m$ image can be represented by its frequency components as follows:

$$f[k, l] = \frac{1}{4\pi^2} \int_{-\pi}^{\pi} \int_{-\pi}^{\pi} F(u, v)\, e^{jku}\, e^{jlv}\, du\, dv \qquad (4.7)$$

where $F(u, v)$ is the *Fourier transform* of the image. The Fourier transform encodes the amplitude and phase of each frequency component and is defined as

$$
\begin{aligned}
F(u, v) &= \mathcal{F}\{f[k, l]\} \\
&= \sum_{k=1}^{n} \sum_{l=1}^{m} f[k, l]\, e^{-jku}\, e^{-jlv}
\end{aligned}
\qquad (4.8)
$$

where $\mathcal{F}$ denotes the Fourier transform operation. The values near the origin of the (u, v) plane are called the low-frequency components of the Fourier transform, and those distant from the origin are the high-frequency components. Note that $F(u, v)$ is a continuous function.

Convolution in the image domain corresponds to multiplication in the spatial frequency domain. Therefore, convolution with large filters, which would normally be an expensive process in the image domain, can be implemented efficiently using the fast Fourier transform. This is an important technique in many image processing applications. In machine vision, however, most algorithms are nonlinear or spatially varying, so the Fourier transform methods cannot be used. In most cases where the vision algorithm can be modeled as a linear, spatially invariant system, the filter sizes are so small that implementing convolution with the fast Fourier transform provides little or no benefit; hence, linear filters, such as the smoothing filters discussed in the following sections, are usually implemented through convolution in the image domain.

4.3   Linear Filters

As mentioned earlier, images are often corrupted by random variations in intensity values, called *noise*. Some common types of noise are *salt and pepper* noise, *impulse* noise, and *Gaussian* noise. Salt and pepper noise contains random occurrences of both black and white intensity values. However, impulse

Figure 4.5: Examples of images corrupted by salt and pepper, impulse, and Gaussian noise. (a) & (b) Original images. (c) Salt and pepper noise. (d) Impulse noise. (e) Gaussian noise.

noise contains only random occurrences of white intensity values. Unlike these, Gaussian noise contains variations in intensity that are drawn from a Gaussian or normal distribution and is a very good model for many kinds of sensor noise, such as the noise due to camera electronics (see Figure 4.5).

Linear smoothing filters are good filters for removing Gaussian noise and, in most cases, the other types of noise as well. A linear filter is implemented using the weighted sum of the pixels in successive windows. Typically, the same pattern of weights is used in each window, which means that the linear filter is spatially invariant and can be implemented using a convolution mask. If different filter weights are used for different parts of the image, but the filter is still implemented as a weighted sum, then the linear filter is spatially varying. Any filter that is not a weighted sum of pixels is a nonlinear filter. Nonlinear filters can be spatially invariant, meaning that the same calculation is performed regardless of the position in the image, or spatially varying. The median filter, presented in Section 4.4, is a spatially invariant, nonlinear filter.

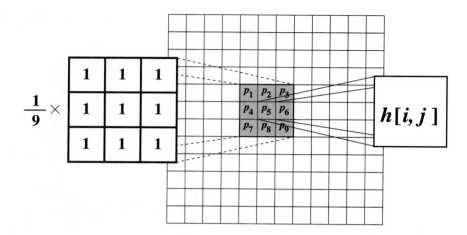

Figure 4.6: An example illustrating the mean filter using a 3 × 3 neighborhood.

Mean Filter

One of the simplest linear filters is implemented by a local averaging operation where the value of each pixel is replaced by the average of all the values in the local neighborhood:

$$h[i,j] = \frac{1}{M} \sum_{(k,l) \in N} f[k,l] \qquad (4.9)$$

where M is the total number of pixels in the neighborhood N. For example, taking a 3 × 3 neighborhood about $[i,j]$ yields:

$$h[i,j] = \frac{1}{9} \sum_{k=i-1}^{i+1} \sum_{l=j-1}^{j+1} f[k,l]. \qquad (4.10)$$

Compare this with Equation 4.6. Now if $g[i,j] = 1/9$ for every $[i,j]$ in the convolution mask, the convolution operation in Equation 4.6 reduces to the local averaging operation shown above. This result shows that a mean filter can be implemented as a convolution operation with equal weights in the convolution mask (see Figure 4.6). In fact, we will see later that many image processing operations can be implemented using convolution.

Figure 4.7: The results of a 3×3, 5×5, and 7×7 mean filter on the noisy images from Figure 4.5.

The size of the neighborhood N controls the amount of filtering. A larger neighborhood, corresponding to a larger convolution mask, will result in a greater degree of filtering. As a trade-off for greater amounts of noise reduction, larger filters also result in a loss of image detail. The results of mean filters of various sizes are shown in Figure 4.7.

When designing linear smoothing filters, the filter weights should be chosen so that the filter has a single peak, called the main lobe, and symmetry in the vertical and horizontal directions. A typical pattern of weights for a 3×3 smoothing filter is

$\frac{1}{16}$	$\frac{1}{8}$	$\frac{1}{16}$
$\frac{1}{8}$	$\frac{1}{4}$	$\frac{1}{8}$
$\frac{1}{16}$	$\frac{1}{8}$	$\frac{1}{16}$

Linear smoothing filters remove high-frequency components, and the sharp detail in the image is lost. For example, step changes will be blurred into gradual changes, and the ability to accurately localize a change will be sacrificed. A spatially varying filter can adjust the weights so that more smoothing is done in a relatively uniform area of the image, and little smoothing is done across sharp changes in the image. The results of a linear smoothing filter using the mask shown above are shown in Figure 4.8.

4.4   Median Filter

The main problem with local averaging operations is that they tend to blur sharp discontinuities in intensity values in an image. An alternative approach is to replace each pixel value with the median of the gray values in the local neighborhood. Filters using this technique are called *median filters*.

Median filters are very effective in removing salt and pepper and impulse noise while retaining image details because they do not depend on values which are significantly different from typical values in the neighborhood. Median filters work in successive image windows in a fashion similar to linear filters. However, the process is no longer a weighted sum. For example, take a 3×3 window and compute the median of the pixels in each window centered around $[i, j]$:

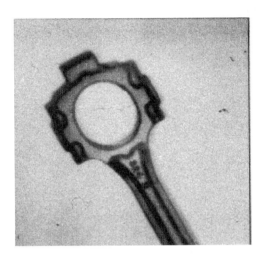 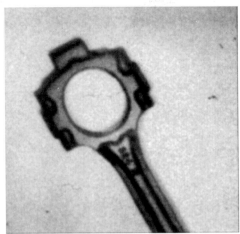

Figure 4.8: The results of a linear smoothing filter on an image corrupted by Gaussian noise. *Left:* Noisy image. *Right:* Smoothed image.

1. Sort the pixels into ascending order by gray level.

2. Select the value of the middle pixel as the new value for pixel $[i, j]$.

This process is illustrated in Figure 4.9. In general, an odd-size neighborhood is used for calculating the median. However, if the number of pixels is even, the median is taken as the average of the middle two pixels after sorting. The results of various sizes of median filters are shown in Figure 4.10.

4.5 Gaussian Smoothing

Gaussian filters are a class of linear smoothing filters with the weights chosen according to the shape of a Gaussian function. The Gaussian smoothing filter is a very good filter for removing noise drawn from a normal distribution.[1] The zero-mean Gaussian function in one dimension is

$$g(x) = e^{-\frac{x^2}{2\sigma^2}}, \tag{4.11}$$

---

[1]The fact that the filter weights are chosen from a Gaussian distribution and that the noise is also distributed as a Gaussian is merely a coincidence.

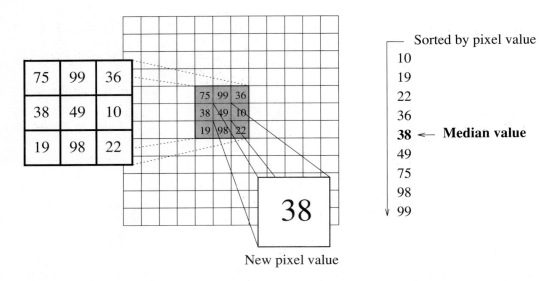

Figure 4.9: An example illustrating the median filter using a 3×3 neighborhood.

where the Gaussian spread parameter σ determines the width of the Gaussian. For image processing, the two-dimensional zero-mean discrete Gaussian function,

$$g[i, j] = e^{-\frac{(i^2 + j^2)}{2\sigma^2}},\qquad(4.12)$$

is used as a smoothing filter. A plot of this function is shown in Figure 4.11.

Gaussian functions have five properties that make them particularly useful in early vision processing. These properties indicate that the Gaussian smoothing filters are effective low-pass filters from the perspective of both the spatial and frequency domains, are efficient to implement, and can be used effectively by engineers in practical vision applications. The five properties are summarized below. Further explanation of the properties is provided later in this section.

1. In two dimensions, Gaussian functions are rotationally symmetric. This means that the amount of smoothing performed by the filter will be the same in all directions. In general, the edges in an image will not be oriented in some particular direction that is known in advance; consequently, there is no reason a priori to smooth more in one direction

Figure 4.10: The results of a 3×3, 5×5, and 7×7 median filter on the noisy images from Figure 4.5.

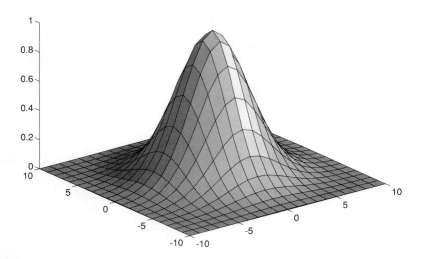

Figure 4.11: The two-dimensional Gaussian function with zero mean.

than in another. The property of rotational symmetry implies that a Gaussian smoothing filter will not bias subsequent edge detection in any particular direction.

2. The Gaussian function has a single lobe. This means that a Gaussian filter smooths by replacing each image pixel with a weighted average of the neighboring pixels such that the weight given to a neighbor decreases monotonically with distance from the central pixel. This property is important since an edge is a local feature in an image, and a smoothing operation that gives more significance to pixels farther away will distort the features.

3. The Fourier transform of a Gaussian has a single lobe in the frequency spectrum. This property is a straightforward corollary of the fact that the Fourier transform of a Gaussian is itself a Gaussian, as will be shown below. Images are often corrupted by undesirable high-frequency signals (noise and fine texture). The desirable image features, such as edges, will have components at *both* low and high frequencies. The single lobe in the Fourier transform of a Gaussian means that the smoothed image will not be corrupted by contributions from unwanted high-frequency signals, while most of the desirable signals will be retained.

4. The width, and hence the degree of smoothing, of a Gaussian filter is parameterized by σ, and the relationship between σ and the degree of smoothing is very simple. A larger σ implies a wider Gaussian filter and greater smoothing. Engineers can adjust the degree of smoothing to achieve a compromise between excessive blur of the desired image features (too much smoothing) and excessive undesired variation in the smoothed image due to noise and fine texture (too little smoothing).

5. Large Gaussian filters can be implemented very efficiently because Gaussian functions are separable. Two-dimensional Gaussian convolution can be performed by convolving the image with a one-dimensional Gaussian and then convolving the result with the same one-dimensional filter oriented orthogonal to the Gaussian used in the first stage. Thus, the amount of computation required for a 2-D Gaussian filter grows linearly in the width of the filter mask instead of growing quadratically.

4.5.1 Rotational Symmetry

The rotational symmetry of the Gaussian function can be shown by converting the function from rectangular to polar coordinates. Remember the two-dimensional Gaussian function

$$g[i,j] = e^{-\frac{(i^2+j^2)}{2\sigma^2}}. \tag{4.13}$$

Since the radius in polar coordinates is given by $r^2 = i^2 + j^2$, it is easy to see that the Gaussian function in polar coordinates,

$$g(r,\theta) = e^{-\frac{r^2}{2\sigma^2}}, \tag{4.14}$$

does not depend on the angle θ and consequently is rotationally symmetric. It is also possible to construct rotationally nonsymmetric Gaussian functions if they are required for an application where it is known in advance that more smoothing must be done in some specified direction. Formulas for rotationally nonsymmetric Gaussian functions are provided by Wozencraft and Jacobs [257, pp. 148–171], where they are used in the probabilistic analysis of communications channels.

4.5.2   Fourier Transform Property

The Gaussian function has the interesting property that its Fourier transform is also a Gaussian function. Since the Fourier transform of a Gaussian is a real function, the Fourier transform is its own magnitude. The Fourier transform of a Gaussian is computed by

$$\mathcal{F}\{g(x)\} = \int_{-\infty}^{\infty} g(x)\, e^{-j\omega x}\, dx \tag{4.15}$$

$$= \int_{-\infty}^{\infty} e^{-\frac{x^2}{2\sigma^2}}\, e^{-j\omega x}\, dx \tag{4.16}$$

$$= \int_{-\infty}^{\infty} e^{-\frac{x^2}{2\sigma^2}} \left(\cos \omega x + j \sin \omega x\right) dx \tag{4.17}$$

$$= \int_{-\infty}^{\infty} e^{-\frac{x^2}{2\sigma^2}} \cos \omega x\, dx + j \int_{-\infty}^{\infty} e^{-\frac{x^2}{2\sigma^2}} \sin \omega x\, dx. \tag{4.18}$$

The Gaussian is a symmetric function and the sine function is antisymmetric, so the integrand in the second integral is antisymmetric. Therefore, the integral must be zero, and the Fourier transform simplifies to:

$$\mathcal{F}\{g(x)\} = \int_{-\infty}^{\infty} e^{-\frac{x^2}{2\sigma^2}} \cos \omega x\, dx \tag{4.19}$$

$$= \sqrt{2\pi}\sigma e^{-\frac{\omega^2}{2\nu^2}}, \qquad \nu^2 = \frac{1}{\sigma^2}. \tag{4.20}$$

The spatial frequency parameter is ω, and the spread of the Gaussian in the frequency domain is controlled by ν, which is the reciprocal of the spread parameter σ in the spatial domain. This means that a narrower Gaussian function in the spatial domain has a wider spectrum, and a wider Gaussian function in the spatial domain has a narrower spectrum. This property relates to the noise suppression ability of a Gaussian filter. A narrow-spatial-domain Gaussian does less smoothing, and in the frequency domain its spectrum has more bandwidth and passes more of the high-frequency noise and texture. As the width of a Gaussian in the spatial domain is increased, the amount of smoothing that the Gaussian performs is increased, and in the frequency domain the Gaussian becomes narrower and passes less high-frequency noise and texture. This simple relationship between spatial-domain Gaussian width and frequency-domain spectral width enhances the ease of use of the Gaussian filter in practical design situations. The Fourier transform duality of Gaussian functions also explains why the single-lobe property in the spatial domain carries over into the frequency domain.

4.5.3 Gaussian Separability

The separability of Gaussian filters is easy to demonstrate:

$$g[i,j] \star f[i,j] = \sum_{k=1}^{m} \sum_{l=1}^{n} g[k,l] f[i-k, j-l] \qquad (4.21)$$

$$= \sum_{k=1}^{m} \sum_{l=1}^{n} e^{-\frac{(k^2+l^2)}{2\sigma^2}} f[i-k, j-l] \qquad (4.22)$$

$$= \sum_{k=1}^{m} e^{-\frac{k^2}{2\sigma^2}} \left\{ \sum_{l=1}^{n} e^{-\frac{l^2}{2\sigma^2}} f[i-k, j-l] \right\}. \qquad (4.23)$$

The summation in brackets is the convolution of the input image $f[i,j]$ with a vertical one-dimensional Gaussian function. The result of this summation is a two-dimensional image, blurred in the vertical dimension, that is then used as the input to a second convolution with a horizontal one-dimensional Gaussian that blurs the image in the horizontal dimension (see Figure 4.12). Since convolution is associative and commutative, the order of the convolutions can be reversed so that the horizontal convolution is performed first and the vertical convolution is performed on the result of the horizontal convolution.

This method can be implemented using the composition of two horizontal convolutions and a single horizontal convolution mask. The input $f[i,j]$ is first convolved with a horizontal Gaussian, and the result is placed in a temporary array in its transposed position. The temporary array is then used as input to the same convolution code so that the vertical convolution is performed by horizontal convolution. The output data from the second convolution is again transposed as the convolution is performed so that the data is restored to its proper (original) orientation. Results of this separable convolution are shown in Figure 4.13.

4.5.4 Cascading Gaussians

A related property of Gaussian filters is that the convolution of a Gaussian with itself yields a scaled Gaussian with larger σ. This is easily shown for

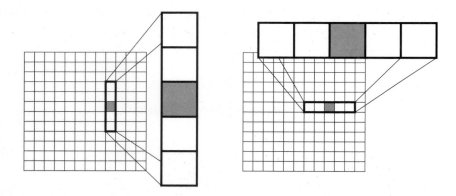

Figure 4.12: An example of the separability of Gaussian convolution. *Left:* Convolution with the vertical mask. *Right:* Convolution with the horizontal mask. Note that the origin of each mask is shaded.

the one-dimensional case:

$$
\begin{aligned}
g(x) \star g(x) &= \int_{-\infty}^{\infty} e^{-\frac{\xi^2}{2\sigma^2}} e^{-\frac{(x-\xi)^2}{2\sigma^2}} \, d\xi \\
&= \int_{-\infty}^{\infty} e^{-\frac{(\frac{x}{2}+\xi)^2}{2\sigma^2}} e^{-\frac{(\frac{x}{2}-\xi)^2}{2\sigma^2}} \, d\xi, \quad \xi \to \xi + \frac{x}{2} \\
&= \int_{-\infty}^{\infty} e^{-\frac{(2\xi^2+\frac{x^2}{2})}{2\sigma^2}} \, d\xi \\
&= e^{-\frac{x^2}{4\sigma^2}} \int_{-\infty}^{\infty} e^{-\frac{\xi^2}{\sigma^2}} \, d\xi \\
&= \sqrt{\pi}\sigma e^{-\frac{x^2}{2(\sqrt{2}\sigma)^2}}.
\end{aligned}
\tag{4.24}
$$

The product of the convolution of two Gaussian functions with spread σ is a Gaussian function with spread $\sqrt{2}\sigma$ scaled by the area of the Gaussian filter. The result holds in two dimensions as well. This means that if an image has been filtered with a Gaussian at a certain spread σ and if the same image must be filtered with a larger Gaussian with spread $\sqrt{2}\sigma$, then instead of filtering the image with the larger Gaussian, the previous result can just be refiltered with the same Gaussian filter of spread σ used to obtain the desired filtered image. This implies a significant reduction in computation in situations where multiple smoothed versions of images must be computed. Similar savings can be obtained when cascading Gaussian filters with different values of σ.

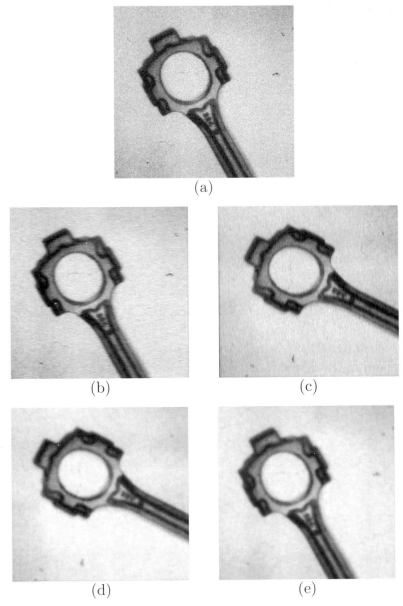

Figure 4.13: The results of separable Gaussian convolution using a single horizontal convolution mask. (a) Original noisy image. (b) Results of convolution with horizontal Gaussian mask. (c) The transposition of (b). (d) The convolution of (c) with the horizontal mask. (e) The transposition of (d). This is the final smoothed image.

4.5.5   Designing Gaussian Filters

An excellent approximation to a Gaussian is provided by the coefficients of the binomial expansion:

$$(1+x)^n = \binom{n}{0} + \binom{n}{1} x + \binom{n}{2} x^2 + \cdots + \binom{n}{n} x^n. \qquad (4.25)$$

In other words, use row n of Pascal's triangle as a one-dimensional, n-point approximation to a Gaussian filter. For example, the five-point approximation is:

1	4	6	4	1

corresponding to the fifth row of Pascal's triangle. This mask is used to smooth an image in the horizontal direction. Remember from Section 4.5.3 that a two-dimensional Gaussian filter can be implemented as the successive convolutions of two one-dimensional Gaussians, one in the horizontal direction and the other in the vertical direction. Also remember that this can be implemented using only the single one-dimensional mask by transposing the image between convolutions and after the final convolution. The results of Gaussian filtering using this approximation are shown in Figure 4.14.

This technique works well for filter sizes up to around $n = 10$. For larger filters, the coefficients in the binomial expansion are too large for most computers; however, arbitrarily large Gaussian filters can be implemented by repeatedly applying a smaller Gaussian filter. The σ of the binomial approximation to a Gaussian filter can be computed by using least-squares to fit a Gaussian function to the binomial coefficients.

Another approach in designing Gaussian filters is to compute the mask weights directly from the discrete Gaussian distribution [146]:

$$g[i,j] = c\, e^{-\frac{(i^2+j^2)}{2\sigma^2}} \qquad (4.26)$$

where c is a normalizing constant. By rewriting this as

$$\frac{g[i,j]}{c} = e^{-\frac{(i^2+j^2)}{2\sigma^2}} \qquad (4.27)$$

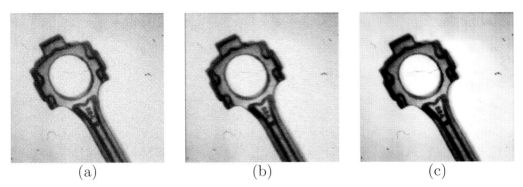

Figure 4.14: The approximation of a Gaussian filter using the fifth row of Pascal's triangle. (a) Original noisy image. (b) Result after smoothing in the horizontal direction. (c) Final result after smoothing in the vertical direction.

and choosing a value for σ^2, we can evaluate it over an $n \times n$ window to obtain a kernel, or mask, for which the value at $[0,0]$ equals 1. For example, choosing $\sigma^2 = 2$ and $n = 7$, the above expression yields the array:

$[i,j]$	-3	-2	-1	0	1	2	3
-3	.011	.039	.082	.105	.082	.039	.011
-2	.039	.135	.287	.368	.287	.135	.039
-1	.082	.287	.606	.779	.606	.287	.082
0	.105	.368	.779	1.000	.779	.368	.105
1	.082	.287	.606	.779	.606	.287	.082
2	.039	.135	.287	.368	.287	.135	.039
3	.011	.039	.082	.105	.082	.039	.011

However, we desire the filter weights to be integer values for ease in computations. Therefore, we take the value at one of the corners in the array, and choose k such that this value becomes 1. Using the above example, we get

$$\frac{g[3,3]}{k} = e^{-\frac{(3^2+3^2)}{2.2}} = 0.011 \implies k = \frac{g[3,3]}{0.011} = \frac{1.0}{0.011} = 91.$$

Now, by multiplying the rest of the weights by k, we obtain

$[i,j]$	-3	-2	-1	0	1	2	3
-3	1	4	7	10	7	4	1
-2	4	12	26	33	26	12	4
-1	7	26	55	71	55	26	7
0	10	33	71	91	71	33	10
1	7	26	55	71	55	26	7
2	4	12	26	33	26	12	4
3	1	4	7	10	7	4	1

This is the resulting convolution mask for the Gaussian filter (also shown in Figure 4.15). However, the weights of the mask do *not* sum to 1. Therefore, when performing the convolution, the output pixel values must be normalized by the sum of the mask weights to ensure that regions of uniform intensity are not affected. From the above example,

$$\sum_{i=-3}^{3} \sum_{j=-3}^{3} g[i,j] = 1115.$$

Therefore,

$$h[i,j] = \frac{1}{1115}(f[i,j] \star g[i,j])$$

where the weights of $g[i,j]$ are all integer values. The results of Gaussian smoothing using the above mask are given in Figure 4.16. Other common Gaussian filter masks are given in Figure 4.17.

4.5.6   Discrete Gaussian Filters

The samples of a Gaussian filter, or the coefficients obtained from the binomial expansion, form a discrete Gaussian filter. When discrete Gaussian filters are convolved, the result is a larger discrete Gaussian filter. If an image is smoothed with an $n \times n$ discrete Gaussian filter, and this intermediate result is smoothed by an $m \times m$ discrete Gaussian filter, then the result is exactly the same as if the original image had been smoothed by an $(n+m-1) \times (n+m-1)$ discrete Gaussian filter. In other words, convolving row n in Pascal's triangle with row m yields row $n+m-1$ in Pascal's triangle.

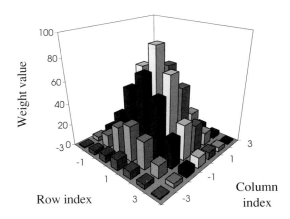

Figure 4.15: A 3-D plot of the 7×7 Gaussian mask.

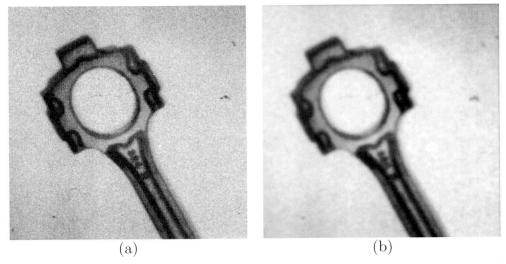

Figure 4.16: The results of smoothing using the 7×7 Gaussian mask. (a) Original image corrupted by Gaussian noise. (b) Smoothed image.

7×7 Gaussian mask

1	1	2	2	2	1	1
1	2	2	4	2	2	1
2	2	4	8	4	2	2
2	4	8	16	8	4	2
2	2	4	8	4	2	2
1	2	2	4	2	2	1
1	1	2	2	2	1	1

15×15 Gaussian mask

2	2	3	4	5	5	6	6	6	5	5	4	3	2	2
2	3	4	5	7	7	8	8	8	7	7	5	4	3	2
3	4	6	7	9	10	10	11	10	10	9	7	6	4	3
4	5	7	9	10	12	13	13	13	12	10	9	7	5	4
5	7	9	11	13	14	15	16	15	14	13	11	9	7	5
5	7	10	12	14	16	17	18	17	16	14	12	10	7	5
6	8	10	13	15	17	19	19	19	17	15	13	10	8	6
6	8	11	13	16	18	19	20	19	18	16	13	11	8	6
6	8	10	13	15	17	19	19	19	17	15	13	10	8	6
5	7	10	12	14	16	17	18	17	16	14	12	10	7	5
5	7	9	11	13	14	15	16	15	14	13	11	9	7	5
4	5	7	9	10	12	13	13	13	12	10	9	7	5	4
3	4	6	7	9	10	10	11	10	10	9	7	6	4	3
2	3	4	5	7	7	8	8	8	7	7	5	4	3	2
2	2	3	4	5	5	6	6	6	5	5	4	3	2	2

Figure 4.17: Other commonly used Gaussian masks [146].

Further Reading

Rosenfeld and Kak [205] provide a brief discussion of histogram modification. Books by Pratt [195] and Gonzalez and Woods [90] on image processing include material on histogram methods, median filters, and linear filters.

There is a discussion of linear systems theory, both continuous and discrete, and linear systems for image processing in the book by Horn [109]. Digital signal processing on two dimensions is covered by Oppenheim and Shafer [190] and by Rabiner and Gold [199]. The relationship between digital filters and numerical methods is discussed by Hamming [96]. More detailed explanations of this property and its use are provided by Crowley and Stern [65] and Burt [52].

Exercises

4.1 A mean filter is a linear filter, but a median filter is not. Why?

4.2 Compare the characteristics of median and mean filters and identify the situations where you will use them.

4.3 Gaussian filtering is usually a preferred averaging method. Why?

4.4 What is the separability property of Gaussian filtering? Why would you want a filtering scheme to be separable?

4.5 In many applications, an image is smoothed by applying Gaussian filters of several sizes. Why would one want to smooth an image using different parameters of the Gaussian?

4.6 What is the cascading property of Gaussian filters? How is it useful in machine vision?

4.7 An image contains a thin vertical line one pixel thick. It has a gray level of 50 and lies on a background of salt and pepper noise having gray values of 0 and 100, where

$$\text{Probability(gray level} = 0) = 0.4$$
$$\text{Probability(gray level} = 100) = 0.6.$$

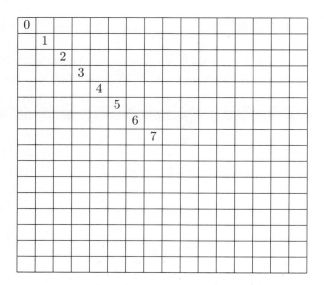

Figure 4.18: Synthetic image for Exercise 4.9.

The gray levels of the background pixels are independent of one another. A horizontal 1×3 operator given by

$$\boxed{-1} \quad \boxed{2} \quad \boxed{-1}$$

is applied to this image. Find the probability distribution of the values of the output when the operator is centered at a pixel:

a. on the line.

b. adjacent to the line.

4.8 An 8×8 image $f[i,j]$ has gray levels given by the following equation:

$$f[i,j] = |i - j|; \quad i,j = 0, 1, 2, 3, 4, 5, 6, 7.$$

Find the output image obtained by applying a 3×3 median filter on the image $f[i,j]$; note that the border pixels remain unchanged.

4.9 Consider the 16×16 image shown in Figure 4.18. The numbers indicate the gray level of that "ring" in the image. For example, the outer ring (border) has a gray value of 0, the next ring has a gray level 1, and so on.

    a. A 3×3 median filter operates on this image. Calculate the values of the central 4×4 pixels in the output image.

    b. Sketch the histogram of an image obtained by adding (pixel by pixel) the original image and its contrast-reversed image. Note that a contrast-reversed image is one in which the gray value l of each pixel is replaced by $(\max l) - l$.

Computer Projects

4.1 Implement the contrast stretching by histogram modification method. Make the range of histogram values $[a, b]$ and $[z_1, z_k]$ as variables. Take a poor contrast image and do experiments to see the effect of contrast enhancement by histogram modification. Implement several different contrast enhancement methods and comment on their performance.

4.2 Implement the Gaussian smoothing filter. Apply this filter to an image by selecting several (at least five) different values of σ. See the amount of smoothing. How would you select the correct σ value for an image?

Chapter 5

Edge Detection

The early stages of vision processing identify features in images that are relevant to estimating the structure and properties of objects in a scene. Edges are one such feature. Edges are significant local changes in the image and are important features for analyzing images. Edges typically occur on the boundary between two different regions in an image. Edge detection is frequently the first step in recovering information from images. Due to its importance, edge detection continues to be an active research area. This chapter covers only the detection and localization of edges. Basic concepts in edge detection will be discussed. Several common edge detectors will be used to illustrate the basic issues in edge detection. Algorithms for combining edges into contours are discussed in Chapter 6.

An edge in an image is a significant local change in the image intensity, usually associated with a discontinuity in either the image intensity or the first derivative of the image intensity. Discontinuities in the image intensity can be either (1) *step* discontinuities, where the image intensity abruptly changes from one value on one side of the discontinuity to a different value on the opposite side, or (2) *line* discontinuities, where the image intensity abruptly changes value but then returns to the starting value within some short distance. However, step and line edges are rare in real images. Because of low-frequency components or the smoothing introduced by most sensing devices, sharp discontinuities rarely exist in real signals. Step edges become *ramp* edges and line edges become *roof* edges, where intensity changes are not instantaneous but occur over a finite distance. Illustrations of these edge profiles are shown in Figure 5.1.

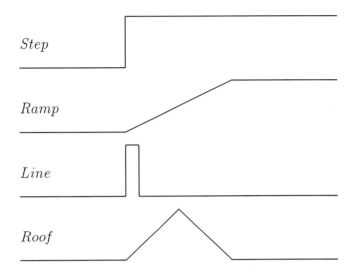

Figure 5.1: One-dimensional edge profiles.

It is also possible for an edge to have both step and line characteristics. For example, a surface that changes orientation from one flat surface to another will produce a step edge; but if the surface has a specular component of reflectance and if the surface corner is rounded, there can be a highlight due to the specular component as the surface orientation of the rounded corner passes the precise angle for specular reflection. The edge profile generated by such a situation looks like a step edge with a superimposed line edge. There are also edges associated with changes in the first derivative of the image intensity. For example, mutual reflection from the sides of a concave corner generate roof edges. Edges are important image features since they may correspond to significant features of objects in the scene. For example, the boundary of an object usually produces step edges because the image intensity of the object is different from the image intensity of the background.

This chapter will deal almost exclusively with step edges, although many of the ideas can be adapted to other types of image intensity changes. The profile of an ideal step edge and the profile of a real image that provides some examples of step edges are displayed in Figure 5.2. By definition, an edge is a significant local change in the image intensity. The plot shows step

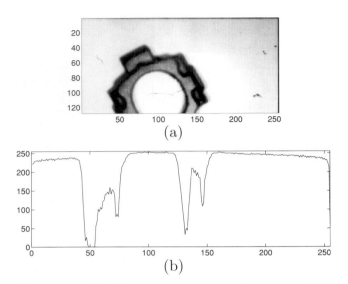

Figure 5.2: (a) The top half of the connecting rod image. (b) The profile of an ideal step change in image intensity is plotted to show that the edges are not perfectly sharp and the image is corrupted by noise. The plot is a horizontal slice through the circular portion of the connecting rod corresponding to the bottom edge of the partial rod image shown above.

changes in image intensity that are edges because the changes are significant and local. The plot also shows changes that are not edges, because they violate part of the definition. The changes due to noise are not edges even though the changes are local, because the changes are not significant. The changes due to shading, such as the ramp on the right side of the plot, are not edges even though the changes are significant, because the changes are not local. Real images are very noisy. It is difficult to develop an edge detection operator that reliably finds step edges and is immune to noise.

Before we discuss important considerations in edge detection operators, some terms must be carefully defined.

Definition 5.1 *An edge point is a point in an image with coordinates $[i, j]$ at the location of a significant local intensity change in the image.*

Definition 5.2 *An edge fragment corresponds to the i and j coordinates of an edge and the edge orientation θ, which may be the gradient angle.*

Definition 5.3 *An edge detector is an algorithm that produces a set of edges (edge points or edge fragments) from an image.*

Definition 5.4 *A contour is a list of edges or the mathematical curve that models the list of edges.*

Definition 5.5 *Edge linking is the process of forming an ordered list of edges from an unordered list. By convention, edges are ordered by traversal in a clockwise direction.*

Definition 5.6 *Edge following is the process of searching the (filtered) image to determine contours.*

The coordinates of an edge point may be the integer row and column indices of the pixel where the edge was detected, or the coordinates of the edge location at subpixel resolution. The edge coordinates may be in the coordinate system of the original image, but more likely are in the coordinate system of the image produced by the edge detection filter since filtering may translate or scale image coordinates. An edge fragment may be conceptualized as a small line segment about the size of a pixel, or as a point with an orientation attribute. The term *edge* is commonly used for either edge points or edge fragments.

The edge set produced by an edge detector can be partitioned into two subsets: correct edges, which correspond to edges in the scene, and false edges, which do not correspond to edges in the scene. A third set of edges can be defined as those edges in the scene that should have been detected. This is the set of missing edges. The false edges are called false positives, and the missing edges are called false negatives.

The difference between edge linking and edge following is that edge linking takes as input an unordered set of edges produced by an edge detector and forms an ordered list of edges. Edge following takes as input an image and produces an ordered list of edges. Edge detection uses local information to decide if a pixel is an edge, while edge following can use global information.

5.1   Gradient

Edge detection is essentially the operation of detecting significant local changes in an image. In one dimension, a step edge is associated with a local peak in the first derivative. The gradient is a measure of change in a function, and an image can be considered to be an array of samples of some continuous function of image intensity. By analogy, significant changes in the gray values in an image can be detected by using a discrete approximation to the gradient. The gradient is the two-dimensional equivalent of the first derivative and is defined as the *vector*

$$\mathbf{G}[f(x,y)] = \begin{bmatrix} G_x \\ G_y \end{bmatrix} = \begin{bmatrix} \frac{\partial f}{\partial x} \\ \frac{\partial f}{\partial y} \end{bmatrix}. \tag{5.1}$$

There are two important properties associated with the gradient: (1) the vector $\mathbf{G}[f(x,y)]$ points in the direction of the maximum rate of increase of the function $f(x,y)$, and (2) the magnitude of the gradient, given by

$$G[f(x,y)] = \sqrt{G_x^2 + G_y^2}, \tag{5.2}$$

equals the maximum rate of increase of $f(x,y)$ per unit distance in the direction G. It is common practice, however, to approximate the gradient magnitude by absolute values:

$$G[f(x,y)] \approx |G_x| + |G_y| \tag{5.3}$$

or

$$G[f(x,y)] \approx \max(|G_x|, |G_y|). \tag{5.4}$$

From vector analysis, the *direction* of the gradient is defined as:

$$\alpha(x,y) = \tan^{-1}\left(\frac{G_y}{G_x}\right) \tag{5.5}$$

where the angle α is measured with respect to the x axis.

Note that the magnitude of the gradient is actually independent of the direction of the edge. Such operators are called *isotropic operators*.

Numerical Approximation

For digital images, the derivatives in Equation 5.1 are approximated by differences. The simplest gradient approximation is

$$G_x \cong f[i, j+1] - f[i, j] \tag{5.6}$$

$$G_y \cong f[i,j] - f[i+1,j]. \tag{5.7}$$

Remember that j corresponds to the x direction and i to the negative y direction. These can be implemented with simple convolution masks as shown below:

$$G_x = \boxed{\begin{array}{|c|c|} -1 & 1 \end{array}} \qquad\qquad G_y = \begin{array}{|c|} \hline 1 \\ \hline -1 \\ \hline \end{array} \tag{5.8}$$

When computing an approximation to the gradient, it is critical that the x and y partial derivatives be computed at exactly the same position in space. However, using the above approximations, G_x is actually the approximation to the gradient at the interpolated point $[i, j + \frac{1}{2}]$ and G_y at $[i + \frac{1}{2}, j]$. For this reason, 2×2 first differences, rather than 2×1 and 1×2 masks, are often used for the x and y partial derivatives:

$$G_x = \begin{array}{|c|c|} \hline -1 & 1 \\ \hline -1 & 1 \\ \hline \end{array} \qquad\qquad G_y = \begin{array}{|c|c|} \hline 1 & 1 \\ \hline -1 & -1 \\ \hline \end{array} \tag{5.9}$$

Now, the positions about which the gradients in the x and y directions are calculated are the same. This point lies between all four pixels in the 2×2 neighborhood at the interpolated point $[i + \frac{1}{2}, j + \frac{1}{2}]$. This fact may lead to some confusion. Therefore, an alternative approach is to use a 3×3 neighborhood and calculate the gradient about the center pixel. These methods are discussed in Section 5.2.

5.2   Steps in Edge Detection

Algorithms for edge detection contain three steps:

Filtering: Since gradient computation based on intensity values of only two points are susceptible to noise and other vagaries in discrete computations, filtering is commonly used to improve the performance of an edge detector with respect to noise. However, there is a trade-off between edge strength and noise reduction. More filtering to reduce noise results in a loss of edge strength.

Enhancement: In order to facilitate the detection of edges, it is essential to determine changes in intensity in the neighborhood of a point. Enhancement emphasizes pixels where there is a significant change in local intensity values and is usually performed by computing the gradient magnitude.

Detection: We only want points with strong edge content. However, many points in an image have a nonzero value for the gradient, and not all of these points are edges for a particular application. Therefore, some method should be used to determine which points are edge points. Frequently, thresholding provides the criterion used for detection.

Examples at the end of this section will clearly illustrate each of these steps using various edge detectors. Many edge detection algorithms include a fourth step:

Localization: The location of the edge can be estimated with subpixel resolution if required for the application. The edge orientation can also be estimated.

It is important to note that detection merely indicates that an edge is present near a pixel in an image, but does not necessarily provide an accurate estimate of edge location or orientation. The errors in edge detection are errors of misclassification: false edges and missing edges. The errors in edge estimation are modeled by probability distributions for the location and orientation estimates. We distinguish between edge detection and estimation because these steps are performed by different calculations and have different error models.

Many edge detectors have been developed in the last two decades. Here we will discuss some commonly used edge detectors. As will be clear, edge detectors differ in use of the computational approach in one or more of the above three steps. We will discuss the implications of these steps after we have discussed the edge detectors.

5.2.1   Roberts Operator

The Roberts cross operator provides a simple approximation to the gradient magnitude:

$$G\left[f[i,j]\right] = |f[i,j] - f[i+1,j+1]| + |f[i+1,j] - f[i,j+1]|. \qquad (5.10)$$

Using convolution masks, this becomes

$$G\left[f[i,j]\right] = |G_x| + |G_y| \tag{5.11}$$

where G_x and G_y are calculated using the following masks:

$$
G_x = \begin{array}{|c|c|} \hline 1 & 0 \\ \hline 0 & -1 \\ \hline \end{array}
\qquad
G_y = \begin{array}{|c|c|} \hline 0 & -1 \\ \hline 1 & 0 \\ \hline \end{array}
\tag{5.12}
$$

As with the previous 2×2 gradient operator, the differences are computed at the interpolated point $[i + \frac{1}{2}, j + \frac{1}{2}]$. The Roberts operator is an approximation to the continuous gradient at that point and not at the point $[i, j]$ as might be expected. The results of Roberts edge detector are shown in the figures at the end of this section.

5.2.2 Sobel Operator

As mentioned previously, a way to avoid having the gradient calculated about an interpolated point between pixels is to use a 3×3 neighborhood for the gradient calculations. Consider the arrangement of pixels about the pixel $[i, j]$ shown in Figure 5.3. The Sobel operator is the magnitude of the gradient computed by

$$M = \sqrt{s_x^2 + s_y^2}, \tag{5.13}$$

where the partial derivatives are computed by

$$
\begin{aligned}
s_x &= (a_2 + ca_3 + a_4) - (a_0 + ca_7 + a_6) \tag{5.14}\\
s_y &= (a_0 + ca_1 + a_2) - (a_6 + ca_5 + a_4) \tag{5.15}
\end{aligned}
$$

with the constant $c = 2$.

Like the other gradient operators, s_x and s_y can be implemented using convolution masks:

$$
s_x = \begin{array}{|c|c|c|} \hline -1 & 0 & 1 \\ \hline -2 & 0 & 2 \\ \hline -1 & 0 & 1 \\ \hline \end{array}
\qquad
s_y = \begin{array}{|c|c|c|} \hline 1 & 2 & 1 \\ \hline 0 & 0 & 0 \\ \hline -1 & -2 & -1 \\ \hline \end{array}
\tag{5.16}
$$

Note that this operator places an emphasis on pixels that are closer to the center of the mask. The figures at the end of this section show the performance of this operator. The Sobel operator is one of the most commonly used edge detectors.

a_0	a_1	a_2
a_7	$[i,j]$	a_3
a_6	a_5	a_4

Figure 5.3: The labeling of neighborhood pixels used to explain the Sobel and Prewitt operators [186].

5.2.3   Prewitt Operator

The Prewitt operator uses the same equations as the Sobel operator, except that the constant $c = 1$. Therefore:

$$s_x = \begin{array}{|c|c|c|} \hline -1 & 0 & 1 \\ \hline -1 & 0 & 1 \\ \hline -1 & 0 & 1 \\ \hline \end{array} \qquad s_y = \begin{array}{|c|c|c|} \hline 1 & 1 & 1 \\ \hline 0 & 0 & 0 \\ \hline -1 & -1 & -1 \\ \hline \end{array} \qquad (5.17)$$

Note that, unlike the Sobel operator, this operator does not place any emphasis on pixels that are closer to the center of the masks. The performance of this edge detector is also shown in the figures at the end of this section.

5.2.4   Comparison

We now compare the different edge detectors discussed so far. The comparisons will be presented according to the first three steps described at the beginning of this section: *filtering, enhancement,* and *detection*. The estimation step will not be shown here. In addition, we will give results of edge detection on noisy images for two specific cases—one utilizing the filtering step and one omitting the filtering step. Results of edge detection using varying amounts of filtering will also be given.

   For each of the following four figures, the sum of the absolute values of the x and y components of the gradient was used as the gradient magnitude (Equation 5.3). The filter used was the 7×7 Gaussian filter described in the previous chapter. The threshold values used for detection are given in the captions.

Figure 5.4 shows results of all the edge detection methods discussed so far, from the simple 1×2 gradient approximation up to the Prewitt operator. Figure 5.5 shows the results of the edge detectors when the filtering step is omitted. The next set of images (Figure 5.6) shows the results of edge detection on the same image now with additive Gaussian noise, $\sigma = 12$. The filter used was the same Gaussian filter as used in the previous figure. The final series of images (Figure 5.7) shows the results of edge detection on the same noisy image. However, for these images the filtering step was again omitted. Note the many false edges detected as a result of the noise.

5.3   Second Derivative Operators

The edge detectors discussed earlier computed the first derivative and, if it was above a threshold, the presence of an edge point was assumed. This results in detection of too many edge points. (Notice the thick lines after thresholding in Figures 5.4 to 5.7.) A better approach would be to find only the points that have local maxima in gradient values and consider them edge points, as shown in Figure 5.8. This means that at edge points, there will be a peak in the first derivative and, equivalently, there will be a zero crossing in the second derivative. Thus, edge points may be detected by finding the zero crossings of the second derivative of the image intensity.

There are two operators in two dimensions that correspond to the second derivative: the Laplacian and second directional derivative.

5.3.1   Laplacian Operator

The second derivative of a smoothed step edge is a function that crosses zero at the location of the edge (see Figure 5.8). The Laplacian is the two-dimensional equivalent of the second derivative. The formula for the Laplacian of a function $f(x, y)$ is

$$\nabla^2 f = \frac{\partial^2 f}{\partial x^2} + \frac{\partial^2 f}{\partial y^2}. \tag{5.18}$$

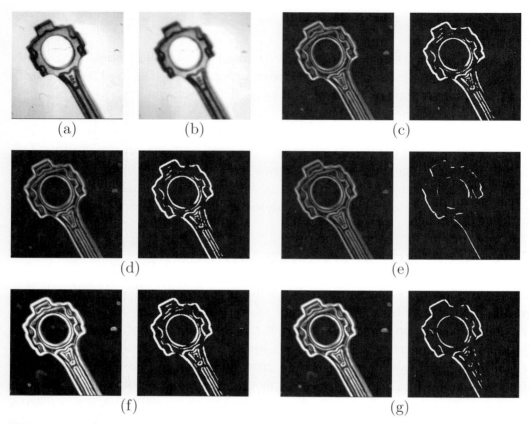

Figure 5.4: A comparison of various edge detectors. (a) Original image. (b) Filtered image. (c) Simple gradient using 1×2 and 2×1 masks, $T = 32$. (d) Gradient using 2×2 masks, $T = 64$. (e) Roberts cross operator, $T = 64$. (f) Sobel operator, $T = 225$. (g) Prewitt operator, $T = 225$.

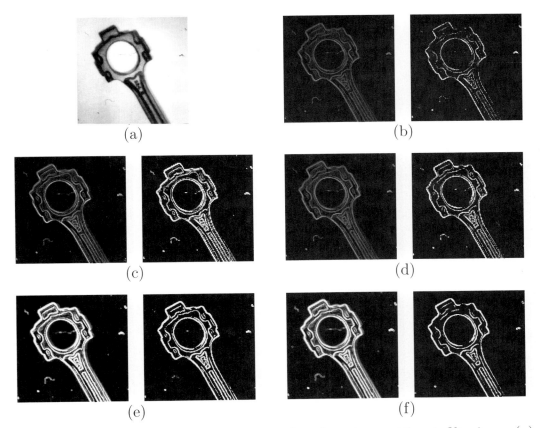

Figure 5.5: A comparison of various edge detectors without filtering. (a) Original image. (b) Simple gradient using 1×2 and 2×1 masks, $T = 64$. (c) Gradient using 2×2 masks, $T = 64$. (d) Roberts cross operator, $T = 64$. (e) Sobel operator, $T = 225$. (f) Prewitt operator, $T = 225$.

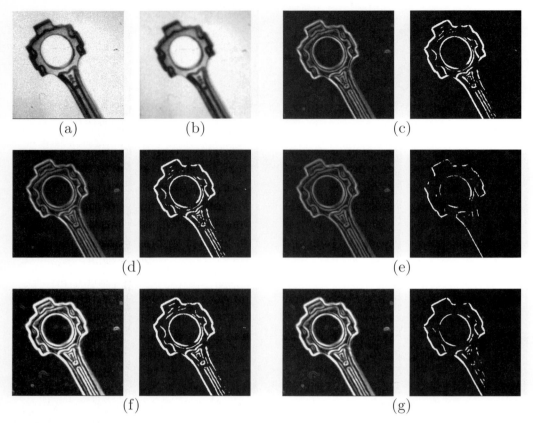

Figure 5.6: A comparison of various edge detectors on a noisy image. (a) Noisy image. (b) Filtered image. (c) Simple gradient using 1×2 and 2×1 masks, $T = 32$. (d) Gradient using 2×2 masks, $T = 64$. (e) Roberts cross operator, $T = 64$. (f) Sobel operator, $T = 225$. (g) Prewitt operator, $T = 225$.

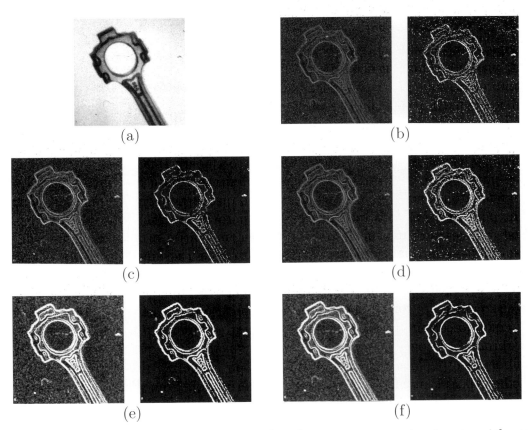

Figure 5.7: A comparison of various edge detectors on a noisy image without filtering. (a) Noisy image. (b) Simple gradient using 1×2 and 2×1 masks, $T = 64$. (c) Gradient using 2×2 masks, $T = 128$. (d) Roberts cross operator, $T = 64$. (e) Sobel operator, $T = 225$. (f) Prewitt operator, $T = 225$.

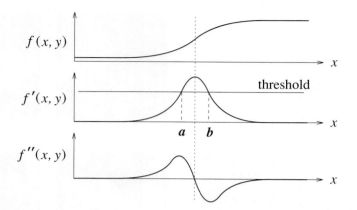

Figure 5.8: If a threshold is used for detection of edges, all points between a and b will be marked as edge pixels. However, by removing points that are *not* a local maximum in the first derivative, edges can be detected more accurately. This local maximum in the first derivative corresponds to a zero crossing in the second derivative.

The second derivatives along the x and y directions are approximated using difference equations:

$$\frac{\partial^2 f}{\partial x^2} = \frac{\partial G_x}{\partial x} \tag{5.19}$$

$$= \frac{\partial \left(f[i, j+1] - f[i, j] \right)}{\partial x} \tag{5.20}$$

$$= \frac{\partial f[i, j+1]}{\partial x} - \frac{\partial f[i, j]}{\partial x} \tag{5.21}$$

$$= (f[i, j+2] - f[i, j+1]) - (f[i, j+1] - f[i, j]) \tag{5.22}$$

$$= f[i, j+2] - 2f[i, j+1] + f[i, j]. \tag{5.23}$$

However, this approximation is centered about the pixel $[i, j+1]$. Therefore, by replacing j with $j-1$, we obtain

$$\frac{\partial^2 f}{\partial x^2} = f[i, j+1] - 2f[i, j] + f[i, j-1], \tag{5.24}$$

which is the desired approximation to the second partial derivative centered about $[i, j]$. Similarly,

$$\frac{\partial^2 f}{\partial y^2} = f[i+1, j] - 2f[i, j] + f[i-1, j]. \tag{5.25}$$

By combining these two equations into a single operator, the following mask can be used to approximate the Laplacian:

$$\nabla^2 \approx \begin{array}{|c|c|c|} \hline 0 & 1 & 0 \\ \hline 1 & -4 & 1 \\ \hline 0 & 1 & 0 \\ \hline \end{array} \tag{5.26}$$

Sometimes it is desired to give more weight to the center pixels in the neighborhood. An approximation to the Laplacian which does this is

$$\nabla^2 \approx \begin{array}{|c|c|c|} \hline 1 & 4 & 1 \\ \hline 4 & -20 & 4 \\ \hline 1 & 4 & 1 \\ \hline \end{array} \tag{5.27}$$

The Laplacian operator signals the presence of an edge when the output of the operator makes a transition through zero. Trivial zeros (uniform zero regions) are ignored. In principle, the zero crossing location can be estimated to subpixel resolution using linear interpolation, but the result may be inaccurate due to noise.

Consider the example shown in Figure 5.9. This figure shows the result of the Laplacian on an image with a simple step edge. A single row of the resulting image is:

$$\begin{array}{|c|c|c|c|c|c|c|c|} \hline 0 & 0 & 0 & 6 & -6 & 0 & 0 & 0 \\ \hline \end{array}$$

In this example, the zero crossing, corresponding to the edge in the original image, lies halfway between the two center pixels. The edge should be marked at either the pixel to the left or the pixel to the right of the edge, as long as it is marked consistently throughout the image. In most cases, however, the zero crossing rarely lies exactly between two pixels, and the actual edge

2	2	2	2	2	8	8	8	8	8
2	2	2	2	2	8	8	8	8	8
2	2	2	2	2	8	8	8	8	8
2	2	2	2	2	8	8	8	8	8
2	2	2	2	2	8	8	8	8	8
2	2	2	2	2	8	8	8	8	8

A sample image containing a vertical step edge.

0	0	0	6	−6	0	0	0
0	0	0	6	−6	0	0	0
0	0	0	6	−6	0	0	0
0	0	0	6	−6	0	0	0

Figure 5.9: The response of the Laplacian to a vertical step edge.

location must be determined by interpolating the pixel values on either side of the zero crossing.

Now consider the example in Figure 5.10. This figure shows the response of the Laplacian to a ramp edge. A single row of the output of the Laplacian is

0	0	0	3	0	−3	0	0

The zero crossing directly corresponds to a pixel in the image. Again, this is an ideal situation, and the actual edge location should be determined by interpolation.

5.3.2   Second Directional Derivative

The second directional derivative is the second derivative computed in the direction of the gradient. The operator is implemented using the formula

$$\frac{\partial^2}{\partial n^2} = \frac{f_x^2 f_{xx} + 2 f_x f_y f_{xy} + f_y^2 f_{yy}}{f_x^2 + f_y^2} \tag{5.28}$$

2	2	2	2	2	5	8	8	8	8
2	2	2	2	2	5	8	8	8	8
2	2	2	2	2	5	8	8	8	8
2	2	2	2	2	5	8	8	8	8
2	2	2	2	2	5	8	8	8	8
2	2	2	2	2	5	8	8	8	8

A sample image containing a vertical ramp edge.

0	0	0	3	0	-3	0	0
0	0	0	3	0	-3	0	0
0	0	0	3	0	-3	0	0
0	0	0	3	0	-3	0	0

Figure 5.10: The response of the Laplacian to a vertical ramp edge.

The Laplacian and second directional derivative operators are not used frequently in machine vision since any operator involving two derivatives is affected by noise more than an operator involving a single derivative. Even very small local peaks in the first derivative will result in zero crossings in the second derivative. To avoid the effect of noise, powerful filtering methods must be used. In the following section, we discuss an approach which combines Gaussian filtering with the second derivative for edge detection.

5.4   Laplacian of Gaussian

As mentioned above, edge points detected by finding the zero crossings of the second derivative of the image intensity are very sensitive to noise. Therefore, it is desirable to filter out the noise before edge enhancement. To do this, the *Laplacian of Gaussian* (LoG), due to Marr and Hildreth [164], combines Gaussian filtering with the Laplacian for edge detection.

The fundamental characteristics of the Laplacian of Gaussian edge detector are

1. The smoothing filter is a Gaussian.

2. The enhancement step is the second derivative (Laplacian in two dimensions).

3. The detection criterion is the presence of a zero crossing in the second derivative with a corresponding large peak in the first derivative.

4. The edge location can be estimated with subpixel resolution using linear interpolation.

In this approach, an image should first be convolved with a Gaussian filter. (We discuss Gaussian filtering in more detail in Section 5.6.) This step smooths an image and reduces noise. Isolated noise points and small structures will be filtered out. Since the smoothing will result in *spreading* of edges, the edge detector considers as edges only those pixels that have locally maximum gradient. This is achieved by using zero crossings of the second derivative. The Laplacian is used as the approximation of the second derivative in 2-D because it is an isotropic operator. To avoid detection of insignificant edges, only the zero crossings whose corresponding first derivative is above some threshold are selected as edge points.

The output of the LoG operator, $h(x, y)$, is obtained by the convolution operation

$$h(x, y) = \nabla^2 \left[(g(x, y) \star f(x, y)) \right]. \tag{5.29}$$

Using the derivative rule for convolution,

$$h(x, y) = \left[\nabla^2 g(x, y) \right] \star f(x, y), \tag{5.30}$$

where

$$\nabla^2 g(x, y) = \left(\frac{x^2 + y^2 - 2\sigma^2}{\sigma^4} \right) e^{-\frac{(x^2 + y^2)}{2\sigma^2}} \tag{5.31}$$

is commonly called the *Mexican hat* operator (shown in Figure 5.11). Thus, the following two methods are mathematically equivalent:

1. Convolve the image with a Gaussian smoothing filter and compute the Laplacian of the result.

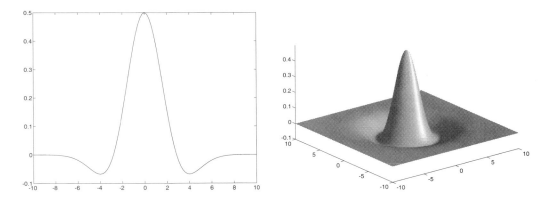

Figure 5.11: The *inverted* Laplacian of Gaussian function, $\sigma = 2$, in one and two dimensions.

2. Convolve the image with the linear filter that is the Laplacian of the Gaussian filter.

If the first method is adopted, Gaussian smoothing masks such as those described in Section 4.5.5 may be used. Typical masks to directly implement the LoG are given in Figure 5.12. In Figure 5.13 we show the result of applying the Laplacian of Gaussian operator and detection of zero crossings. For a discussion on efficient methods to implement the Laplacian of Gaussian, see [117].

At the beginning of Section 5.2, we stated that filtering (usually smoothing), enhancement, and detection were the three steps in edge detection. This is still true for edge detection using the Laplacian of Gaussian. Smoothing is performed with a Gaussian filter, enhancement is done by transforming edges into zero crossings, and detection is done by detecting the zero crossings.

It can be shown that the slope of the zero crossing depends on the contrast of the change in image intensity across the edge. The problem of combining edges obtained by applying different-size operators to images remains. In the above approach, edges at a particular resolution are obtained. To obtain real edges in an image, it may be necessary to combine information from operators at several filter sizes.

5 × 5 Laplacian of Gaussian mask

0	0	−1	0	0
0	−1	−2	−1	0
−1	−2	16	−2	−1
0	−1	−2	−1	0
0	0	−1	0	0

17 × 17 Laplacian of Gaussian mask

0	0	0	0	0	0	−1	−1	−1	−1	−1	0	0	0	0	0	0
0	0	0	0	−1	−1	−1	−1	−1	−1	−1	−1	−1	0	0	0	0
0	0	−1	−1	−1	−2	−3	−3	−3	−3	−3	−2	−1	−1	−1	0	0
0	0	−1	−1	−2	−3	−3	−3	−3	−3	−3	−3	−2	−1	−1	0	0
0	−1	−1	−2	−3	−3	−3	−2	−3	−2	−3	−3	−3	−2	−1	−1	0
0	−1	−2	−3	−3	−3	0	2	4	2	0	−3	−3	−3	−2	−1	0
−1	−1	−3	−3	−3	0	4	10	12	10	4	0	−3	−3	−3	−1	−1
−1	−1	−3	−3	−2	2	10	18	21	18	10	2	−2	−3	−3	−1	−1
−1	−1	−3	−3	−3	4	12	21	24	21	12	4	−3	−3	−3	−1	−1
−1	−1	−3	−3	−2	2	10	18	21	18	10	2	−2	−3	−3	−1	−1
−1	−1	−3	−3	−3	0	4	10	12	10	4	0	−3	−3	−3	−1	−1
0	−1	−2	−3	−3	−3	0	2	4	2	0	−3	−3	−3	−2	−1	0
0	−1	−1	−2	−3	−3	−3	−2	−3	−2	−3	−3	−3	−2	−1	−1	0
0	0	−1	−1	−2	−3	−3	−3	−3	−3	−3	−3	−2	−1	−1	0	0
0	0	−1	−1	−1	−2	−3	−3	−3	−3	−3	−2	−1	−1	−1	0	0
0	0	0	0	−1	−1	−1	−1	−1	−1	−1	−1	−1	0	0	0	0
0	0	0	0	0	0	−1	−1	−1	−1	−1	0	0	0	0	0	0

Figure 5.12: Some useful Laplacian of Gaussian masks [146].

Figure 5.13: Results of the Laplacian of Gaussian edge detector.

Scale Space

The Gaussian smoothing operation results in the blurring of edges and other sharp discontinuities in an image. The amount of blurring depends on the value of σ. A larger σ results in better noise filtering but at the same time loses important edge information, which affects the performance of an edge detector. If a small filter is used, there is likely to be more noise due to insufficient averaging. For large filters, edges which are close to each other may get merged by smoothing and may be detected as only a single edge. In general, small filters result in too many noise points and large filters result in dislocation of edges and even false edges. The exact size of the filter cannot be determined without knowing the size and location of objects in an image.

Many approaches are being developed which apply filtering masks of multiple sizes and then analyze the behavior of edges at these different *scales* of filtering. The basic idea in these approaches is to exploit the fact that at higher scales, larger filtering masks result in robust but displaced edges. The location of these edges can be determined at smaller scales.

5.5   Image Approximation

An image is an array of samples of a continuous function. Most ideas about images are discussed in the continuous domain and then the desired properties are computed using discrete approximations. If we can estimate the continuous function from which the image samples were taken, then we may be able to compute image properties from the estimated function. This may allow the computation of edge location to subpixel precision.

Let

$$z = f(x, y) \tag{5.32}$$

be a continuous image intensity function like the one shown in Figure 5.14. The task is to reconstruct this continuous function from the sampled gray values. For complex images, the continuous image intensity function may contain extremely high powers of x and y. This makes reconstruction of the original function extremely difficult, if not impossible. Therefore, we try to model the image as a simple piecewise analytical function. Now the task becomes the reconstruction of the individual piecewise functions, or *facets*. In other words, try to find simple functions which best approximate the intensity values *only* in the local neighborhood of each pixel. This is illustrated in Figure 5.15. This approximation is called the *facet model* [102]. Figure 5.16 shows the coordinate system for the facet model using 5×5 neighborhoods.

The continuous image intensity function is approximated locally at every pixel in the image. For an $n \times m$ image, you will obtain $n \cdot m$ approximating functions, each valid only about a specific pixel in the image. These functions, and not the pixel values, are used to locate edges in the image.

A variety of analytical functions of varying complexity can be used to approximate image intensities. Often for simple images, piecewise constant or piecewise bilinear functions are adequate approximations of the intensity values. However, for images with more complex regions, biquadratic, bicubic, and even higher-power functions are used. For this example, we will model the image neighborhood as the following bicubic polynomial:

$$\begin{aligned} f(x, y) = \; & k_1 + k_2 x + k_3 y + k_4 x^2 + k_5 xy + k_6 y^2 \\ & + k_7 x^3 + k_8 x^2 y + k_9 xy^2 + k_{10} y^3 \end{aligned} \tag{5.33}$$

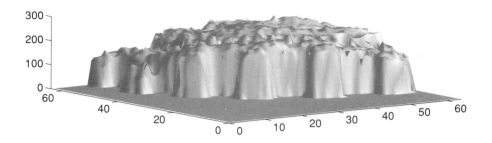

Figure 5.14: A graphical representation of the continuous image intensity function.

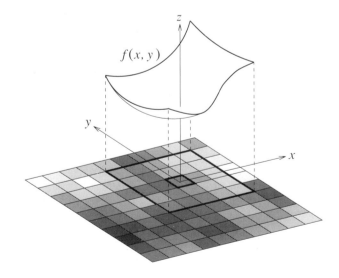

Figure 5.15: An illustration of an approximated function within a 5×5 neighborhood.

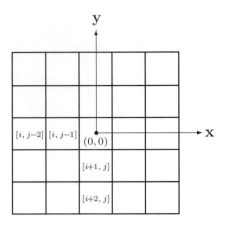

Figure 5.16: An example of the coordinate system for the facet model using a 5 × 5 neighborhood. The continuous intensity function is approximated at every pixel location *only* within this neighborhood. The array indices are marked on the pixels for clarity. Note that pixel $[i, j]$ lies in the center of the neighborhood.

where x and y are coordinates with respect to $(0, 0)$, the center of the image plane neighborhood that is being approximated (see Figure 5.16).

The goal now is to calculate the coefficients k_i in Equation 5.33 for each approximating function. Use least-squares methods to compute the coefficients k_i using singular-value decomposition, or, if you are using a 5 × 5 neighborhood, use the masks shown in Figure 5.17 to directly compute the coefficients for the bicubic approximation.[1]

To detect edges, use the fact that edge points occur at relative extrema in the first directional derivative of the function approximating the image intensity in the neighborhood of a pixel. The presence of relative extrema in the first derivative will result in a zero crossing in the second derivative in the direction of the first derivative.

The first derivative in the direction θ is given by

$$f'_\theta(x, y) = \frac{\partial f}{\partial x} \cos \theta + \frac{\partial f}{\partial y} \sin \theta. \qquad (5.34)$$

---

[1]For a complete discussion of the construction of these masks, refer to [103, chap. 8].

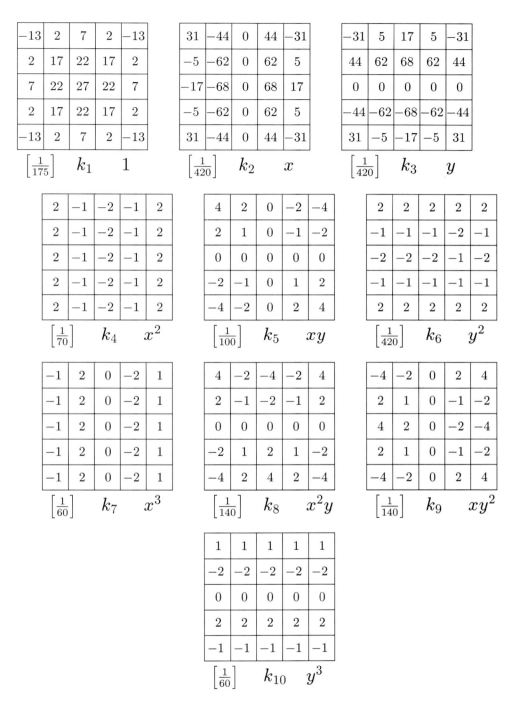

Figure 5.17: Masks for computing the coefficients of the bicubic approximation [103].

The second directional derivative in the direction θ is given by

$$f_\theta''(x, y) = \frac{\partial^2 f}{\partial x^2} \cos^2 \theta + 2\frac{\partial^2 f}{\partial x \partial y} \cos \theta \sin \theta + \frac{\partial^2 f}{\partial y^2} \sin^2 \theta. \tag{5.35}$$

Since the local image intensity was approximated by a bicubic polynomial, the angle θ may be chosen to be the angle of the approximating plane. This will result in

$$\sin \theta = \frac{k_3}{\sqrt{k_2^2 + k_3^2}} \tag{5.36}$$

$$\cos \theta = \frac{k_2}{\sqrt{k_2^2 + k_3^2}}. \tag{5.37}$$

At a point (x_0, y_0), the second directional derivative in the direction θ is given by

$$\begin{aligned} f_\theta''(x_0, y_0) = {} & 2\left(3k_7 \cos^2 \theta + 2k_8 \sin \theta \cos \theta + k_9 \sin^2 \theta\right) x_0 \\ & + 2\left(k_8 \cos^2 \theta + 2k_9 \sin \theta \cos \theta + 3k_{10} \sin^2 \theta\right) y_0 \\ & + 2\left(k_4 \cos^2 \theta + k_5 \sin \theta \cos \theta + k_6 \sin^2 \theta\right). \end{aligned} \tag{5.38}$$

Since we are considering points only on the line in the direction θ, $x_0 = \rho \cos \theta$ and $y_0 = \rho \sin \theta$. Substituting this in the above, we get

$$\begin{aligned} f_\theta''(x_0, y_0) = {} & 6\left(k_{10} \sin^3 \theta + k_9 \sin^2 \theta \cos \theta + k_8 \sin \theta \cos^2 \theta + k_7 \cos^3 \theta\right) \rho \\ & + 2\left(k_6 \sin^2 \theta + k_5 \sin \theta \cos \theta + k_4 \cos^2 \theta\right) \tag{5.39} \\ = {} & A\rho + B. \tag{5.40} \end{aligned}$$

Thus, there is an edge at (x_0, y_0) in the image if for some ρ, $|\rho| < \rho_0$ where ρ_0 is the length of the side of a pixel,

$$f_\theta''(x_0, y_0; \rho) = 0 \tag{5.41}$$

and

$$f_\theta'(x_0, y_0; \rho) \neq 0. \tag{5.42}$$

In other words, mark the pixel as an edge pixel if the location of the edge falls within the boundaries of the pixel (see Figure 5.18). However, do *not* mark the pixel as an edge pixel if the point lies outside the pixel boundaries as shown in Figure 5.19. The results of this operator are shown in Figure 5.20.

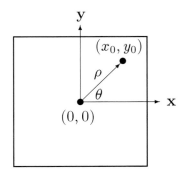

Figure 5.18: An enlarged view of the pixel at the center of the approximated function. (x_0, y_0) is the location of the edge as determined by Equations 5.41 and 5.42. This pixel will be marked as an edge pixel since the location of the edge falls within its boundaries.

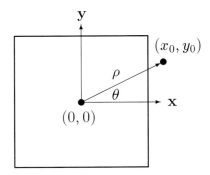

Figure 5.19: An enlarged view of the pixel at the center of the approximated function. This pixel will *not* be marked as an edge pixel since the location of the edge does not fall within the pixel boundaries.

Figure 5.20: Edges obtained with facet model edge detector.

5.6   Gaussian Edge Detection

The essential idea in detecting step edges is to find points in the sampled image that have locally large gradient magnitudes. Much of the research work in step edge detection is devoted to finding numerical approximations to the gradient that are suitable for use with real images. The step edges in real images are not perfectly sharp since the edges are smoothed by the low-pass filtering inherent in the optics of the camera lens and the bandwidth limitations in the camera electronics. The images are also severely corrupted by noise from the camera and unwanted detail in the scene. An approximation to the image gradient must be able to satisfy two conflicting requirements: (1) the approximation must suppress the effects of noise, and (2) the approximation must locate the edge as accurately as possible. There is a trade-off between noise suppression and localization. An edge detection operator can reduce noise by smoothing the image, but this will add uncertainty to the location of the edge; or the operator can have greater sensitivity to the presence of edges, but this will increase the sensitivity of the operator to noise. The type of linear operator that provides the best compromise between noise immunity and localization, while retaining the advantages of Gaussian filtering, is the first derivative of a Gaussian. This operator corresponds to smoothing an image with a Gaussian function and then computing the gradient. The gradient can be numerically approximated by using the standard finite-difference approximation for the first partial derivatives in the x and y directions listed in Section 5.1. The operator that is the combination of a Gaussian smoothing filter and a gradient approximation is not rotationally symmetric. The operator is symmetric along the edge and antisymmetric perpendicular to the edge (along the line of the gradient). This means that the operator is sensitive to the edge in the direction of steepest change, but is insensitive to the edge and acts as a smoothing operator in the direction along the edge.

5.6.1 Canny Edge Detector

The Canny edge detector is the first derivative of a Gaussian and closely approximates the operator that optimizes the product of signal-to-noise ratio and localization. The Canny edge detection algorithm is summarized by the following notation. Let $I[i,j]$ denote the image. The result from convolving the image with a Gaussian smoothing filter using separable filtering is an array of smoothed data,

$$S[i,j] = G[i,j;\sigma] \star I[i,j], \tag{5.43}$$

where σ is the spread of the Gaussian and controls the degree of smoothing.

The gradient of the smoothed array $S[i,j]$ can be computed using the 2×2 first-difference approximations (Section 5.1) to produce two arrays $P[i,j]$ and $Q[i,j]$ for the x and y partial derivatives:

$$
\begin{aligned}
P[i,j] &\approx (S[i,j+1] - S[i,j] \\
&\quad + S[i+1,j+1] - S[i+1,j])/2
\end{aligned}
\tag{5.44}
$$

$$
\begin{aligned}
Q[i,j] &\approx (S[i,j] - S[i+1,j] \\
&\quad + S[i,j+1] - S[i+1,j+1])/2.
\end{aligned}
\tag{5.45}
$$

The finite differences are averaged over the 2×2 square so that the x and y partial derivatives are computed at the same point in the image. The magnitude and orientation of the gradient can be computed from the standard formulas for rectangular-to-polar conversion:

$$M[i,j] = \sqrt{P[i,j]^2 + Q[i,j]^2} \tag{5.46}$$

$$\theta[i,j] = \arctan(Q[i,j], P[i,j]), \tag{5.47}$$

where the arctan function takes two arguments and generates an angle over the entire circle of possible directions. These functions must be computed efficiently, preferably without using floating-point arithmetic. It is possible to compute the gradient magnitude and orientation from the partial derivatives by table lookup. The arctangent can be computed using mostly fixed-point arithmetic[2] with a few essential floating-point calculations performed in software using integer and fixed-point arithmetic [59, chap. 11]. Sedgewick [218, p. 353] provides an algorithm for an integer approximation to the gradient angle that may be good enough for many applications.

Nonmaxima Suppression

The magnitude image array $M[i, j]$ will have large values where the image gradient is large, but this is not sufficient to identify the edges, since the problem of finding locations in the image array where there is rapid change has merely been transformed into the problem of finding locations in the magnitude array $M[i, j]$ that are local maxima. To identify edges, the broad ridges in the magnitude array must be thinned so that only the magnitudes at the points of greatest local change remain. This process is called nonmaxima suppression, which in this case results in thinned edges.

Nonmaxima suppression thins the ridges of gradient magnitude in $M[i, j]$ by suppressing all values along the line of the gradient that are not peak values of a ridge. The algorithm begins by reducing the angle of the gradient $\theta[i, j]$ to one of the four sectors shown in Figure 5.21,

$$\zeta[i, j] = \text{Sector}(\theta[i, j]). \tag{5.48}$$

The algorithm passes a 3×3 neighborhood across the magnitude array $M[i, j]$. At each point, the center element $M[i, j]$ of the neighborhood is compared with its two neighbors along the line of the gradient given by the sector value $\zeta[i, j]$ at the center of the neighborhood. If the magnitude array value $M[i, j]$ at the center is not greater than both of the neighbor magnitudes along the gradient line, then $M[i, j]$ is set to zero. This process thins the broad ridges of gradient magnitude in $M[i, j]$ into ridges that are only one pixel wide. The

---

[2]In this context, fixed-point arithmetic is like integer arithmetic except that the number carries an implicit scale factor that assumes that the binary point is to the left of the number. Fixed-point arithmetic can be implemented using integer arithmetic on many machines.

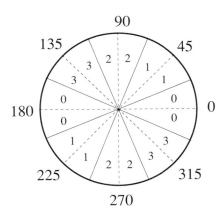

Figure 5.21: The partition of the possible gradient orientations into sectors for nonmaxima suppression is shown. There are four sectors, numbered 0 to 3, corresponding to the four possible combinations of elements in a 3×3 neighborhood that a line must pass through as it passes through the center of the neighborhood. The divisions of the circle of possible gradient line orientations are labeled in degrees.

values for the height of the ridge are retained in the nonmaxima-suppressed magnitude.

Let

$$N[i,j] = \text{nms}(M[i,j], \zeta[i,j]) \qquad (5.49)$$

denote the process of nonmaxima suppression. The nonzero values in $N[i,j]$ correspond to the amount of contrast at a step change in the image intensity. In spite of the smoothing performed as the first step in edge detection, the nonmaxima-suppressed magnitude image $N[i,j]$ will contain many false edge fragments caused by noise and fine texture. The contrast of the false edge fragments is small.

Thresholding

The typical procedure used to reduce the number of false edge fragments in the nonmaxima-suppressed gradient magnitude is to apply a threshold to $N[i,j]$. All values below the threshold are changed to zero. The result of applying a threshold to the nonmaxima-suppressed magnitude is an array of the edges detected in the image $I[i, j]$. There will still be some false edges because the threshold τ was too low (false positives), and portions of actual contours may be missing (false negatives) due to softening of the edge contrast by shadows or because the threshold τ was too high. Selecting the proper threshold is difficult and involves some trial and error. A more effective thresholding scheme uses two thresholds.

The double thresholding algorithm takes the nonmaxima-suppressed image, $N[i, j]$, and applies two thresholds τ_1 and τ_2, with $\tau_2 \approx 2\tau_1$, to produce two thresholded edge images $T_1[i, j]$ and $T_2[i, j]$. Since image T_2 was formed with a higher threshold, it will contain fewer false edges; but T_2 may have gaps in the contours (too many false negatives). The double thresholding algorithm links the edges in T_2 into contours. When it reaches the end of a contour, the algorithm looks in T_1 at the locations of the 8-neighbors for edges that can be linked to the contour. The algorithm continues to gather edges from T_1 until the gap has been bridged to an edge in T_2. The algorithm performs edge linking as a by-product of thresholding and resolves some of the problems with choosing a threshold. The Canny edge detection algorithm is outlined in Algorithm 5.1.

The edge detection algorithm presented in this section has been run on several test images. Figure 5.22 shows the image of a connecting rod. Figures 5.23 and 5.24 present the results of applying the edge detection algorithm summarized in this section to the test image in Figure 5.22. In Figure 5.23, a 7×7 Gaussian filter was used to smooth the image before computing the gradient; in Figure 5.24, a 31×31 Gaussian filter was used. The nonmaxima-suppressed gradient magnitude for the smaller filter size exhibits excellent fine detail in the edges but suffers from excessive unwanted edge fragments

Algorithm 5.1 Canny Edge Detection

1. *Smooth the image with a Gaussian filter.*

2. *Compute the gradient magnitude and orientation using finite-difference approximations for the partial derivatives.*

3. *Apply nonmaxima suppression to the gradient magnitude.*

4. *Use the double thresholding algorithm to detect and link edges.*

Figure 5.22: A test image of a connecting rod. The image was acquired by a Reticon 256×256 area CCD array camera.

due to noise and fine texture. For the larger filter size, there are fewer unwanted edge fragments, but much of the detail in the edges has been lost. This illustrates the trade-off between edge localization and noise immunity.

5.7 Subpixel Location Estimation

In many applications, it is necessary to estimate the location of an edge to better than the spacing between pixels (subpixel resolution). The methods for obtaining subpixel resolution for gradient and second-order edge detection algorithms are very different and will be considered separately.

First, consider the output of a second-order edge detector such as the Laplacian of Gaussian. The edge is signaled by a zero crossing between pixels. In principle, the edge position can be computed to subpixel resolution by using linear interpolation. In practice, the output of second-order edge detection schemes, even with Gaussian presmoothing, is too noisy to allow any simple interpolation method to provide accurate results.

Obtaining subpixel resolution in edge location after edge detection with a gradient-based scheme is both practical and efficient. The result of applying a Gaussian smoothing filter and first derivative to an ideal step edge is a profile that is exactly the same shape as the Gaussian filter used for smoothing. If

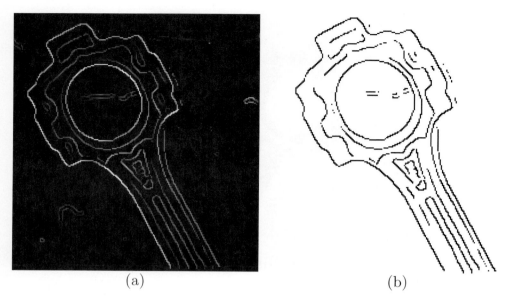

(a)   (b)

Figure 5.23: The result of edge detection applied to the test image from Figure 5.22 with a 7×7 Gaussian smoothing filter. The picture in part (a) is a gray level image of the result after smoothing with a Gaussian smoothing filter, computing the gradient approximation, and suppressing nonmaxima. The weak edge fragments due to noise and fine texture do not show up clearly. Part (b) is a plot of the image from part (a) with every pixel that was greater than zero drawn in black. This plot shows the weak edge fragments that are present in the result of edge detection but do not show up clearly in a gray level display of the results.

the step edge is not ideal but makes a gradual transition from one level to another, then the result of Gaussian smoothing and a first derivative can be approximated by a broader Gaussian.

Consider a set of measurements drawn from a normal distribution. The center of the bell-shaped curve corresponds to the mean value of the normal distribution, which can be estimated by averaging the measurements. Now suppose that a histogram of the measurements, rather than the raw measurements themselves, is all the information that is available. The mean can be estimated by dividing the sum of the values for the centers of the histogram buckets, weighted by the number of entries in each bucket, by the area of the histogram. By analogy, the location of an edge can be estimated to subpixel resolution by averaging along the profile in the output of the Gaussian edge

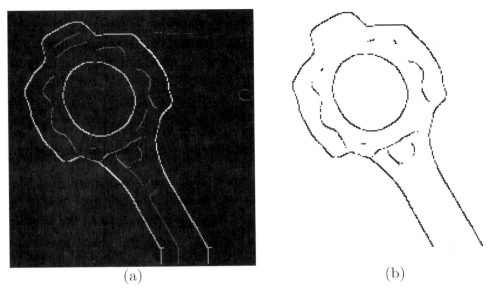

Figure 5.24: The result of edge detection applied to the test image from Figure 5.22. A 31 × 31 Gaussian filter was used to smooth the image, followed by computation of the gradient approximation and then nonmaxima suppression. Part (a) displays the nonmaxima-suppressed result as a gray level image. Note how the large smoothing filter has rounded the edges. Part (b) plots the results to show all of the edge fragments.

detector. To compute the edge location to subpixel resolution, take samples of the magnitude of the Gaussian edge detector output (without nonmaxima suppression) along the gradient direction to either side of the edge until the gradient magnitude falls below a threshold. Use the samples of the gradient magnitude g_i as weights to compute the weighted sum of the position d_i along the gradient. The subpixel correction to the position of the edge along the gradient direction is given by

$$\delta d = \frac{\sum_{i=1}^{n} g_i d_i}{\sum_{i=1}^{n} g_i},$$ (5.50)

where d_i is the distance of a pixel along the gradient from the pixel where the edge was detected, and g_i is the gradient magnitude.

A simpler algorithm that is probably as effective is to compute the position $(\delta x, \delta y)$ of the edge to subpixel resolution, relative to the pixel where the edge was detected, by applying a first moment calculation to the magnitude of the gradient of the Gaussian edge detector. The correction can be added to the coordinates of the pixel to provide a more accurate estimate of edge location.

The algorithm for computing the average along the profile of gradient magnitude, though more complex, has the advantage that the profile can be compared with the ideal profile using statistical techniques and the results of this comparison can be used as the criterion for edge detection. If the profile is not close to a Gaussian, the edge does not correspond to the ideal step model. It may not be possible in this case to accurately estimate the edge location using the techniques presented in this section.

In Chapter 12, we will present methods for calibrating the coordinate system of the image. The integer coordinates of a pixel $[i, j]$ can be mapped to coordinates (x, y) in the image plane of the camera. The correction to the edge location provided by either of the methods for estimating edge location to subpixel resolution described above is added to the (x, y) coordinates of the pixel to obtain the precise location of the edge in the image plane. This precise, calibrated value is what is needed for measuring feature dimensions.

5.8   Edge Detector Performance

Measures for evaluating the performance of edge detectors have been formulated by Abdou and Pratt [1] and DeMicheli, Caprile, Ottonello, and Torre [66]. The criteria to consider in evaluating the performance of an edge detector include

1. Probability of false edges

2. Probability of missing edges

3. Error in estimation of the edge angle

4. Mean square distance of the edge estimate from the true edge

5. Tolerance to distorted edges and other features such as corners and junctions

The first two criteria concern the performance of an algorithm as a detector of edges. The second two criteria concern the performance of an algorithm as an estimator of the edge location and orientation. The last criterion concerns the tolerance of the edge algorithm to edges that depart from the ideal model used to formulate the algorithm.

5.8.1 Methods for Evaluating Performance

The performance of an edge detector can be evaluated in two stages: count the number of false and missing edges and measure the variance (or error distribution) for the estimated location and orientation.

For a test case, select a synthetic image where the true edges are known to lie along a contour that can be modeled by a curve with a simple mathematical formula—for example, a filled rectangle where the boundary contour can be modeled by line segments or two filled rectangles where the gap between them is known. Count the number of correct, missing, and false edges by comparing the results of the edge detector with the original (synthetic) image. This is a harder task than it appears to be. The results vary with the threshold, smoothing filter size, interactions between edges, and other factors. If you run an edge detector over the test image with no added noise, no smoothing, and no interactions between edges, then you should get a perfect set of edges (no missing or false edges). Use this set of edges as the standard for comparison.

Now consider the edges obtained from a test case that has added noise, or other distortions in the image that create missing or false edges. Compute a one-to-one match of edges in the test image to edges in the standard, based on the criterion of Euclidean distance. Ideally, we should use a proper matching algorithm such as the method for the disparity analysis of images presented in Section 14.3. Edges too far from the edges in the standard are false edges; edges that pair closely with one edge in the standard are correct. After this procedure, the edges in the standard that are not paired with one edge in the test case are missing edges.

This procedure tests an edge detector based only on its ability to indicate the presence or absence of edges, but says nothing about how accurately the edge locations or orientations are estimated. Compare the locations and orientations of edges in the set of correct edges (computed above) with the original test image. This comparison requires that the model of the test case be available. For the filled rectangle, the model is the line segments that make up the sides of the rectangle. The edge locations and orientations must be compared with a mathematical description of the model of the scene contours. For each edge with location (x, y), how far is this location from the true location? What is the difference between the orientation of the edge and the orientation of the true curve? The edge location (x, y) could correspond to any point along the contour, but the closest point along the contour is used as the corresponding point, and the distance between the edge point and the closest point is computed. For a line segment, use the formulas in Section 6.4. Estimate the error distribution from a histogram of location errors, or tabulate the sum of the squared error and divide by $n - 1$, where n is the number of edges, to estimate the variance (refer to the formula in Appendix B). The orientation error of an edge is measured by comparing the orientation of the edge fragment with the angle of the normal to the curve that models the scene contour, evaluated at the closest point to the edge point.

5.8.2   Figure of Merit

One method to judge the performance of edge detectors is to look at an edge image and subjectively evaluate the performance. However, this does not provide an objective measure of performance. To quantitatively evaluate the performance of various edge detectors, we should formulate a criterion that may help in judging the relative performance under controlled conditions. We observe that in the response of an edge detector, there can be three types of errors:

- Missing valid edges

- Errors in localizing edges

- Classification of noise as edges

A *figure of merit* for an edge detector should consider these three errors. One such figure of merit, called Pratt's figure of merit [196], is:

$$FM = \frac{1}{\max(I_A, I_I)} \sum_{i=1}^{I_A} \frac{1}{1 + \alpha d_i^2} \tag{5.51}$$

where I_A, I_I, d, and α are detected edges, the ideal edges, the distance between the actual and ideal edges, and a design constant used to penalize displaced edges, respectively.

Note that since this figure involves missing edge points, location of edge points, and false edge points, it can be applied only to a limited class of images. One may generate known objects of controlled contrast at known location and then use the above figure of merit. It is a common practice to evaluate the performance for synthetic images by introducing random noise in images. A plot of signal-to-noise ratio against the figure of merit gives the degradation in the performance of the detector.

5.9 Sequential Methods

All the edge detectors described above are parallel in nature: they can be applied at a single pixel, using local information, independent of the result at other pixels. In practice, the performance of such edge detectors may not be acceptable due to too many missing edges. The edges detected by such detectors have to be linked to form boundaries of objects. Missing edges result in breaks in the boundaries. Several methods have been suggested to improve the performance of edge detection and linking. Such approaches include edge following, relaxation, and boundary tracking. We will discuss relaxation in a later chapter.

Edge following tries to use information from the neighborhood of an edge point to improve the performance of an edge detector. This approach may also be used to thin edges from a simple edge detector response. The basic approach is that an edge point looks at its neighbors. If the direction of an edge is compatible with that of its neighbors, the edges are linked; incompatible edges are removed and, by looking at a large neighborhood, one may fill in missing edges.

The edge following algorithm scans an image to find a strong edge used as a starting point for following the boundary of an object. Depending on the direction of the edge, the edge detector is applied to extend the edge in the proper direction. One may implement the tracking operation, shown in Figure 5.25, either without backtracking or with backtracking. Using this approach, even very weak edge segments of an object may be detected.

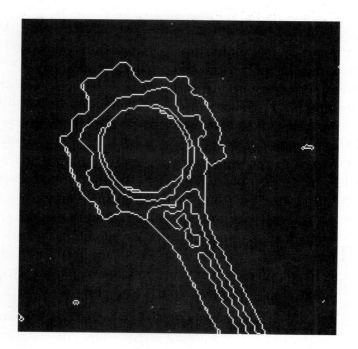

Figure 5.25: An illustration of edge following.

5.10   Line Detection

Up to this point, this chapter has discussed only how to detect step edges in images. Lines are also very important features that may have to be detected. A line can be modeled as two edges with opposite polarity that are close together, separated by less distance than the width of the smoothing filter.

Edge detection filters are designed to respond to step edges and do not provide a meaningful response to lines. A separate algorithm must be used to detect lines, and this line detection algorithm will not provide a meaningful response to step edges. One outstanding problem in machine vision is how to combine edge and line detection in a single system.

Lines can be detected with a modified version of the Canny algorithm: perform nonmaxima suppression on the smoothed image instead of the gradient magnitude. A line is the derivative of a step, so the derivative step in the Canny algorithm is unnecessary.

Further Reading

Edge detection has been one of the most popular research areas since the early days of computer vision. Roberts [202] and Sobel [225] present two early edge detectors that are still commonly used. Some other edge detectors that were popular in the early days were by Prewitt [198], Hueckel [116], and Frei and Chen [84]. Many statistical [258] and several filtering approaches have also been used for edge detection. Algorithms based on the Laplacian of Gaussian [164] and based on the gradient of Gaussian [53] were very popular in the 1980s. The Laplacian of Gaussian edge detection scheme [164] is still dominant in models of biological edge detection. Haralick [102] presented an edge detection scheme based on the second directional derivative. His scheme incorporated a form of image smoothing based on approximating the image with local surface patches.

Although edge detection using Guassian filters of different scales was introduced earlier, it began to receive considerable attention after the paper by Witkin [253]. The papers by Yuille and Poggio [260, 261] provided key theoretical results in this area. Another paper, by Hummel [118], provided additional results. Shah, Sood, and Jain [220] and Lu and Jain [159, 160] studied interaction among edges at different scales and developed a reasoning methodology for detecting edges in scale space. Edge focusing is another approach to edge detection in scale space [25].

Another class of edge detection algorithms searches the image or a filtered version of the image for patterns of image intensity that may be edges. These algorithms combine edge detection with edge linking. The analysis of the patterns of image intensity can be very elaborate, and these algorithms are usually used only in situations where it is necessary to find edges in images with very poor quality. A good treatment of this is provided in [74].

Though many edge detectors have been developed, there is still no well-defined metric to help in selecting the appropriate edge detector for an application. Lack of a performance measure makes judicious selection of an edge detector for a particular application a difficult problem. Some discussion of this may be found in [103, 196].

Exercises

5.1 What is an edge? How does it relate to the boundary of an object? How does it relate to the boundary of a region?

5.2 How can an edge be modeled in an image? Which is the most commonly used model in edge detection? Why?

5.3 What is an isotropic edge detector? How can you implement it?

5.4 What is a directional edge detector? How can you implement it? Where will you use a directional edge detector? Give edge detector masks for detecting edges oriented at 45° and −45°.

5.5 Name all the steps required in edge detection. Can you implement an edge detector by skipping one or more of these steps? How?

5.6 Why is the Laplacian not a good edge operator?

5.7 Describe the Laplacian of Gaussian edge detector. Considering different steps in edge detection, show how the Laplacian is not a good edge operator, but the Laplacian of Gaussian is.

5.8 How can you select the correct size of the operator in the LoG operators? What factors should you consider in selecting the proper size of the operator? Can you have an automatic selection algorithm?

5.9 What is the facet model of an image? How can you use it for edge detection? Can you use this model for region growing also?

5.10 Compare the Gaussian edge detector with the Laplacian of Gaussian. Use all steps in edge detection and compare what the two operators do at these steps. Where is the difference? Do you think that their performances will differ significantly? Explain clearly the difference and the effect of the difference in edge detection.

5.11 Can edges be located at subpixel resolution? How? Is there any particular approach that will be more suitable for this? Consider subpixel edge location estimation for the gradient, Laplacian, and facet models. Compare the different estimation problems.

5.12 To select a particular edge detector in a machine vision application, we should know the comparative performance of edge detectors in the type of application that we are faced with. How can we compare performance of various edge detectors in an objective manner? List all important factors in edge detection and then define a performance measure that can be effectively evaluated.

5.13 What is edge tracking? What factors must be considered in edge tracking?

5.14 The equation for the sloped planar facet model is obtained by setting all terms above k_3 to zero, yielding

$$g[i, j] = k_3 i + k_2 j + k_1.$$

The error in this approximation is given by

$$\epsilon^2 = \sum_{i=-l}^{l} \sum_{j=-l}^{l} \left(\hat{k_3} i + \hat{k_2} j + \hat{k_1} - g[i, j] \right)^2.$$

For the case $l = 1$, estimate the parameters k_1, k_2, and k_3 for the sloped facet model that best approximates the gray levels in the 3×3 neighborhood given below.

	j −1	0	1
i −1	5	7	9
0	3	7	7
1	1	3	5

What is the magnitude of the gradient?

5.15 Suppose that an image is smoothed with an $n \times n$ Gaussian filter. During smoothing, the square filter window is moved across the image. The pixel at position $[i, j]$ in the upper left corner of the window is replaced by the smoothed value. After smoothing, the gradient magnitude is computed using the approximations in Section 5.1. As the 2×2 operators are moved across the smoothed image, the pixel at position $[i, j]$

in the upper left corner of the window is replaced by the gradient magnitude. After edge detection, the edge location (x_{ij}, y_{ij}) for each edge pixel is computed to subpixel resolution. Where is the edge location in the coordinate system of the original (unsmoothed) image?

Computer Projects

5.1 Implement the Roberts, Sobel, and Prewitt operators. Apply these to various images. Do you get edges where you expect them? Manually identify several edge segments and note their locations. Compare the locations of edge segments given by your program with manually marked edges. Are you satisfied? Can you improve your program to do better? Change the threshold values for your detection step and see the results.

5.2 Generate a synthetic image that contains one or more known objects with clear, known intensity discontinuities. You should know the locations of these discontinuities precisely. The rest of the image should be without edges. Use any graphics technique or cut and paste using image processing programs. Use this image to do systematic experiments with different edge detectors. Apply the edge detector to the image and obtain all edges.

Define a measure to evaluate the performance of an edge detector and use that in all experiments. What factors would you consider in defining the performance measure? Repeat your experiment by adding noise. You can systematically add random noise using a random number generator. You may also want to systematically change the intensity values of the objects and background. Change the threshold values for your detection step and see the results.

5.3 Develop a test-bed to generate or acquire images for evaluating the performance measure that you defined in the above exercise. Apply each edge detector that you studied in this chapter, and plot the performance of various edge detectors by varying the parameters controlling the quality of images. These curves represent performance characteristics of edge detectors and may be useful in selecting a suitable edge detector for your application.

5.4 In many applications computational time should also be considered in selecting the proper edge detector. Compute the time requirement, using theoretical and experimental analysis, for each edge detector. Based on the performance measure you defined and determined and the time requirement, develop a general guideline for selecting an edge detector in different applications.

5.5 Facet model edge detection:

a. Using the gradient-based edge detection method with *sloped planar* facets (see Exercise 5.14), detect edges in an image of your choice. Specifically, you should do the following:

- Calculate k_1, k_2, and k_3 at each pixel.
- Calculate the gradient at each pixel.
- Identify pixels at which the gradient is above some threshold.

b. Add Gaussian noise to the above images and detect the edges as in part **a**.

5.6 Detect edges using the *cubic* facet approximation. In particular, you should do the following:

- Find the cubic fit coefficients k_1 to k_{10} at each pixel.
- Find the gradient angle α.
- Find the subpixel deviation ρ at which the second derivative is zero.
- Find if the zero crossing occurs within the boundary of the pixel.
- Confirm that the first derivative is nonzero and the third derivative is negative.
- Mark all such pixels as 255 and reset the others to zero.

Chapter 6

Contours

Edges must be linked into a representation for a region boundary. This representation is called a contour. The contour can be open or closed. Closed contours correspond to region boundaries, and the pixels in the region may be found by a filling algorithm. An open contour may be part of a region boundary. Gaps can occur in a region boundary because the contrast between regions may not be enough to allow the edges along the boundary to be found by an edge detector. The edge detection threshold may have been set too high, or the contrast along some portion of the boundary may be so weak relative to other areas of the image that no single threshold works everywhere in the image. Open contours also occur when line fragments are linked together—for example, when line fragments are linked along a stroke in a drawing or sample of handwriting.

A contour may be represented as an ordered list of edges or by a curve. A curve is a mathematical model for a contour. Examples of curves include line segments and cubic splines. There are several criteria for a good contour representation:

Efficiency: The contour should be a simple, compact representation.

Accuracy: The contour should accurately fit the image features.

Effectiveness: The contour should be suitable for the operations to be performed in later stages of the application.

The accuracy of the contour representation is determined by the form of curve used to model the contour, by the performance of the curve fitting algorithm, and by the accuracy of the estimates of edge location. The simplest representation of a contour is an ordered list of its edges. This representation is as accurate as the location estimates for the edges, but is the least compact representation and may not provide an effective representation for subsequent image analysis. Fitting the appropriate curve model to the edges increases accuracy, since errors in edge location are reduced through averaging, and it increases efficiency by providing a more appropriate and more compact representation for subsequent operations. For example, a set of edges that lie along a line can be represented most efficiently by fitting a line to the edges. This representation simplifies later calculations, such as determining the orientation or length of the line, and increases the accuracy, since the mean squared error between the estimated line and the true line will be smaller than the error between the true line and any of the edges.

The first section in this chapter presents the elementary differential geometry of curves in the plane. The second section gives a collection of techniques for calculating contour properties such as length, tangent, and curvature from the list of edges, without fitting a curve model to the edges. The remaining sections cover curve models and techniques for fitting the models to contours.

Before proceeding, some terms must be defined. A curve *interpolates* a list of points if the curve passes through the points. *Approximation* is fitting a curve to a list of points with the curve passing close to the points, but not necessarily passing exactly through the points. In the following sections, we will begin by assuming that the edges provided by an edge detection algorithm are exact and will fit curves to the edge points using interpolation methods. The edges provided by edge detection applied to real images will not be exact. There will be some error in the estimated location of the edge. Later sections will present methods for curve approximation.

Definition 6.1 *An edge list is an ordered set of edge points or fragments.*

Definition 6.2 *A contour is an edge list or the curve that has been used to represent the edge list.*

Definition 6.3 *A boundary is the closed contour that surrounds a region.*

In this chapter, the term *edges* will usually refer to edge points. The edge orientation is not used by most curve fitting algorithms. In the few cases where the algorithm does use the edge orientation, it will be clear from the context that the term *edges* refers to edge fragments.

6.1   Geometry of Curves

Planar curves can be represented in three different ways: the explicit form $y = f(x)$, the implicit form $f(x, y) = 0$, or the parametric form $(x(u), y(u))$ for some parameter u. The explicit form is rarely used in machine vision since a curve in the x–y plane can twist around in such a way that there can be more than one point on the curve for a given x.

The parametric form of a curve uses two functions, $x(u)$ and $y(u)$, of a parameter u to specify the point along the curve from the starting point of the curve at $\mathbf{p}_1 = (x(u_1), y(u_1))$ to the end point $\mathbf{p}_2 = (x(u_2), y(u_2))$. The length of a curve is given by the arc length:

$$\int_{u_1}^{u_2} \sqrt{\left(\frac{dx}{du}\right)^2 + \left(\frac{dy}{du}\right)^2} \, du. \tag{6.1}$$

The unit tangent vector is

$$\mathbf{t}(u) = \frac{\mathbf{p}'(u)}{|\mathbf{p}'(u)|}, \tag{6.2}$$

where $\mathbf{p}(u) = (x(u), y(u))$. The curvature of the curve is the derivative of the tangent: $\mathbf{n}(u) = \mathbf{p}''(u)$.

Consider three points along the curve: $\mathbf{p}(u + \Delta)$, $\mathbf{p}(u)$, and $\mathbf{p}(u - \Delta)$. Imagine a circle passing through these three points, which uniquely determine the circle. In the limit as $\Delta \to 0$, this circle is the osculating circle. The osculating circle touches the curve at $\mathbf{p}(u)$, and the center of the circle lies along the line containing the normal to the curve. The curvature is the inverse of the radius of the osculating circle.

6.2   Digital Curves

In this section, we present a set of algorithms for computing the elements of curve geometry, such as contour length, tangent orientation, and curvature, from the list of edge points. Slope and curvature are difficult to compute precisely in the digital domain, since the angle between neighboring pixels is quantized to 45° increments.

The basic idea is to estimate the tangent orientation using edge points that are not adjacent in the edge list. This allows a larger set of possible tangent orientations. Let $\mathbf{p}_i = (x_i, y_i)$ be the coordinates of edge i in the edge list. The k-slope is the (angle) direction vector between points that are k edges apart. The left k-slope is the direction from $\mathbf{p}_{i-k}$ to $\mathbf{p}_i$, and the right k-slope is the direction from $\mathbf{p}_i$ to $\mathbf{p}_{i+k}$. The k-curvature is the difference between the left and right k-slopes.

Suppose that there are n edge points $(x_1, y_1), \ldots, (x_n, y_n)$ in the edge list. The length of a digital curve can be approximated by adding the lengths of the individual segments between pixels:

$$S = \sum_{i=2}^{n} \sqrt{(x_i - x_{i-1})^2 + (y_i - y_{i-1})^2}. \tag{6.3}$$

A good approximation is obtained by traversing the edge list and adding 2 along sides and 3 along diagonals, and dividing the final sum by 2. The distance between end points of a contour is

$$D = \sqrt{(y_n - y_1)^2 + (x_n - x_1)^2}. \tag{6.4}$$

6.2.1 Chain Codes

Chain codes are a notation for recording the list of edge points along a contour. The chain code specifies the direction of a contour at each edge in the edge list. Directions are quantized into one of eight directions, as shown in Figure 6.1. Starting at the first edge in the list and going clockwise around the contour, the direction to the next edge is specified using one of the eight chain codes. The direction is the chain code for the 8-neighbor of the edge. The chain code represents an edge list by the coordinates of the first edge and the list of chain codes leading to subsequent edges. A curve and its chain code are shown in Figure 6.2.

The chain code has some attractive properties. Rotation of an object by $45°$ can be easily implemented. If an object is rotated by $n \times 45°$, then the code for the rotated object is obtained by adding $n \bmod 8$ to the original code. The derivative of the chain code, also called *difference code*, obtained by using first difference, is a rotation-invariant boundary description. Some other characteristics of a region, such as area and corners, may be directly computed using the chain code. The limitation of this representation is

2	3	4
1	·	5
8	7	6

Figure 6.1: The chain codes for representing the directions between linked edge points.

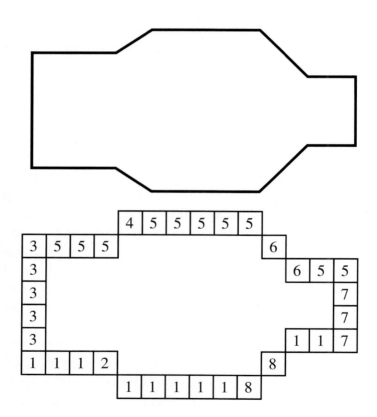

Figure 6.2: A curve and its chain code.

the limited set of directions used to represent the tangent at a point. This limitation can be removed by using one of the curve representations presented in the following sections. Once a curve has been fitted to the list of edges, any of the geometric quantities presented in Section 6.1 can be computed from the mathematical formula for the curve.

6.2.2 Slope Representation

The slope representation of a contour, also called the Ψ-s plot, is like a continuous version of the chain code. We want to represent a contour using arbitrary tangent directions, rather than the limited set of tangent directions allowed by the chain code. Suppose that we start at the beginning of the edge list and compute the tangent and arc length using the formulas presented for digital curves. We may plot the tangent Ψ versus arc length s to obtain a representation for the contour in the Ψ-s space. The Ψ-s plot is a representation of the shape of the contour. For example, a contour that consists of line segments and circular arcs will look like a sequence of line segments in the Ψ-s plot. Horizontal line segments in the Ψ-s plot correspond to line segments in the contour; line segments at other orientations in the Ψ-s plot correspond to circular arcs. Portions of the Ψ-s plot that are not straight lines correspond to other curve primitives.

The contour may be split into straight lines and circular arcs by segmenting the Ψ-s plot into straight lines. This method has been used by many researchers, and there are several versions of this approach for splitting a contour into segments.

One may use the Ψ-s plot as a compact description of the shape of the original contour. In Figure 6.3, we show a contour and its Ψ-s plot. For a closed contour, the Ψ-s plot is periodic.

6.2.3 Slope Density Function

The slope density function is the histogram of the slopes (tangent angles) along a contour. This can be a useful descriptor for recognition. Correlating the slope density function of a model contour with the slope density function for a contour extracted from an image allows the orientation of the object to be determined. This also provides a means for object recognition.

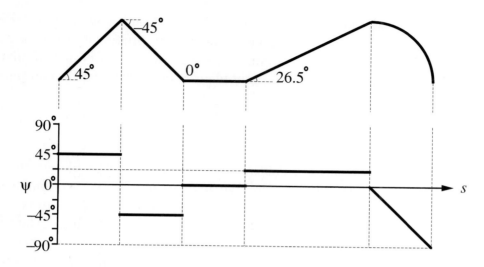

Figure 6.3: Slope representations of a contour.

6.3   Curve Fitting

The rest of this chapter will cover four curve models and the methods for fitting the models to edge points. The models include:

- Line segments

- Circular arcs

- Conic sections

- Cubic splines

Any fitting algorithm must address two questions:

1. What method is used to fit the curve to the edges?

2. How is the closeness of the fit measured?

Sections 6.4 through 6.7 will cover techniques for fitting curve models to edges with the assumption that the edge locations are sufficiently accurate that selected edge points can be used to determine the fit. Section 6.8 will present successively more powerful methods that can handle errors in the edge locations.

Let d_i be the distance of edge point i from a line. There are several measures of the goodness of fit of a curve to the candidate edge points. All of them depend on the error between the fitted curve and the candidate points forming the curve. Some commonly used methods follow.

Maximum absolute error measures how much the points deviate from the curve in the worst case:

$$\text{MAE} = \max_i |d_i| \qquad (6.5)$$

Mean squared error gives an overall measure of the deviation of the curve from the edge points:

$$\text{MSE} = \frac{1}{n} \sum_{i=1}^{n} d_i^2 \qquad (6.6)$$

Normalized maximum error is the ratio of the maximum absolute error to the length of the curve:

$$\varepsilon = \frac{\max_i |d_i|}{S} \qquad (6.7)$$

Number of sign changes in the error is a good indicator of the appropriateness of the curve as a model for the edges in the contour.

Ratio of curve length to end point distance is a good measure of the complexity of the curve.

The normalized maximum error provides a unitless measure of error independent of the length of the curve. In other words, a given amount of deviation from a curve may be equally significant, in some applications, as twice as much deviation from a curve that is twice as long. If the curve model is a line segment, then it is not necessary to compute the arc length; the distance D between the end points can be used:

$$D = \sqrt{(y_n - y_1)^2 + (x_n - x_1)^2}. \qquad (6.8)$$

Sign changes are a very useful indication of goodness of fit. Fit a list of edge points with a straight line and examine the number of sign changes. One sign change indicates that the list of edges may be modeled by a line segment, two sign changes indicate that the edges should be modeled by a quadratic curve, three sign changes indicate a cubic curve, and so on. Numerous sign changes indicate that a small increase in the complexity of the curve will not improve the fit significantly. A good fit has a random pattern to the sign changes. Runs of errors of the same sign indicate a systematic error in fitting; possibly due to the wrong curve model.

In the following sections, we will use simple curve fitting methods to illustrate the use of the polyline, circular arc, conic section, and cubic spline models. Section 6.8 will present more powerful curve fitting methods, using polylines as the primary example; but, in principle, any of the models presented in the following sections could be used with any of the curve fitting methods presented in Section 6.8.

The choice of curve fitting model must be guided by the application. The use of straight line segments (polylines) is appropriate if the scene consists of straight lines and is the starting point for fitting other models. Circular arcs are a useful representation for estimating curvature, since the curve is segmented into sections with piecewise constant curvature. Conic sections provide a convenient way to represent sequences of line segments and circular arcs, as well as elliptic and hyperbolic arcs, and explicitly represent inflection points. Cubic splines are good for modeling smooth curves and do not force estimates of tangent vectors and curvature to be piecewise constant.

6.4   Polyline Representation

A polyline is a sequence of line segments joined end to end. The polyline representation for a contour fits the edge list with a sequence of line segments. The polyline interpolates a selected subset of the edge points in the edge list. The ends of each line segment are edge points in the original edge list. Each line segment models the run of contiguous edges between its end points. The points where line segments are joined are called vertices. Note that polylines are two-dimensional curves in the image plane, as are all of the curves discussed in this chapter, and the vertices are points in the image plane.

The polyline algorithm takes as input an ordered list of edge points $\{(x_1, y_1), (x_2, y_2), \ldots, (x_n, y_n)\}$. The edge point coordinates may be computed to subpixel resolution (see Section 5.7). Since line segments are fit between two edge points selected as vertices, only the coordinates of the edges that are selected as vertices need to be computed precisely.

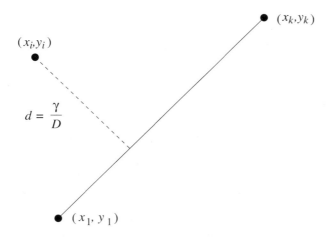

Figure 6.4: Diagram showing the perpendicular distance of a point from a line segment. The value γ is computed by plugging the coordinates (x_i, y_i) of the point into the equation for the line segment.

The formula for a line segment that approximates a list of edge points and joins the first and last edge points (x_1, y_1) and (x_k, y_k) can be derived by noting that the slope of the line between the end points is the same as the slope of the line between the first point and an arbitrary point along the line:

$$\frac{y - y_1}{x - x_1} = \frac{y_k - y_1}{x_k - x_1}. \tag{6.9}$$

Multiplying out and rearranging terms gives the implicit form for a line segment, parameterized by the coordinates of the end points:

$$x(y_1 - y_k) + y(x_k - x_1) + y_k x_1 - y_1 x_k = 0. \tag{6.10}$$

The distance of any point (x_i, y_i) from the line is $d = r/D$, where r is computed by plugging the coordinates of the point into the equation for the line segment,

$$r = x_i(y_1 - y_k) + y_i(x_k - x_1) + y_k x_1 - y_1 x_k, \tag{6.11}$$

and D is the distance between the end points. (Refer to Figure 6.4.) The sign of r can be used to compute the number of sign changes C. The normalized distance is d/D. The normalized maximum absolute error is

$$\varepsilon = \frac{\max_i |d_i|}{D}, \tag{6.12}$$

where d_i is the distance between the line and the position of the ith edge
in the edge list. The normalized maximum error is frequently used as the
measure for the goodness of fit of a line segment to a set of edges. All of
these formulas assume that the perpendicular projection of a point onto a
line is within the line segment; that is, both on the line and between the end
points of the line segment. This is the case for the situations throughout this
chapter, but in other cases the formulas may need to be modified to compute
the distance of the point from the nearest end point of the line segment.

There are two approaches to fitting polylines: top-down splitting and
bottom-up merging.

6.4.1   Polyline Splitting

The top-down splitting algorithm recursively adds vertices, starting with an
initial curve. Consider the curve shown in Figure 6.5. The initial curve is the
line segment between the first and last edge points, labeled A and B. The
point in the edge list that is farthest from the straight line is found. If the
normalized maximum error is above a threshold, then a vertex is inserted at
the edge point farthest from the line segment, labeled as point C in Figure 6.5.
The splitting algorithm is recursively applied to the two new line segments
and the edge list. The edge list is partitioned into two lists corresponding
to the two line segments. The edge points in the list that are farthest from
each segment are found, and new vertices are introduced if the points are
too far from the line segments. The polyline splitting algorithm terminates
when the normalized maximum error, for all edge points along the polyline,
is below the threshold. This recursive procedure is very efficient. Segment
splitting is also called *recursive subdivision*.

6.4.2   Segment Merging

In segment merging, edge points are added to line segments as the edge list
is traversed. New segments are started when the edge points deviate too

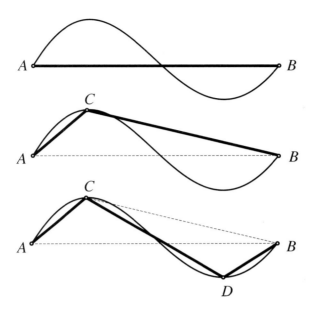

Figure 6.5: Splitting method for polylines.

far from the line segment. The merge approach is also called the bottom-up approach to polyline fitting.

There are several measures that can be used to determine if an edge point is too far from the line segment that is being formed. One method is to use sequential least-squares, which performs a least-squares fit of the line segment to the edge points and updates the parameters of the line segment incrementally as each new edge point is processed. The fitting algorithm calculates the squared residual between the line segment and the edge points. When the error exceeds a threshold, a vertex is introduced and a new segment is started from the end point of the last segment.

The tolerance band algorithm uses a different method for determining the placement of vertices. Two line segments that are parallel to the line segment approximating the edge points at a distance ϵ from the center line segment are computed. (See Figure 6.6.) The value of ϵ represents the absolute amount of deviation from the fitted line that is tolerated. Edges are added to the current line segment as long as the new edges are inside the tolerance band. The parameters of the line segment may be recomputed as new edges are added to the segment. The approximating line segment does

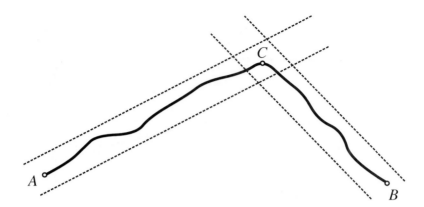

Figure 6.6: Tolerance band for fitting line segments.

not have to remain parallel to the sides of the tolerance band. The vertex at the end of the segment is the starting point for the next segment. This approach usually results in too many segments. Corner locations and angles are not accurately estimated since a vertex is not created until the algorithm has processed edges up to the boundary of the tolerance band.

6.4.3   Split and Merge

The top-down method of recursive subdivision and the bottom-up method of merging can be combined as the split and merge algorithm. Splitting and merging methods are only partially successful when used by themselves, but the accuracy of line segment approximations to a list of edges can be improved by interleaving merge and split operations. Figure 6.7 shows an example where a split followed by a merge can repair a badly placed vertex.

The basic idea is to interleave split and merge passes. After recursive subdivision, allow adjacent segments to be replaced by a single segment between the first and last end points if the new segment fits the edges with less normalized error. Note that it is necessary to use normalized error since multiple line segments will always fit a list of edges with less error than for a single line segment. After segment merging, the new segment may be split at a different point. Alternate applications of split and merge continue until no segments are merged or split.

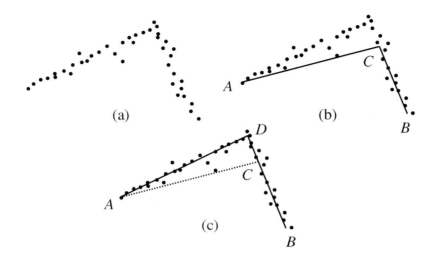

Figure 6.7: A bad corner estimate produced by a bottom-up edge merge that missed the true corner location can be repaired by split and merge passes that split the first segment at a point closer to the true corner and then merge the two segments into a single line segment.

6.4.4 Hop-Along Algorithm

The hop-along algorithm approximates a contour by a sequence of line segments, like the split and merge method described above, but works on short sublists of edges. The algorithm starts at one end of a list of edge points, grabs some fixed number of edges, and fits a line segment between the first and last edge points. If the fit is bad, the algorithm does a split at the point of maximum error and repeats with the segment closest to the beginning of the run. In other words, the algorithm falls back until it finds a good line segment approximation to some initial sequence of edges. The algorithm makes the current segment the previous segment, and continues with the remaining edge points. The algorithm also checks to see if the current segment can be merged with the previous segment. The algorithm is like a split and merge algorithm, but it does not start with the entire list of edges and does not waste time doing lots of splits. The algorithm hops along, working on modest-sized runs of edges. The algorithm is given as Algorithm 6.1.

Algorithm 6.1   Hop-Along Algorithm for Polyline Fitting

1. *Start with the first k edges from the list.*

2. *Fit a line segment between the first and last edges in the sublist.*

3. *If the normalized maximum error is too large, shorten the sublist to the point of maximum error. Return to step 2.*

4. *If the line fit succeeds, compare the orientation of the current line segment with that of the previous line segment. If the lines have similar orientations, replace the two line segments with a single line segment.*

5. *Make the current line segment the previous line segment and advance the window of edges so that there are k edges in the sublist. Return to step 2.*

The algorithm hops along, advancing the window of edges by a constant k. If the fit of a line segment to the edges is not good enough, the algorithm falls back to the point of maximum error. Since the algorithm considers only a short run of edges, it is more efficient than pure recursive subdivision or the split and merge algorithm, which would start with the entire list of edges and waste a lot of time splitting the edge list into manageable pieces.

6.5   Circular Arcs

After a list of edges is approximated by line segments, subsequences of the line segments can be replaced by circular arcs if desired. Replacing line segments by circular arcs involves fitting circular arcs through the end points of two or more line segments. In other words, circular arc fitting is done on the vertices in the polyline. Representing the contour as a sequence of line segments and circular arcs breaks the contour into sections with piecewise constant curvature. Many image analysis algorithms use curvature information.

Just as we derived the implicit formula for the line segment between two points, we need to derive the implicit formula for a circle through three points. The implicit equation for a circle with radius r and center (x_0, y_0) is

$$(x - x_0)^2 + (y - y_0)^2 = r^2. \tag{6.13}$$

Consider three points $\mathbf{p}_1 = (x_1, y_1)$, $\mathbf{p}_2 = (x_2, y_2)$, and $\mathbf{p}_3 = (x_3, y_3)$. Transform the origin of the coordinate system to point $\mathbf{p}_1$. In the new coodinate system,

$$x' = x - x_1 \tag{6.14}$$
$$y' = y - y_1 \tag{6.15}$$

and the equation for the circle is

$$(x' - x_0')^2 + (y' - y_0')^2 = r^2. \tag{6.16}$$

Substitute the coordinates in the x'–y' space for the points p_1, p_2, and p_3 in the implicit equation for a circle:

$$x_0'^2 + y_0'^2 - r^2 = 0 \tag{6.17}$$
$$x_2'^2 - 2x_2'x_0' + x_0'^2 + y_2'^2 - 2y_2'y_0' + y_0'^2 - r^2 = 0 \tag{6.18}$$
$$x_3'^2 - 2x_3'x_0' + x_0'^2 + y_3'^2 - 2y_3'y_0' + y_0'^2 - r^2 = 0 \tag{6.19}$$

This yields three nonlinear equations for the three unknowns x_0', y_0', and r. Subtract the first equation from the second and third equations:

$$2x_2'x_0' + 2y_2'y_0' = x_2'^2 + y_2'^2 \tag{6.20}$$
$$2x_3'x_0' + 2y_3'y_0' = x_3'^2 + y_3'^2 \tag{6.21}$$

This yields two linear equations in the two unknowns x_0' and y_0', which are the coordinates of the center of the circle in the x'–y' space. Add (x_1, y_1) to (x_0', y_0') to get the center of the circle in the original coordinate system. Compute the radius of the circle from $r^2 = x_0'^2 + y_0'^2$.

To calculate the error in fitting a circular arc, define the distance of point Q from the circle as the distance of Q from the circle along a line passing through the center of the circle. Let the radius of the circle be r. Compute the distance q with coordinates (x_i, y_i) from point Q to the center (x_0, y_0) of the circle:

$$q = \sqrt{(x_i - x_0)^2 + (y_i - y_0)^2} \tag{6.22}$$

The distance from point Q to the circular arc is

$$d = q - r \tag{6.23}$$

Now that we have a formula for fitting a circular arc to three points, we need a method for evaluating the goodness of fit so we can determine whether or not the circular arc is a better approximation to the edges than the line segments. If the ratio of the length of the contour to the distance between the first and last end points is more than a threshold, then it may be possible to replace the line segments with a circular arc. The circular arc is fit between the first and last end points and one other point. There are several methods for fitting a circular arc to a sequence of polylines, depending on how the middle point is chosen:

1. Use the polyline vertex that is farthest from the line joining the first and last end points.

2. Use the edge point that is farthest from the line joining the first and last end points.

3. Use the polyline vertex that is in the middle of the sequence of vertices between the first and last end points.

4. Use the edge point that is in the middle of the list of edges between the first and last end points.

Calculate the signed distance between all edge points and the circular arc. Compute the maximum absolute error and the number of sign changes. If the normalized maximum error is below a threshold and the number of sign changes is large, then accept the circular arc; otherwise, retain the polyline approximation. The algorithm for replacing line segments with circular arcs is outlined in Algorithm 6.2.

Algorithm 6.2 Replacing Line Segments with Circular Arcs

1. *Initialize the window of vertices to the three end points of the first two line segments in the polyline.*

2. *Compute the ratio of the length of the part of the contour corresponding to the two line segments to the distance between the end points. If the ratio is small, then leave the first line segment unchanged, advance the window of vertices by one vertex, and repeat this step.*

3. *Fit a circle through the three vertices.*

4. *Calculate the normalized maximum error and number of sign changes.*

5. *If the normalized maximum error is too large or the number of sign changes is too small, then leave the first line segment unchanged, advance the window of vertices, and return to step 2.*

6. *If the circle fit succeeds, then try to include the next line segment in the circular arc. Repeat this step until no more line segments can be subsumed by this circular arc.*

7. *Advance the window to the next three polyline vertices after the end of the circular arc and return to step 2.*

After running Algorithm 6.2 over the polyline, the contour will be represented by a sequence of line segments and circular arcs. It may be inconvenient to have two different curve primitives in the representation. In the next section, we will present conic sections, which allow line segments, circular arcs, and other primitives to coexist in the same representation. Conic sections also provide smooth transitions between sections, if desired, as well as the explicit representation of corners.

6.6   Conic Sections

This section describes how to approximate lists of edge points with conic sections. As with circular arcs, the method assumes that the edge points are first approximated by a polyline and replaces subsequences of line segments by conics.

The implicit (algebraic) form of a conic is

$$f(x, y) = ax^2 + 2hxy + by^2 + 2ex + 2gy + c = 0. \tag{6.24}$$

There are three types of conic sections: hyperbolas, parabolas, and ellipses. Circles are a special case of ellipses. Geometrically, conic sections are defined by intersecting a cone with a plane as shown in Figure 6.8.

Conic sections can be fit between three vertices in the polyline approximation to a contour. The locations where conic sections are joined are called knots. Conic splines are a sequence of conic sections that are joined end to

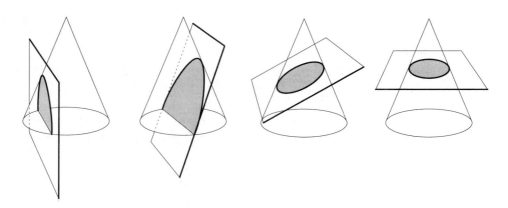

Figure 6.8: Conic sections are defined by intersecting a cone with a plane.

end, with equal tangents at the knots to provide a smooth transition between adjacent sections of the curve. Let the polyline vertices be V_i. The
conic approximation is shown in Figure 6.9.

Each conic section in a conic spline is defined by two end points, two
tangents, and one additional point. The knots K_i can be located between
the vertices of the polyline:

$$K_i = (1 - \nu_i)V_i + \nu_i V_{i+1} \qquad (6.25)$$

where ν_i is between 0 and 1. The tangents are defined by the triangle with
vertices V_i, V_{i+1}, and V_{i+2}. The additional point is

$$Z_i = \gamma_i V_{i+1} + (1 - \gamma_i)\frac{K_i + K_{i+1}}{2}, \qquad (6.26)$$

as shown in Figure 6.10.

There are several special cases of the conic section that can be handled
in a uniform way by this representation. If $\nu_{i+1} = 0$, then the ith section of
the conic spline is the line segment from K_i to V_{i+1}. If $\nu_i = 1$ and $\nu_{i+1} = 0$,
then K_i, K_{i+1}, and V_{i+1} collapse to the same point and there is a corner in
the sequence of conic sections. These special properties allow line segments
and corners to be represented explicitly in a conic spline, without resorting
to different primitives or special flags.

The algorithm presented here for computing conic splines uses the guided
form of a conic section, which represents a conic section using three lines that

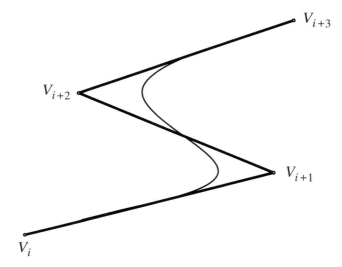

Figure 6.9: Conic sections are approximations defined between three points.

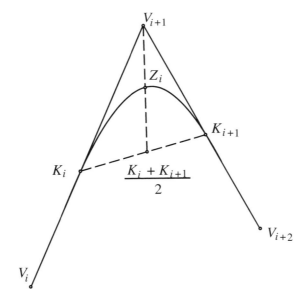

Figure 6.10: A conic section is defined by the two end points and tangents obtained from three vertices of the polyline approximation, plus one additional point.

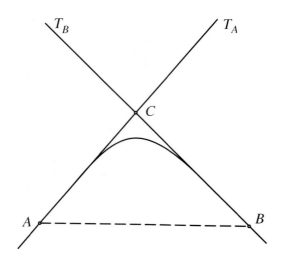

Figure 6.11: The guided form for a conic.

bound the conic. (See Figure 6.11.) The equation of a line is

$$a_0 + a_1 x + a_2 y = 0. \tag{6.27}$$

Let the first and last vertices in a polyline be A and B, and let point C be an intermediate vertex in the polyline. The first and last vertices are joined by the chord AB. The guided form of conic is the family of conics with end points at A and B and tangents AC and BC defined by the equation

$$(a_0 + a_1 x + a_2 y)(b_0 + b_1 x + b_2 y) = \rho(u_0 + u_1 x + u_2 y)^2, \tag{6.28}$$

where

$$a_0 + a_1 x + a_2 y = 0 \tag{6.29}$$

is the line containing the line segment AC,

$$b_0 + b_1 x + b_2 y = 0 \tag{6.30}$$

is the line containing the line segment BC, and

$$u_0 + u_1 x + u_2 y = 0 \tag{6.31}$$

is the line containing the chord AB. The family of conic sections is parameterized by ρ.

The algorithm for fitting a conic section to a list of edge points starts with a polyline and classifies the vertices as corners, soft vertices, or knots. Soft vertices have angles near 180°, and the adjacent line segments are nearly collinear and may be replaced with a conic section. A sequence of soft vertices corresponds to a sequence of line segments with gradually changing orientation that most likely were fitted to edge points sampled along a smooth curve. Corners have vertex angles above $180° + T_1$ or below $180° - T_1$, where T_1 is a threshold, and are unlikely to be part of the conic. Knots are placed along a line segment that has soft vertices at either end that are angled in opposite directions. A conic section cannot have an inflection, so two conic sections must be joined at the knot. The placement of the knot along the line segment is determined by the relative angles of the soft vertices at the ends of the line segment. Let the angles of the two soft vertices V_i and V_{i+1} be A_i and A_{i+1}, respectively. If $A_i = A_{i+1}$, then the knot is placed halfway between the vertices, which means that $\nu = 1/2$ in Equation 6.25. If the angles are not the same, then the knot location should be biased away from the vertex with the larger angle, since the conic may not bend away from the line segment fast enough to follow the corner. The value for ν in Equation 6.25 can be set using the formula

$$\nu_i = \frac{A_1}{A_1 + A_2}. \tag{6.32}$$

Each sequence of line segments joined by soft vertices is replaced by a guided conic through the first and last vertices (or knots). The tangents are defined by the orientation of the first and last line segments. The tangents and end points determine four of the five degrees of freedom for the conic. The conic is fully specified by having it pass through the soft vertex in the middle of the sequence.

6.7 Spline Curves

The term *spline* refers to a function represented using piecewise polynomials. Splines occur in many applications. In data analysis, splines are used to fit a set of data points when no function model is available [245]. In computer graphics and computer-aided design, splines are used to represent free-form curves. In machine vision, splines provide a general-purpose representation for curves when no simpler model is adequate.

A spline can be made from any class of functions joined end to end. The most common form of spline is the cubic spline, which is a sequence of piecewise cubic polynomials. The curve representations presented in previous sections, such as sequences of line segments, circular arcs, and conic sections, are other examples of splines. Cubic splines allow more complex curves to be represented using fewer spline segments. Cubic splines are widely used in computer drawing programs for free-form curves and for representing character outlines in fonts. Since cubic splines are so widely used, it may be necessary for a machine vision program to fit this curve model to an edge list. Since interactive graphics interfaces for manipulating cubic spline curves are well known, a contour represented as a cubic spline can be modified manually by the user if necessary. This is a very important consideration, since the results of fitting a curve to edges may never be perfect.

One point to make clear is the difference between geometric and parametric equivalence. Two curves are geometrically equivalent if they trace the same set of points. In other words, the two curves are geometrically equivalent if they correspond to the same shape (or set of points) in space. Two curves are parametrically equivalent if their equations are identical. In other words, two curves are parametrically equivalent if their representation uses the same formula with the same parameters. Parametric equivalence is stronger than geometric equivalence.

Two curves can be geometrically equivalent but have different parametric representations. This is an important concept for fitting curves in machine vision. A machine vision system might produce a representation based on cubic splines that is very close (geometrically) to the true representation of an object boundary, but the representation may not be at all similar in a parametric sense. In applications such as object recognition or comparing the image of an industrial part with its model, it is not possible to compare the parametric forms of the cubic spline curves. The comparison must be based on geometric equivalence.

Cubic splines have enough degrees of freedom to allow the orientation of edge fragments to be used in the approximation. Recall that most edge detection algorithms can provide estimates of edge orientation (gradient angle) as well the position of the edge. Only the positions of edges were used in the algorithms for fitting line segments, circular arcs, and conic sections. With cubic splines, we can introduce an example of how to use the orientation information produced by an edge detector.

The equation for a cubic curve in the plane is

$$\mathbf{p}(u) = (x(u), y(u)) = \mathbf{a}_0 + \mathbf{a}_1 u + \mathbf{a}_2 u^2 + \mathbf{a}_3 u^3, \qquad (6.33)$$

where the coefficients $\mathbf{a}_0$, $\mathbf{a}_1$, $\mathbf{a}_2$, and $\mathbf{a}_3$ are two-element vectors (points in the image plane) and the parameter u covers the interval $[0, 1]$. The cubic curve begins at point $\mathbf{p}(0) = (x(0), y(0))$ and ends at point $\mathbf{p}(1) = (x(1), y(1))$. The cubic spline is a sequence of cubic curves $\mathbf{p}_1(u), \mathbf{p}_2(u), \ldots, \mathbf{p}_n(u)$, defined over successive intervals $[0, 1], [1, 2], \ldots, [n-1, n]$ and joined at the end points so that $\mathbf{p}_i(i) = \mathbf{p}_{i+1}(i)$. Each of the cubic curves in the spline is called a spline segment, and the edge points where the segments are joined are called knots.

As with the curve fitting algorithms presented in previous sections, the sequence of edge points is partitioned into subsequences and a spline segment is fit to each subsequence. Each cubic curve segment in the spline requires eight parameters. The positions of the first and last edge points in the subsequence provide four constraints. First-order continuity (equal tangent vectors) at the knots provides two more constraints. The orientation of the edges at the knots provides only one additional constraint on each segment, since the edge is shared by adjacent segments. Second-order continuity (equal curvature) at the knots would provide two more constraints, but then there would be too many equations for the eight parameters of each cubic spline segment.

It is important for the spline segments to be joined smoothly at the knots, and this is achieved in computer graphics by requiring second-order continuity. Requiring second-order continuity would overconstrain each spline segment, since the segments are already constrained to pass through selected edges with the orientation (tangent angle) constrained by the orientation of the edge; but one additional constraint can be provided by minimizing the magnitude of the second-order discontinuity at the knot. In other words, minimize the difference in curvature at the knots.

For the entire cubic spline curve, minimize the sum of the squared magnitude of the difference in the second derivative at the $n-1$ knots:

$$\chi^2 = \sum_{i=1}^{n-1} (\Delta \ddot{\mathbf{p}}), \qquad (6.34)$$

where the difference in the second derivatives of two spline segments at their common knot is

$$\Delta\ddot{\mathbf{p}} = \ddot{\mathbf{p}}_{i-1}(1) - \ddot{\mathbf{p}}_i(0) \tag{6.35}$$

$$= 2\left(\mathbf{t}_{i-1} + 4\mathbf{t}_i + \mathbf{t}_{i+1} + 3(\mathbf{p}_{i-1} - \mathbf{p}_i)\right). \tag{6.36}$$

The variable $\mathbf{t}_i$ is the tangent vector at knot i. The tangent vector has an orientation $\hat{\mathbf{t}}_i$ given by the edge orientation (gradient angle) and a signed magnitude γ_i which is unknown:

$$\mathbf{t}_i = \gamma_i\hat{\mathbf{t}}_i. \tag{6.37}$$

In other words, the orientation of an edge at the knot is modeled as a unit tangent vector, but the cubic spline requires a tangent with sign and magnitude to indicate from which direction the curve should pass through the knot and at what speed. The algorithm solves a system of linear equations for the n unknowns γ_i which provide the missing information for constructing the cubic spline segments between the knots.

This algorithm does not have any additional parameters or thresholds, but, as with the algorithms for fitting polylines, circular arcs, and conics presented in previous sections, the knots must be chosen from the edge list. The knot locations can be determined by using any of the polyline algorithms described above to compute a polyline approximation to the contour. The polyline vertices can be used as the knot locations. The number and placement of knots can be adjusted to improve the fit of the cubic spline to the entire set of edge points.

The cubic spline fitting algorithm is very efficient, since the solution only requires solving a small system of linear equations for the tangent signs and magnitudes. There are many interactive graphics interfaces that allow the user to easily adjust the cubic spline curve, if necessary.

6.8   Curve Approximation

The curve fitting methods described in previous sections interpolated the curve through a subset of the edges. Higher accuracy can be obtained by computing an approximation that is not forced to pass through particular edges.

This section presents methods for approximating a curve. There are several ways to approximate curves, depending on the reliability with which edge points can be grouped into contours. If it is certain that all of the edge points linked into a contour actually belong to the contour, then total least-squares regression can be used to fit a curve to the edge points. If some grouping errors are present, then robust regression methods can be used for computing the curve approximation. Finally, if the grouping of edges into contours is very unreliable, or if the edges are so scattered that grouping cannot be easily done using the edge linking or following methods discussed previously, then cluster analysis techniques must be used to perform grouping and curve fitting simultaneously. An excellent example of an algorithm for grouping and fitting scattered edge points is the Hough transform. All of these methods will be presented in the following sections.

The methods for fitting line segments, circular arcs, conic sections, and cubic splines to edge points presented in Sections 6.4 through 6.7 are trivial regression problems that fit curve segments between end points. These algorithms assume that the edge locations can be accurately computed, possibly using subpixel methods. Edge points in between the end points were not used in the regression. The accuracy of the curve approximation is determined by the accuracy of the location of the edge points chosen as the segment end points. The methods presented in this section will use all of the edge points to calculate the best approximation of the curve to the edge points.

The general curve fitting problem is a regression problem with the curve modeled by the implicit equation

$$f(x, y; a_1, a_2, \ldots, a_p) = 0 \qquad (6.38)$$

with p parameters. The curve estimation problem is to fit the curve model to a set of edge points $\{(x_1, y_1), (x_2, y_2), \ldots, (x_n, y_n)\}$.

In the noise-free case, one can use p observations to formulate p equations for the p unknown curve parameters. Unfortunately, in most applications this direct approach is not suitable due to noise. Real applications usually require the best estimate of the parameter values using all of the information in the edge list.

The next section will cover least-squares regression as it is used for curve fitting in machine vision. Least-squares methods are appropriate when the errors are normally distributed. Section 6.8.3 will present robust methods for curve fitting that are useful when some of the edge points have been incorrectly linked into the contour. These incorrectly assigned points are called *outliers*.

6.8.1   Total Regression

Classical linear regression minimizes the difference between a data point and the model in only one dimension, the dimension of the dependent variable. For example, a functional model of the form

$$y = f(x, a_1, \ldots, a_p) \tag{6.39}$$

relating the dependent variable y to the independent variable x, with the p model parameters a_1 through a_p, assumes that there are no errors in the independent variable x. In machine vision, errors in the x and y coordinates of location are equally likely and the curve model may be a vertical line, for instance, which cannot be represented in functional form. In machine vision, lines and other curve models are fitted to edges using total regression, which minimizes the sum of the squares of the perpendicular distances of the data points from the regression model. The advantage of this technique is that it compensates for errors in both the x and y directions. Total regression has actually already been presented in Chapter 2 where it was used to derive the equations for determining the orientation of a blob, although the term *total regression* was not used at the time.

To avoid problems when the line is vertical, represent the equation for a line by using polar coordinates:

$$x \cos \theta + y \sin \theta - \rho = 0. \tag{6.40}$$

Minimize the sum of the squared perpendicular distances of points (x_i, y_i) from the line:

$$\chi^2 = \sum_i (x_i \cos \theta + y_i \sin \theta - \rho)^2. \tag{6.41}$$

The solution to the total regression problem is

$$\rho = \bar{x} \cos \theta + \bar{y} \sin \theta \tag{6.42}$$

with

$$\bar{x} = \frac{1}{n}\sum_{i=1}^{n} x_i \tag{6.43}$$

$$\bar{y} = \frac{1}{n}\sum_{i=1}^{n} y_i. \tag{6.44}$$

The orientation of the total regression line is θ, given by

$$\tan 2\theta = \frac{a}{b}, \tag{6.45}$$

with

$$a = 2\sum_{i=1}^{n} x_i' y_i' \tag{6.46}$$

$$b = \sum_{i=1}^{n} x_i'^2 - \sum_{i=1}^{n} y_i'^2 \tag{6.47}$$

and

$$x_i' = x_i - \bar{x} \tag{6.48}$$

$$y_i' = y_i - \bar{y}. \tag{6.49}$$

Total regression uses a least-squares error norm that is optimal if the errors are from a normal distribution, but is not suitable if there are outliers present in the data. In the case of fitting a curve model to edge data, outliers would occur if the edge linking procedure incorporated one or more edges from other contours into the edge list for a contour. Outliers can occur even if the edge linking procedure performs flawlessly. For example, consider a list of edges from two adjacent sides of a rectangle. The corner must be identified in order to segment the edges into the two sides before fitting a line to the edges. If the corner point is not identified correctly, some edges may be assigned to the wrong side, and these edges are outliers.

In general, errors in classification introduce errors into the regression problem that are not normally distributed. In such a case, the errors may be modeled by a mixture distribution that combines a Gaussian distribution for modeling the normal errors with a broad-tailed distribution for modeling the outliers due to imperfect classification.

6.8.2   Estimating Corners

The best method for estimating corners is to use one of the methods for fitting a line to edge points and then compute the intersection of the lines. This method compensates for the error introduced by edge detection operators that round off the corners and is more accurate than using a corner detector which only uses local information.

Given the implicit equations for two lines,

$$a_1 x + b_1 y + c_1 = 0 \tag{6.50}$$
$$a_2 x + b_2 y + c_2 = 0, \tag{6.51}$$

the location of the intersection is

$$y = \frac{c_1 a_2 - c_2 a_1}{a_1 b_2 - a_2 b_1} \tag{6.52}$$

$$x = \frac{c_2 b_1 - c_1 b_2}{a_1 b_2 - a_2 b_1}. \tag{6.53}$$

If $a_1 b_2 - a_2 b_1$ is close to zero, then the lines are nearly parallel and cannot be intersected.

A good method for detecting corners is to try to fit pairs of lines over successive sublists of $2n + m$ edge points along the contour. The parameter n is the number of edge points required for an accurate line fit, and the parameter m is the number of edge points to skip between the sides of the corner. The gap skips over the edge points in the rounded part of the corner. A corner is detected by testing the magnitude of $a_1 b_2 - a_2 b_1$ against a threshold.

6.8.3   Robust Regression

If the errors are not from a normal distribution, then least-squares is not the best fitting criterion. Figure 6.12 shows an example of the problems encountered by least-squares regression when the data set contains outliers. Even a single outlier is enough to pull the regression line far away from its correct location. Robust regression methods try various subsets of the data points and choose the subset that provides the best fit.

The physical analogy shown in Figure 6.13 may make this discussion more clear. Imagine that you want to find the center of mass of a set of points

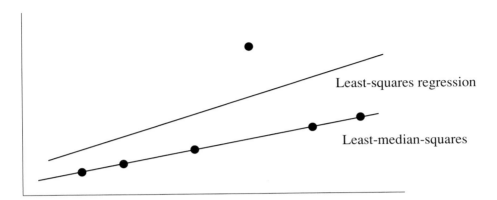

Figure 6.12: Illustration of the difference between fitting a curve using least-squares regression and fitting a curve using robust methods to a data set that contains outliers.

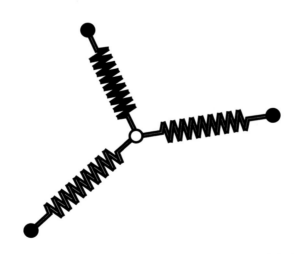

Figure 6.13: A physical analogy that illustrates the sensitivity of least-squares methods to outliers. Even a single outlier renders a least-squares solution useless.

in the plane. Attach springs with equal spring constants to the fixed points and to a small object that can move freely. The object will be pulled to the average of the locations of the points. The springs implement a least-squares norm through the spring equation for potential energy. This physical analogy corresponds to the derivation of the calculation of an average from the criterion that the sum of the squares of the residuals, the differences between each point and the average, should be minimized. Now suppose that one of the points can be moved. Call this point a leverage point. It is possible to force the location of the average to be shifted to any arbitrary point by pulling the leverage point far enough away. This illustrates the extreme sensitivity of estimators based on least-squares criteria to outliers. Even a single outlier can ruin an estimate. Ideally, one would like to break the spring connected to an outlier so that the estimate remains unharmed. Changing the spring constants so that points that are far away exert little influence on the estimate corresponds to the implementation of robust estimators based on influence functions. Breaking the springs attached to outliers corresponds to resampling schemes where a consistent subset of samples is determined. Resampling plans repeatedly draw random subsets and choose the subset that yields the best estimate. Examples of resampling algorithms include random sample consensus, least-median-squares regression, and other computer-intensive methods in regression.

The spring analogy also extends to linear regression with the same conclusions: even a single outlier will distort the regression estimate. A linear, multivariate model of order n is represented by the equation

$$\hat{y}_i = \hat{\theta}_1 x_{i1} + \hat{\theta}_2 x_{i2} + \cdots + \hat{\theta}_n x_{in} \tag{6.54}$$

for the ith data point, where $\hat{\theta}_i$ are the estimates of the model parameters θ_i. The residual for each data point (the deviation of the data point from the estimated model) is $r_i = y_i - \hat{y}_i$. In least-squares regression, the estimates of the model parameters are given by minimizing the sum of the squares of the residuals:

$$\min_{\hat{\theta}} \sum_{i=1}^{n} r_i^2. \tag{6.55}$$

As demonstrated by the spring analogy described above, the model parameters can be arbitrary if only one of the data points is an outlier.

Often, the noise and outliers can be modeled as a mixture distribution: a linear combination of a normal distribution to model the noise and a broad-tailed distribution to account for outliers. In this case, it makes sense to formulate an estimator with a norm that resembles a least-squares norm for small errors but is insensitive to large errors so that outliers are ignored. This is called the influence function approach.

The breakdown point is the smallest percentage of data points that can be incorrect to an arbitrary degree and not cause the estimation algorithm to produce an arbitrarily wrong estimate [207]. Let Z be a set of n data points. Suppose that the set Z' is a version of set Z with m points replaced with arbitrary values. Let a regression estimator be denoted by $\hat{\theta} = T(Z)$. The bias in an estimate due to outliers is given by

$$Bias = \sup_{Z'} \parallel T(Z') - T(Z) \parallel .\tag{6.56}$$

The idea behind the breakdown point is to consider what happens to the bias as the number of outliers m as a percentage of the number of data points n is increased. Since the data points can be replaced with arbitrary values, for some ratio of m to n the bias can potentially be unbounded. This is the breakdown point. Below the breakdown point, the regression estimator may be able to totally reject outliers, or the outliers may have only some small effect on the estimate. Beyond the breakdown point, the outliers can drive the estimator to produce an arbitrary answer in the sense that the answer will depend on the outliers and not on the legitimate data. In other words, the result provided by the estimator is unpredictable. The breakdown point is defined as

$$\epsilon_n^\star = \min \left\{ \frac{m}{n} : Bias(m; T, Z) \text{ is infinite} \right\} .\tag{6.57}$$

For least-squares regression, $\epsilon_n^\star = 1/n$, and in the limit as the number of data points becomes large, $\epsilon_\infty^\star = 0$. In other words, least-squares regression has no immunity to outliers; a single outlier can completely ruin the result.

Least-median-squares regression is a very simple technique to implement and has proven to be very powerful in solving regression problems when there is a large percentage of outliers. Least-median-squares regression can tolerate up to 50 percent outliers in a data set. What this means is that up to half of the data points in a data set can be arbitrary without significantly affecting the regression result.

In least-median-squares regression, the estimates of the model parameters

Algorithm 6.3   Least-Median-Squares Regression

Assume that there are n data points and p parameters in the linear model.

1. *Choose p points at random from the set of n data points.*

2. *Compute the fit of the model to the p points.*

3. *Compute the median of the square of the residuals.*

The fitting procedure is repeated until a fit is found with sufficiently small median of squared residuals or up to some predetermined number of resampling steps.

are given by minimizing the median of the squares of the residuals:

$$\min_{\hat{\theta}} \operatorname*{med}_{i} r_i^2. \tag{6.58}$$

The least-median-squares algorithm is described in Algorithm 6.3.

The median has a 50 percent breakdown point, and this property carries over to least-median-squares regression [207]. In other words, even if as many as half of the data points are outliers, the regression estimate is not seriously affected. If more than 50 percent of the data points are outliers, then least-median-squares regression may not work well, and more powerful techniques, such as the Hough transform, must be used.

6.8.4   Hough Transform

The last few years have seen increasing use of parameter estimation techniques that use a voting mechanism. One of the most popular voting methods is the Hough transform. In the Hough transform, each point on a curve votes for several combinations of parameters; the parameters that win a majority of votes are declared the winners. Let us consider this approach for fitting a straight line to data. Consider the equation of a straight line:

$$y = mx + c. \tag{6.59}$$

In the above equation, x and y are observed values, and m and c represent the parameters. If the values of the parameters are given, the relationship

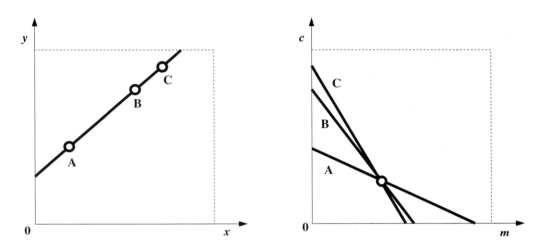

Figure 6.14: Image–to–parameter space mapping of a point in the Hough transform.

between the coordinates of the point is clearly specified. Let us rewrite the above equation as

$$c = -xm + y. \tag{6.60}$$

Now, in the above equation, let us assume that m and c are variables of interest, and x and y are constants. The equation above represents a straight line in the m–c space. The slope and intercept of this line are determined by x and y. A point (x, y) corresponds to a straight line in m–c space. This mapping is shown in Figure 6.14. It should be mentioned here that the shape of the curve in the parameter space depends on the original function used to represent the curve. In practice, the polar form of the line

$$\rho = x \cos \theta + y \sin \theta \tag{6.61}$$

is used rather than the explicit form to avoid problems with lines that are nearly vertical. Edge points (x, y) are mapped into the (ρ, θ) parameter space.

In the case of a straight line, as represented above, if there are n points lying on this straight line, then these points will correspond to a family of straight lines in the parameter space, as shown in Figure 6.14. All these lines will pass through the point (m, c) in the parameter space. This point gives the parameters of the original straight line.

Algorithm 6.4   Hough Transform Algorithm

1. *Quantize the parameter space appropriately.*

2. *Assume that each cell in the parameter space is an accumulator. Initialize all cells to zero.*

3. *For each point (x, y) in the image space, increment by 1 each of the accumulators that satisfy the equation.*

4. *Maxima in the accumulator array correspond to the parameters of model instances.*

If we are interested in finding the straight line that best fits n points in an image, then we can use the above mapping from the image space to the parameter space. In this approach, called the Hough transform, we represent the parameter space as an array of accumulators, representing discrete parameter values. Each point in the image votes for several parameters, according to the transformation equation. To find parameters that characterize the line, we should detect peaks in the parameter space. This general idea is highlighted in Algorithm 6.4.

The Hough transform does not require prior grouping or linking of the edge points, and the edge points that lie along the curve of interest may constitute a small fraction of the edges in the image. In particular, the number of edges that actually lie along the curve could be less than half of the number of edges in the scene, which would rule out most robust regression methods. The assumption behind the Hough transform is that in the presence of large amounts of noise, the best that can be done is to find the point in the parameter space that satisfies the maximum number of edges in the image. If the peak in the parameter space covers more than one accumulator, then the centroid of the region that contains the peak provides an estimate of the parameters.

If there are several curves in the image that are matched by the model, then there will be several peaks in the parameter space. It is possible to detect each peak, remove the edges associated with the curve instance corresponding to the peak, and continue to detect the remaining curves, until the peaks are not significant. However, it can be difficult to determine whether a peak is significant.

Another problem with the Hough transform is that the size of the discrete parameter space increases very quickly as the number of parameters increases. For a circular arc, the parameter space has three dimensions; for other curves the dimensionality may be even higher. Since the number of accumulators increases exponentially with the dimension of the space, the Hough transform may be computationally very inefficient for complex models. Several methods have been suggested to improve the performance of the Hough transform. One method uses gradient information for boundaries to reduce work in the parameter space. Suppose that the curve model is a circle. This model has three parameters: two parameters for the center of the circle and one parameter for the radius of the circle. If the gradient angle for edges is available, then this provides a constraint that reduces the number of degrees of freedom and hence the required size of the parameter space. The direction of the vector from the center of the circle to each edge is determined by the gradient angle, leaving the value of the radius as the only unknown parameter. There are other methods that may be used to reduce the size of the parameter space.

The details of using the gradient angle to reduce the size of the parameter space are explained for circle fitting. The algorithm is detailed in Algorithm 6.5. The implicit equation for a circle is

$$(x - a)^2 + (y - b)^2 = r^2. \tag{6.62}$$

The parametric equations for a circle in polar coordinates are

$$x = a + r \cos \theta \tag{6.63}$$
$$y = b + r \sin \theta. \tag{6.64}$$

Solve for the parameters of the circle to obtain the equations:

$$a = x - r \cos \theta \tag{6.65}$$
$$b = y - r \sin \theta. \tag{6.66}$$

Given the gradient angle θ at an edge point (x, y), compute $\cos \theta$ and $\sin \theta$. Note that these quantities may already be available as a by-product of edge detection. Eliminate the radius from the pair of equations above to yield

$$b = a \tan \theta - x \tan \theta + y. \tag{6.67}$$

Algorithm 6.5   Circle Fitting Algorithm

1. *Quantize the parameter space for the parameters a and b.*

2. *Zero the accumulator array $M(a,b)$.*

3. *Compute the gradient magnitude $G(x,y)$ and angle $\theta(x,y)$.*

4. *For each edge point in $G(x,y)$, increment all points in the accumulator array $M(a,b)$ along the line*

$$b = a \tan\theta - x \tan\theta + y. \qquad (6.68)$$

5. *Local maxima in the accumulator array correspond to centers of circles in the image.*

This is the equation for updating the accumulators in the parameter space. For each edge point at position (x,y) with edge orientation θ, increment the accumulators along the line given by Equation 6.67 in the (a,b) parameter space.

If the radius is known, then it is only necessary to increment the accumulator for the point (a,b) given by

$$a = x - r\cos\theta \qquad (6.69)$$
$$b = y - r\sin\theta. \qquad (6.70)$$

It is not necessary that the curves to be detected by the Hough transform be described by a parametric equation. The Hough transform can be generalized into a voting algorithm (see Algorithm 6.6) that implements template matching efficiently.

Algorithm 6.6 encodes the shape of the object boundary in a table for efficient access. One point on the object is chosen as the reference point. By definition, the location of the reference point in the image is the location of the object. For each image gradient point at (x,y) with gradient angle θ, the possible locations of the reference point are given by

$$a = x - r(\theta)\cos(\alpha(\theta)) \qquad (6.71)$$
$$b = y - r(\theta)\sin(\alpha(\theta)). \qquad (6.72)$$

Algorithm 6.6 Generalized Hough Transform

1. *Pick a reference point on the object.*

2. *Compute the gradient angles θ_i along the object boundary.*

3. *For each gradient point θ_i, store the distance r_i and angle α_i from the reference point.*

Each possible reference point location is incremented. The location of the peak in the parameter space is the estimate for the location of the object. It is not easy to generalize this technique to incorporate changes in scale or rotation.

6.9   Fourier Descriptors

Since the position along a closed contour is a periodic function, Fourier series may be used to approximate the contour. The resolution of the contour approximation is determined by the number of terms in the Fourier series.

Suppose that the boundary of an object is expressed as a sequence of coordinates $\mathbf{u}(n) = [x(n), y(n)]$, for $n = 0, 1, 2, \ldots, N - 1$. We can represent each coordinate pair as a complex number so that

$$u(n) = x(n) + j\, y(n) \tag{6.73}$$

for $n = 0, 1, 2, \ldots, N - 1$. In other words, the x axis is treated as the real axis, and the y axis is treated as the imaginary axis of a series of complex numbers. Note that for a closed boundary, this sequence is periodic with period N and that now the boundary is represented in one dimension.

The discrete Fourier transform (DFT) representation of a one-dimensional sequence $u(n)$ is defined as

$$u(n) = \sum_{k=0}^{N-1} a(k) e^{\frac{j 2\pi k n}{N}}, \qquad 0 \le n \le N - 1 \tag{6.74}$$

$$a(k) = \frac{1}{N} \sum_{n=0}^{N-1} u(n) e^{\frac{-j 2\pi k n}{N}}, \qquad 0 \le k \le N - 1. \tag{6.75}$$

The complex coefficients $a(k)$ are called the *Fourier descriptors* of the boundary.

Fourier descriptors are compact representations for closed contours. However, low-resolution approximations, using only the low-order terms in the series, can be used as an even more compact representation. If only the first M coefficients are used, which is equivalent to setting $a(k) = 0$ for $k > M-1$, the following approximation to $u(n)$ is obtained:

$$\hat{u}(n) = \sum_{k=0}^{M-1} a(k)e^{\frac{j2\pi kn}{N}}, \qquad 0 \le n \le N - 1. \qquad (6.76)$$

Although only M terms are used to obtain each component of the boundary $u(n)$, n still ranges from 0 to $N - 1$. In other words, the same number of points are in the approximated boundary, but not as many terms are used in reconstructing each point.

Simple geometric transformations of a boundary, such as translation, rotation, and scale, are related to simple operations of the boundary's Fourier descriptors (see Exercise 6.18). This makes the use of Fourier descriptors attractive for boundary matching. However, Fourier descriptors do have problems with occluded shapes. There are other methods for obtaining similar descriptors, using other boundary representations.

Further Reading

The Hough transform is an efficient method for detecting lines and other features from imperfect edges. A discussion of generalized Hough transform methods is given in the paper by Ballard [18]. Generalizing the Hough transform to detect arbitrary shapes, Stockman used Hough transform techniques for pose clustering in 2-D and 3-D problems in object recognition and localization [228]. Asada and Brady [45, 10] present a rigorous approach to contour and region descriptors. Work on measuring the deviation from roundness of a circle has been reported by Van-Ban and Lee [243].

The hop-along algorithm approximates a contour by a sequence of line segments but works on short subsequences of edges and is given in [193]. Curve fitting using line segments (polylines) and circular arcs is adapted from Pavlidis [193]. The method for fitting cubic splines to edge points using orientation was published by Tehrani, Weymouth, and Schunck [231].

Algorithms for fitting a circle to three points have been developed in computer graphics [204].

A discussion of robust regression using M-estimators is provided in numerical recipes [197, pp. 558–565]. Another robust regression technique is least-median-squares regression [207]. The influence function approach was pioneered by Huber [115, 97]. For resampling algorithms including random sample consensus, see [38, 80]. Least-median squares regression is discussed by Rousseeuw and Leroy [207], and another good source for information on computer-intensive methods in regression is a review article by Efron [71].

The Fourier descriptor technique has been applied to medical imagery by Staib and Duncan [227].

Exercises

6.1 What is a contour? How is it related to a region? What does an open contour represent?

6.2 List the criterion you will consider in selecting a contour representation. Discuss the implications of these factors for object recognition.

6.3 What is the difference between interpolation and approximation methods? Which one is better?

6.4 To implement rotation by $n \times 22.5°$, one may use a chain code using 16 directions. How can you implement such a code? Why is the 8-direction chain code almost always used?

6.5 The D_8 distance with respect to the origin is defined as $\max(x, y)$. Using this measure, find the signature of the contour in Figure 6.15 by plotting the D_8 distance as a function of pixel number. In addition, find the 8-direction chain code and difference code measured in the counterclockwise direction for the following object. Note that the starting point is an empty circle as opposed to a solid dot.

6.6 What are the criteria for choosing the threshold for normalized maximum error? (*Hint:* Consider the variance in the estimate of edge location.)

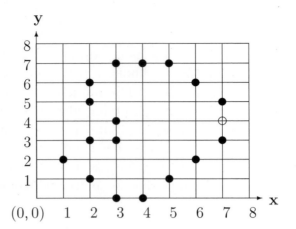

Figure 6.15: Contour for Exercise 6.5.

6.7 Consider the method for estimating the location of a corner presented in Section 6.8.2. What is the value of $a_1 b_2 - a_2 b_1$ when the lines are at right angles? What is the value when the lines meet at an angle θ? This is the formula to use for setting the threshold for corner detection.

6.8 Consider the method for estimating the location of a corner presented in Section 6.8.2. Assume that the error in the x and y coordinates of the edge locations has a normal distribution with variance σ^2. What is the error distribution for the location of the corner? How is the error affected by the angle of the corner?

6.9 Why is the Ψ-s representation considered a continuous chain code? What are its most attractive features?

6.10 How can you compare two objects using their Ψ-s representations?

6.11 Discuss different error measures that you can use in approximation. What is the role of an error measure in an approximation technique?

6.12 What is a conic section? How many types of conic sections are possible? Are there mathematical conditions that define types of a conic section?

6.13 What is a spline? Where can it be used? Why are cubic splines considered more powerful representations than polylines and conic sections?

6.14 Discuss the difference between geometric equivalence and parametric equivalence.

6.15 Why is the least-squares measure used in total regression? List its advantages and disadvantages.

6.16 How does robust regression overcome limitations of the total regression? Why is robust regression not very popular in approximation?

6.17 What are the strengths and weaknesses of Fourier descriptors for approximating and representing closed contours?

6.18 Several geometric transformations of object boundaries are related to simple operations on the Fourier descriptors as follows:

Transformation	Boundary	Fourier descriptor
Identity	$u(n)$	$a(k)$
Translation	$\tilde{u}(n) = u(n) + u_0$	$\tilde{a}(k) = a(k) + u_0\delta(k)$
Scaling or zooming	$\tilde{u}(n) = \alpha u(n)$	$\tilde{a}(k) = \alpha a(k)$
Starting point	$\tilde{u}(n) = u(n - n_0)$	$\tilde{a}(k) = a(k)e^{\frac{-j2\pi n_0 k}{N}}$
Rotation	$\tilde{u}(n) = u(n)e^{j\theta_0}$	$\tilde{a}(k) = a(k)e^{j\theta_0}$
Reflection	$\tilde{u}(n) = u^*(n)e^{j2\theta} + 2\gamma$	$\tilde{a}(k) = a^*(-k)e^{j2\theta} + 2\gamma\delta(k)$

Consider a simple square object with coordinates of boundary points $A = (0,0), B = (0,1), C = (1,1),$ and $D = (1,0)$.

a. Find its Fourier descriptors when the starting point is A and the boundary is traversed in the order A, B, C, D.

b. Find the descriptors when the object translates such that A is at $(2,3)$.

c. Find the descriptors of the translated object if the length of the sides of the square is changed to 2.

d. Find the descriptors of the translated and scaled object if the starting point is changed to B.

6.19 The Fourier descriptors of a simple square object with coordinates of boundary points given by $A = (-0.5, -0.5), B = (-0.5, 0.5), C = (0.5, 0.5), D = (0.5, -0.5)$ are given by

$$a(0) = a(1) = a(2) = 0 \tag{6.77}$$
$$a(3) = -0.5 - j\,0.5. \tag{6.78}$$

Starting with this data and using the properties of the Fourier descriptors given in the above problem, find the Fourier descriptors for the object given by $P = (0, 1), Q = (-1, 2), R = (0, 3), S = (1, 2)$.

6.20 What is the Hough transform? Is it related to robust regression? How?

6.21 Can you extend the Hough transform to detect an arbitrary shape? How will you develop a Hough transform that detects an object in its rotated and scaled versions?

6.22 Find the Hough transform of the lines enclosing an object with vertices $A = (2, 0), B = (2, 2)$, and $C = (0, 2)$. Sketch the modified object enclosed by lines obtained by replacing (ρ, θ) of the object lines by $(\rho^2, \theta + 90°)$. Calculate the area of the modified object.

6.23 The two-dimensional Hough transform for line detection can be generalized to the 3-D case to detect planes. The Hough domain parameters are then specified by the three variables ρ, θ, and ϕ, where the angles are measured as shown in Figure 6.16. Consider a unit cube whose diagonal corners, A and G, are located at $(1, 1, 1)$ and $(2, 2, 2)$ as shown. Find the Hough transform of the plane passing through the vertices C, H, and F.

Computer Projects

6.1 Develop an algorithm to find the chain code of a given curve. Can you determine corners using chain code? If so, implement this algorithm and test it on several images.

6.2 Develop a program to start with an image and find the slope–arc–length plot of the largest object in the image. Segment this object in its linear and circular segments.

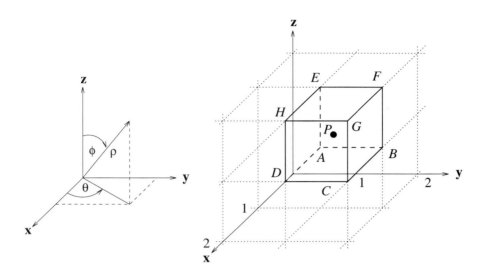

Figure 6.16: Diagram for Exercise 6.23.

6.3 Develop an algorithm to accept given points and provide a cubic spline. Apply this algorithm to the output of an edge detector.

6.4 Develop a Hough transform algorithm to detect straight lines in images. Use this to approximate an image by finding all lines that are above a fixed size, say 20 points.

6.5 Four binary objects are shown in Figure 6.17(a). Scaled versions of the same objects are shown in Figure 6.17(b), and scaled and rotated versions are shown in Figure 6.17(c). Consider various contour representation methods and comment on their suitability for matching scaled and scaled-rotated versions of objects with their corresponding original images. Implement one of the methods as a computer program.

6.6 Create a synthetic image of a rectangle with uniform intensity against a uniform background. Compute the edges using any edge detector from Chapter 5. Fit polylines to the edges. How close are the vertices to the true corner locations? Is the error consistently biased in one direction?

6.7 Create a synthetic image of a rectangle with rounded corners, modeled as quarter circles. Use edge detection and polyline fitting, and then replace runs of polylines with circular arcs using the algorithm presented

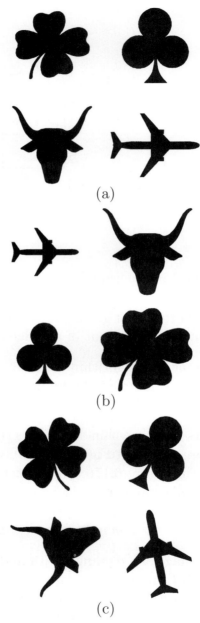

(a)

(b)

(c)

Figure 6.17: Binary objects used in Computer Project 6.5. (a) Four reference binary objects. (b) Scaled versions of the objects in (a). (c) Scaled and rotated versions of the objects in (a).

in this chapter for fitting circular arcs. Measure the error in the computed end points of the circles. Are the errors symmetrical, or are the corners unevenly distorted so that the rectangle appears skewed?

6.8 Experiment with the corner fitting algorithm described in Section 6.8.2. Create a synthetic image of a rectangle, adding noise from a normal distribution and scattered edges to simulate false positives in edge detection. Plot the error in the corner estimates versus the noise level.

6.9 Implement least-median-squares regression for fitting a line to a list of edge points. Add false edges to the list to simulate grouping errors. Plot the maximum distance of the estimated line from the true line versus the number of false edges.

6.10 Consider an object that is symmetrical about a vertical axis. Suppose that two lists of edge fragments along the left and right sides of the object are available. Adapt the algorithm for fitting cubic splines to edges to enforce the symmetry constraint. Generalize the algorithm so that the axis can be at any orientation.

6.11 Suppose that the list of edge points along the contour of an object is available and that the object is symmetric about some axis. Process the edge list into a sequence of line segments and circular arcs. Develop an algorithm for matching line segments and circular arcs to detect that the object is symmetric and estimate the axis of symmetry. After the axis of symmetry has been determined, this information can be used to improve the estimates of the line segments and circular arcs. Develop an algorithm for refining the contour representation. Experiment with an iterative algorithm that detects the axis of symmetry, uses this information to refine the contour representation, and then uses the improved contour representation to refine the axis estimate, repeating these steps until the axis and contour representations converge.

6.12 Consider an image of two rectangles. The task is to measure the gap between the two rectangles. The facing sides of the two rectangles are parallel, so the width of the gap is constant along its length. Explore different algorithms for measuring the gap.

a. Develop a formula for the average distance between two line segments that are nearly parallel and approximately of the same length. (*Hint:* Represent each line segment in parametric form on the interval $[0, 1]$.)

b. Compute the polyline representation for the rectangles. Are the line segments that bound the gap skewed due to poor estimates of the corner locations? Use the formula from part **a** to estimate the width of the gap.

c. Modify the algorithm to use improved techniques for estimating the locations of the corners. Measure the improvement in the estimate of the gap width.

d. Implement least-median-squares regression and repeat the measurements of the gap width, comparing the results using least median squares regression with the previous techniques.

e. Repeat the experiment using lines estimated with the Hough transform.

Synthetic noise, including noise from a normal distribution and scattered edges that model false positives in edge detection, can be added to the image to test the accuracy of gap measurement at various noise levels. Prepare a plot of the error in gap width measurement versus the noise level for each of the techniques.

6.13 Consider a line of text on a printed page that has been scanned into the computer. The task is to fit a line through the baseline of the text. The estimated baseline can guide algorithms for character recognition. Note that the baseline is not necessarily horizontal since the page may not be exactly vertical when scanned.

a. Suppose that edge detection has reduced the line of text to a list of edge points. Compare line fitting with least-squares, least-median-squares, and the Hough transform.

b. Consider multiple lines of text, reduced to multiple edge lists. Assume that the baseline separation is constant and that the baselines are parallel. Use these constraints to formulate a more accurate algorithm for determining the baselines.

    c. Suppose that the baselines are parallel, but the baseline separation varies between lines. How does this change the algorithm?

6.14 Consider a list of edge points that form a circle. The task is to estimate the position and radius of the circle. Use nonlinear regression to fit a circle to the edge points, assuming that the radius of the circle is known. Modify the algorithm to determine both the position and radius. Finally, modify the algorithm to determine both the position and radius of the circle, but include a penalty term for variation in the estimated radius from a nominal value.

Chapter 7

Texture

7.1 Introduction

Texture plays an important role in many machine vision tasks such as surface inspection, scene classification, and surface orientation and shape determination. For example, surface texture features are used in the inspection of semiconductor wafers, gray-level distribution features of homogeneous textured regions are used in the classification of aerial imagery, and variations in texture patterns due to perspective projection are used to determine three-dimensional shapes of objects.

Texture is characterized by the spatial distribution of gray levels in a neighborhood. Thus, texture cannot be defined for a point. The resolution at which an image is observed determines the scale at which the texture is perceived. For example, in observing an image of a tiled floor from a large distance we observe the texture formed by the placement of tiles, but the patterns within the tiles are not perceived. When the same scene is observed from a closer distance, so that only a few tiles are within the field of view, we begin to perceive the texture formed by the placement of detailed patterns composing each tile. For our purposes, we can define texture as repeating patterns of local variations in image intensity which are too fine to be distinguished as separate objects at the observed resolution. Thus, a connected set of pixels satisfying a given gray-level property which occur repeatedly in an image region constitutes a textured region. A simple example is a repeated pattern of dots on a white background. Text printed on white paper such as this page also constitutes texture. Here, each gray-level primitive is formed by the connected set of pixels representing each character. The process of placing the characters on lines and placing lines in turn as elements of the page results in an ordered texture. There are three primary issues in texture analysis: texture classification, texture segmentation, and shape recovery from texture.

In texture classification, the problem is identifying the given textured region from a given set of texture classes. For example, a particular region in an aerial image may belong to agricultural land, forest region, or an urban area. Each of these regions has unique texture characteristics. The texture analysis algorithms extract distinguishing features from each region to facilitate classification of such patterns. Implicit in this is the assumption that the boundaries between regions have already been determined. Statistical methods are extensively used in texture classification. Properties such as gray-level co-occurrence, contrast, entropy, and homogeneity are computed from image gray levels to facilitate classification. These are discussed in Section 7.2. The statistical methods are particularly useful when the texture primitives are small, resulting in *microtextures*. On the other hand, when the size of the texture primitive is large, it becomes necessary to first determine the shape and properties of the basic primitive and then determine the rules which govern the placement of these primitives, forming *macrotextures*. Such structural methods are briefly discussed in Section 7.3. As an alternative to the bottom-up analysis of image pixels to determine texture properties for classification, model-based methods to synthesize texture have been studied. In these methods a model for texture is first assumed and its parameters are then estimated from the image region such that an image synthesized using the model closely resembles the input image region. The parameters are then useful as discriminating features to classify the region. These are discussed in Section 7.4.

As opposed to texture classification, in which the class label of a single homogenous region is determined using properties computed from the region, texture segmentation is concerned with automatically determining the boundaries between various textured regions in an image. Although quantitative texture measures, once determined, are useful in segmentation, most of the statistical methods for determining the texture features do not provide accurate measures unless the computations are limited to a single texture region. Both region-based methods and boundary-based methods have been attempted to segment textured images. These methods are analogous to those used for object-background separation methods discussed in earlier chapters. Texture segmentation is still an active area of research, and numerous methods, each designed for a particular application, have been proposed in the literature. However, there are no general methods which are useful in a wide variety of situations. Thus, we do not cover texture segmentation methods in this book.

Image plane variations in the texture properties, such as density, size, and orientation of texture primitives, are the cues exploited by shape-from-texture algorithms. For example, the texture gradient, defined as the magnitude and direction of maximum change in the primitive size of the texture elements, determines the orientation of the surface. Quantifying the changes in the shape of texture elements (e.g., circles appearing as ellipses) is also useful to determine surface orientation. These are discussed in Section 7.5.

7.2   Statistical Methods of Texture Analysis

Since texture is a spatial property, a simple one-dimensional histogram is not useful in characterizing texture (for example, an image in which pixels alternate from black to white in a checkerboard fashion will have the same histogram as an image in which the top half is black and the bottom half is white). In order to capture the spatial dependence of gray-level values which contribute to the perception of texture, a two-dimensional dependence matrix known as a *gray-level co-occurrence* matrix is extensively used in texture analysis. Another measure that has been used extensively is the *autocorrelation* function. These are discussed briefly in this section.

Gray-Level Co-occurrence Matrix

The gray-level co-occurrence matrix $P[i, j]$ is defined by first specifying a displacement vector $\mathbf{d} = (dx, dy)$ and counting all pairs of pixels separated by $\mathbf{d}$ having gray levels i and j. For example, consider the simple 5×5 image having gray levels 0, 1, and 2 as shown in Figure 7.1(a). Since there are only three gray levels, $P[i, j]$ is a 3×3 matrix. Let the position operator be specified as $(1, 1)$, which has the interpretation: one pixel to the right and one pixel below. In a 5×5 image there are 16 pairs of pixels which satisfy this spatial separation. We now count all pairs of pixels in which the first

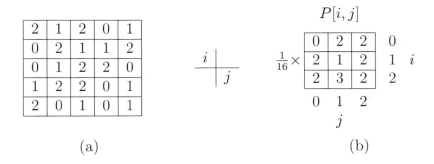

Figure 7.1: (a) A 5×5 image with three gray levels 0, 1, and 2. (b) The gray-level co-occurrence matrix for $\mathbf{d} = (1, 1)$.

pixel has a value of i and its matching pair displaced from the first pixel by $\mathbf{d}$ has a value of j, and we enter this count in the ith row and jth column of the matrix $P[i, j]$. For example, there are three pairs of pixels having values $[2, 1]$ which are separated by the specified distance, and hence the entry $P[2, 1]$ has a value of 3. The complete matrix $P[i, j]$ is shown in Figure 7.1(b).

Note that $P[i, j]$ is not symmetric since the number of pairs of pixels having gray levels $[i, j]$ does not necessarily equal the number of pixel pairs having gray levels $[j, i]$. The elements of $P[i, j]$ are normalized by dividing each entry by the total number of pixel pairs. In our example, each entry is divided by 16. This normalized $P[i, j]$ is then treated as a probability mass function since the entries now add up to 1.

It is easy to illustrate that the gray-level co-occurrence matrix captures the spatial distribution of gray levels with the following simple example. Consider the 8×8 binary image of a checkerboard shown in Figure 7.2(a), where each square corresponds to a single pixel. Since there are only two gray levels, $P[i, j]$ is a 2×2 matrix. If we define $\mathbf{d} = (1, 1)$ as before, we get the normalized $P[i, j]$ shown in Figure 7.2(b). Notice that the only pairs that occur are $[1, 1]$ and $[0, 0]$ because of the well-defined structure of pixels; the off-diagonal elements are zero. Similarly, if the vector $\mathbf{d}$ is defined as $(1, 0)$, the only entries will be those corresponding to $[0, 1]$ and $[1, 0]$, as shown in Figure 7.2(c), and the diagonal elements are zero.

In the above example, if the black pixels are randomly distributed throughout the image with no fixed structure, the gray-level co-occurrence matrix

Figure 7.2: (a) An 8×8 checkerboard image. (b) The gray-level co-occurrence matrix for $\mathbf{d} = (1,1)$. (c) The gray-level co-occurrence matrix for $\mathbf{d} = (1,0)$.

will not have any preferred set of gray-level pairs. In such a case the matrix is expected to be uniformly populated. Thus, a feature which measures the randomness of gray-level distribution is the *entropy*, defined as

$$\text{Entropy} = -\sum_i \sum_j P[i,j] \log P[i,j]. \tag{7.1}$$

Note that the entropy is highest when all entries in $P[i,j]$ are equal; such a matrix corresponds to an image in which there are no preferred gray-level pairs for the specified distance vector $\mathbf{d}$. The features of *energy*, *contrast*, and *homogeneity* are also defined using the gray-level co-occurrence matrix as given below:

$$\text{Energy} = \sum_i \sum_j P^2[i,j] \tag{7.2}$$

$$\text{Contrast} = \sum_i \sum_j (i-j)^2 P[i,j] \tag{7.3}$$

$$\text{Homogeneity} = \sum_i \sum_j \frac{P[i,j]}{1 + |i-j|} \tag{7.4}$$

The choice of the displacement vector $\mathbf{d}$ is an important parameter in the definition of the gray-level co-occurrence matrix. Occasionally, the co-occurrence matrix is computed for several values of **d** and the one which maximizes a statistical measure computed from $P[i, j]$ is used. The gray-level co-occurrence matrix approach is particularly suitable for describing microtextures. It is not suitable for textures comprising large area primitives since it does not capture shape properties. Gray-level co-occurrence matrices have been used extensively in remote sensing applications for land-use classification.

Autocorrelation

The autocorrelation function $p[k, l]$ for an $N \times N$ image is defined as follows:

$$p[k, l] = \frac{\frac{1}{(N-k)(N-l)} \sum_{i=1}^{(N-k)} \sum_{j=1}^{(N-l)} f[i, j] \, f[i+k, j+l]}{\frac{1}{N^2} \sum_{i=1}^{N} \sum_{j=1}^{N} f^2[i, j]}, \quad 0 \leq k, l \leq N{-}1. \quad (7.5)$$

For images comprising repetitive texture patterns the autocorrelation function exhibits periodic behavior with a period equal to the spacing between adjacent texture primitives. When the texture is coarse, the autocorrelation function drops off slowly, whereas for fine textures it drops off rapidly. The autocorrelation function is used as a measure of periodicity of texture as well as a measure of the scale of the texture primitives.

7.3 Structural Analysis of Ordered Texture

When the texture primitive is large enough to be individually segmented and described, then structural methods which describe the primitives and their placement rules are useful. For example, consider a simple texture formed by the repeated placement of homogeneous gray-level discs in a regular grid pattern as shown in Figure 7.3(a). Such a texture can be described by first segmenting the discs using a simple method such as connected component labeling, described earlier, and then determining the regular structure formed by the centroids of these connected components. For more general binary images the primitives can be first extracted using morphological methods and then their placement rules determined. Such morphological methods are particularly useful when the image is corrupted by noise or other nonrepeating random patterns which would be difficult to separate in a simple connected component method. For example, when the image shown in Figure 7.3(a) is

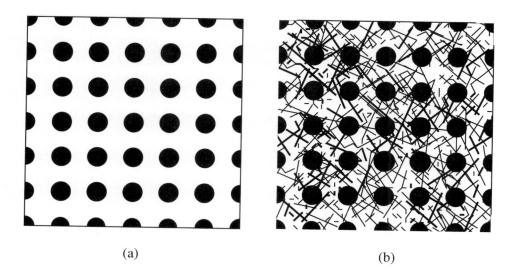

(a)   (b)

Figure 7.3: (a) A simple texture formed by repeated placement of discs on a regular grid. (b) Texture in (a) corrupted by random streaks of lines.

corrupted by noise resulting in random streaks of lines as shown in Figure 7.3(b), morphological techniques (see Chapter 2) can be used to locate all discs.

For gray-scale images we can define a predicate which is satisfied by all pixels within each blob corresponding to a primitive. A commonly used predicate is the gray-level homogeneity predicate. The image is initially processed using a Laplacian of Gaussian filter (see Chapter 5). Primitive regions are then identified by grouping all those pixels which are not on or near edge pixels. For homogeneous blobs properties such as size, elongation, and orientation are useful features. Measures based on co-occurrence of these primitives obtained by analyzing their spatial relationship are then used to characterize the texture.

7.4   Model-Based Methods for Texture Analysis

An approach to characterize texture is to determine an analytical model of the textured image being analyzed. Such models have a set of parameters.

The values of these parameters determine the properties of the texture, which may be synthesized by applying the model. The challenge in texture analysis is to estimate these model parameters so that the synthesized texture is visually similar to the texture being analyzed.

Markov random fields (MRFs) have been studied extensively as a model for texture. In the discrete Gauss-Markov random field model, the gray level at any pixel is modeled as a linear combination of gray levels of its neighbors plus an additive noise term as given by the following equation:

$$f[i,j] = \sum_{[k,l]} f[i-k, j-l] \, h[k,l] + n[i,j] \tag{7.6}$$

Here the summation is carried out over a specified set of pixels which are neighbors to the pixel $[i,j]$. The parameters of this model are the weights $h[k,l]$. These parameters are computed from the given texture image using least-squares method. These estimated parameters are then compared with those of the known texture classes to determine the class of the particular texture being analyzed.

When patterns forming texture have the property of self-similarity at different scales, fractal-based models may be used. A set is said to have the property of self-similarity if it can be decomposed as a nonoverlapping union of N copies of itself scaled down by a factor r. Such a texture is characterized by its fractal dimension D, given by the equation

$$D = \frac{\log N}{\log(\frac{1}{r})} \tag{7.7}$$

The fractal dimension is a useful feature for texture characterization. Estimation of D from an image is rather difficult because natural textures do not strictly follow the deterministic repetitive model of fractals assumed above, but have statistical variations.

7.5   Shape from Texture

Variations in the size, shape, and density of texture primitives provide clues for estimation of surface shape and orientation. These are exploited in *shape-from-texture* methods to recover 3-D information from 2-D images. As an illustration, consider the regular ordered texture shown in Figure 7.3(a) slanted at an angle α such that the top of the surface is farther away from the camera than the bottom; for simplicity, let us assume that all points along a given horizontal line are at the same depth from camera (i.e., there is no tilt). This is illustrated in Figure 7.4(a). The corresponding image captured is shown in Figure 7.5. Note that the discs now appear as ellipses, which is a clue that the surface is not parallel to the image plane. The sizes of these ellipses decrease as a function of y' in the image plane. In other words, there are more ellipses for a unit area in the image plane near the top of the image compared with the center, resulting in a density gradient. Furthermore, the aspect ratio (ratio of minor to major diameters of an ellipse) does not remain constant, resulting in an aspect ratio gradient [36]. To show this, we first derive an expression for the major and minor diameters of the ellipse as a function of the slant angle and the position of the ellipse in the image plane.

Let the diameter of the disc be d. Consider the disc at the image center. The major diameter of the ellipse in the image plane corresponding to this disc is given by the perspective projection equation

$$d_{\mathrm{major}}(0,0) = \frac{d\,f}{z} \qquad\qquad (7.8)$$

where z is the distance of the disc from the camera center and f is the focal length of the camera. The minor diameter of this ellipse is affected not only by the perspective projection but also by the foreshortening effect due to the slant angle α. This is given by the equation

$$d_{\mathrm{minor}}(0,0) = \frac{d\,f}{z} \cos \alpha. \qquad\qquad (7.9)$$

Thus, the aspect ratio of the ellipse at the center of the image plane is equal to $\cos \alpha$. All ellipses along the same horizontal line in the image plane will have the same aspect ratio.

Now consider an ellipse with its center at $(0, y')$ in the image plane. The disc corresponding to this ellipse is at an angle with respect to the optical axis as shown in Figure 7.4(b) and (c), where $\tan \theta = y'/f$. The major diameter of the ellipse is now given by Equation 7.8 with a slight modification. Since the disc is now located at a distance S away from the camera center, z must be replaced with S. From Figure 7.4(b) and (c):

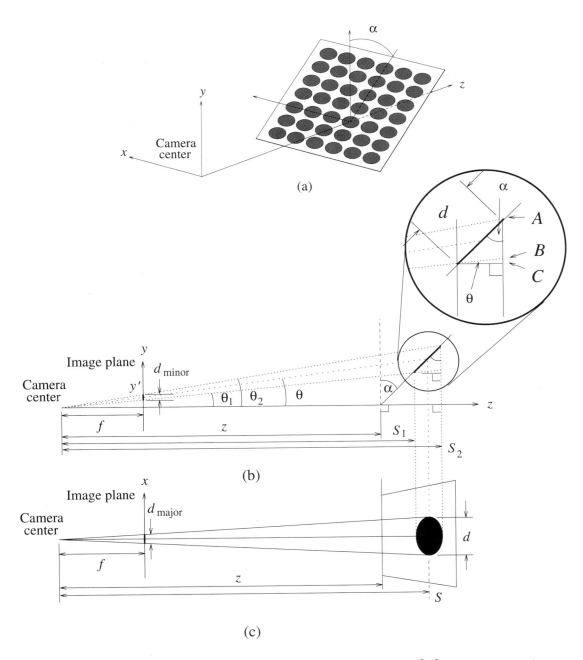

Figure 7.4: (a) The three-dimensional representation of the camera system with slanted texture plane. (b) The y–z view of (a). Note that the x axis is going into the paper. (c) The x–z vew of (a). Note that the y axis is coming out of the paper.

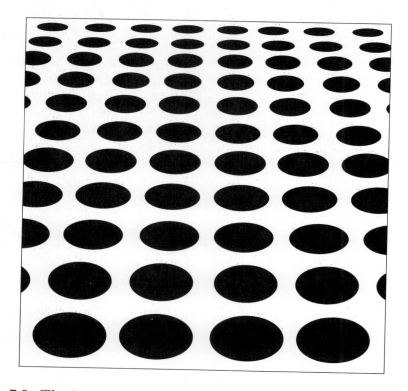

Figure 7.5: The image captured from the camera system in Figure 7.4.

$$\frac{S - z}{\tan \alpha} = S \tan \theta \qquad (7.10)$$

$$S(1 - \tan \theta \tan \alpha) = z \qquad (7.11)$$

$$S = \frac{z}{1 - \tan \theta \tan \alpha}. \qquad (7.12)$$

Therefore,

$$d_{\mathrm{major}}(0, y') = \frac{d\,f}{z}(1 - \tan \theta \tan \alpha). \qquad (7.13)$$

The derivation of $d_{\mathrm{minor}}(0, y')$ is a bit more involved. First, from Figure 7.4(b),

$$S_1 = \frac{z}{1 - \tan \theta_1 \tan \alpha} \qquad (7.14)$$

and
$$S_2 = \frac{z}{1 - \tan\theta_2 \tan\alpha}. \tag{7.15}$$

Now if we assume that the diameter of the disk d is very small so that it subtends a small angle at the camera center, we can approximate

$$\theta_1 \approx \theta_2 \approx \theta. \tag{7.16}$$

Therefore,

$$S_1 \approx S_2 \approx S \tag{7.17}$$
$$\approx \frac{z}{1 - \tan\theta\,\tan\alpha}. \tag{7.18}$$

Now from Figure 7.4(b) we know that

$$AC = d\cos\alpha. \tag{7.19}$$

However, we need to find the distance AB.

Noting that
$$S_1 - S_2 = d\sin\alpha, \tag{7.20}$$

we find
$$BC = d\sin\alpha\,\tan\theta. \tag{7.21}$$

Therefore,

$$AB = d(\cos\alpha - \sin\alpha\,\tan\theta) \tag{7.22}$$
$$= d\cos\alpha(1 - \tan\alpha\,\tan\theta). \tag{7.23}$$

Now, by the perspective projection,

$$\frac{d_{\text{minor}}(0, y')}{f} = \frac{d\cos\alpha(1 - \tan\alpha\,\tan\theta)}{\frac{z}{1 - \tan\theta\,\tan\alpha}} \tag{7.24}$$

Therefore, the minor diameter of the ellipse at $(0, y')$ is given by

$$d_{\text{minor}}(0, y') = \frac{d\,f}{z}\cos\alpha\,(1 - \tan\theta\,\tan\alpha)^2 \tag{7.25}$$

Thus, the aspect ratio given by $\cos\alpha(1 - \tan\theta\,\tan\alpha)$ decreases as θ increases, resulting in an *aspect ratio gradient*.

Variations in the image plane features such as size, shape, density, and aspect ratio of texture primitives can be exploited to recover the surface orientation (and ultimately the 3-D surface shape) of scene objects. To do this, however, we must have accurate methods to delineate each primitive in the image plane. For simple binary primitives such as the disc used in our illustration, fairly accurate segmentation of individual image plane primitives for measurement is possible. However, for more complex gray-level textures corrupted by noise, it is difficult to accurately estimate the image plane features.

7.6   Further Reading

There is a wealth of literature describing methods for texture modeling, synthesis, analysis, and segmentation. More details and references to many of the topics covered in this chapter may be found in [103, 239, 200]. Haralick, Shanmugam, and Dinstein [100] and Conners and Harlow [61] have used gray-level co-occurrence matrices to analyze textures in satellite images. Pentland [194] describes textures using fractals. Cross and Jain [64] describe a Markov random field model for texture. Methods for estimating the model parameters may be found in [138]. The shape-from-texture methods described in this chapter can be seen in more detail in [36]. Other approaches are given in [13, 252, 135]. Rao [200] includes a taxonomy for texture description and identification.

Exercises

7.1 What is texture? How is it defined for images?

7.2 How can you classify texture? Give some major classes and the characteristics that are used to define each class.

7.3 How are the following used in texture recognition:

- Fractals

- Co-occurrence matrices

- Fourier transforms

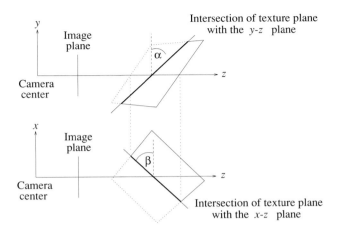

Figure 7.6: Diagram illustrating a textured plane with an arbitrary slant and tilt with respect to the camera axes.

- Markov random fields

Discuss their application and mention the types of texture where the above are most suitable.

7.4 How can images be segmented using texture features? Give an approach for image segmentation using texture.

7.5 How can the shape of a surface be determined by using its texture characteristics? Give details of an algorithm to determine shape from texture.

7.6 Mention three applications of a machine vision system where texture plays a prominent role.

7.7 For Figure 7.1(a), find the gray-level co-occurrence matrix for $\mathbf{d} = (0, 2), (2, 0), (2, 2)$. Repeat for Figure 7.2(a).

7.8 Derive the expressions for d_{major} and d_{minor} when the plane containing the texture pattern is tilted in addition to slant (i.e., oriented in any arbitrary direction as long as it is visible to the camera) as shown in Figure 7.6.

Computer Projects

7.1 Take an image of a textured plane (no tilt, only slant) and measure the major and minor diameters to verify their expressions.

7.2 Similar to above, now with unknown slant angle. Estimate α from the measured values of d_{major} and d_{minor}.

7.3 Same as above, now with both slant and tilt unknown.

7.4 Develop a fractal-based algorithm to discriminate among textures of different kind. Implement this algorithm and test its performance on several images. Relate the performance of this algorithm to judgments by humans about the similarity and ordering of texture.

7.5 Implement a texture recognition algorithm based on co-occurrence matrices. Apply it to several texture images to study its discrimination power.

7.6 Implement a texture segmentation algorithm. Test it on several images containing several regions of different types of texture.

7.7 Take a picture of a brick wall from the side so that the wall is at about 45 degrees from the optical axis of the camera. Develop a shape-from-texture algorithm and apply it to recover the structure of the surface.

Chapter 8

Optics

This chapter covers the essentials of geometrical optics. Radiometry is covered in Chapter 9. Machine vision relies on the pinhole camera model, which models the geometry of perspective projection but omits the effects of depth of field, based on the fact that only those points within a certain depth range are in focus on the image plane. Perspective projection assumes that the view volume is an infinite pyramid, limited only by the top, bottom, and sides of the viewable rectangle on the image plane. Optical effects such as depth of field and light attenuation limit the view volume by introducing limits on the distances to viewable objects. Objects may not be seen clearly if they are too close or too far away.

The pinhole camera model assumes that the perspective rays pass through an infinitesimal aperture at the front of the camera. In practice, the aperture must be larger to admit more light. Lenses are placed in the aperture to focus the bundle of rays from each point in the scene onto the corresponding point in the image plane as indicated by the geometry of perspective projection. The scene point, the projected point on the image plane, and the center of the aperture are on the ray for the line of sight in perspective projection. A lens gathers more light, allowing the camera to work with less ambient illumination or with a faster shutter speed, but the depth of field is limited. With the exception of limited depth of field and the camera calibration problems discussed in Section 12.9, lenses do not introduce any effects that violate the assumptions of ideal perspective projection, and most machine vision algorithms do not depend on the design of the optical system for image formation.

It may be necessary to calculate the depth of field in order to select a camera lens for a particular machine vision application. This chapter will cover the lens equation and present the equations for calculating the depth of field.

8.1   Lens Equation

The lens equation relates the distance z' of the image plane from the center of the lens (optical origin), the distance z to the point in the scene, and the focal length f of the lens:

$$\frac{1}{z'} - \frac{1}{z} = \frac{1}{f}. \tag{8.1}$$

When $z \rightarrow \infty$, the distance z' of the image plane from the optical origin is equal to the focal length f. The focal length is the distance of the image plane from the optical origin when parallel rays are focused to a single point in the image plane. In photogrammetry, the optical origin is called the center of projection and z' is called the camera constant (see Chapter 12 on camera calibration). When the lens is focused on points that are not at infinity (points at close range), $z' < f$, so using f to approximate z' overestimates the camera constant. The distance z' of the image plane from the optical origin approaches f as the camera focus is changed to accommodate points in the scene that are farther away.

8.2   Image Resolution

Spatial resolution is determined by pixel spacing, lens aberrations, diffraction, and depth of field. For most machine vision applications, lens distortions and diffraction are not the limiting effects. Spatial resolution is determined by the interplay between pixel spacing and depth of field.

Suppose that the spacing between pixels in the image plane is Δ. The resolution limit is 2Δ because that is the separation between features that allows two features to be perceived as distinct. In other words, perceiving some separation between features requires at least one imaging element between the features. Resolution is often listed as resolving power in units of lines per inch or lines per millimeter:

$$\text{RP} = \frac{1}{\text{RL}} \text{ lines/mm} = \frac{1}{2\Delta} \text{ lines/mm}. \tag{8.2}$$

In electronic cameras using photodiodes, the resolution of an image is limited by the spacing between the imaging elements (pixels) in a charge-coupled imaging device.

Photographic film has grains (crystals) of silver halide suspended in clear gelatin. When a ray of light hits a grain of silver halide, it changes to metallic silver. The unexposed grains are washed away during the developing process. The resolution of film is limited by the average spacing between grains of silver halide. A typical spacing between grains is 5 μm, with each grain about 0.5 μm^2 in cross section. The sensitivity (film speed) increases with grain size.

In the human vision system, the spatial resolution is determined by the spacing between cones in the fovea. A typical spacing is 30 seconds of arc or $1/120°$. This provides a resolution limit of one minute of arc, or $1/60°$, or 0.3×10^{-3} radians. Since this visual angle is small, the distance from the eye can be multiplied by the resolution limit in radians to calculate the spacing between features that corresponds to the resolution limit of human vision. For example, the length of your arm is around 40 cm. At this distance the resolution limit is

$$0.3 \times 10^{-3} \times 40 \text{ cm} = 120 \ \mu\text{m}. \tag{8.3}$$

The spacing between dots placed on a page by a laser printer is 300 dots per inch or 85 μm. The human vision system can perceive seperations below the resolution limit by interpolating feature locations to subpixel resolution, so the jaggedness of printed text and line drawings is just barely noticeable when reading a page held at arm's length.

8.3 Depth of Field

The purpose of a lens is to allow a larger aperture to be used in the construction of the camera so more light can enter the camera, allowing the camera to function with less ambient illumination or a faster shutter speed. A larger aperture comes with the penalty of reduced depth of field. There is a trade-off between aperture size and depth of field: a smaller aperture provides more depth of field but admits less light; a larger aperture admits more light but reduces the depth of field.

For a particular setting of the distance z' of the image plane from the center of the lens, only points on a plane at distance z, obtained by solving the lens Equation 8.1, are in perfect focus. In practice, the depth of field is determined by the spatial resolution of the imaging device. Some amount of defocusing below the resolution of the imaging device can be tolerated. There is a range of image plane distances z' with an acceptable level of defocusing and a corresponding range of scene distances z, called the depth of field, with scene points that are in focus to an acceptable degree.

When a scene point is out of focus, it creates a circle of image intensity on the image plane instead of a single point. If the diameter of this circle is below the resolution of the imaging device, then the amount of defocusing is not significant. Suppose that the diameter of the circle is b, the diameter of the lens aperture is d, the focal length is f, and the correct setting of the image plane distance from the center of the lens is z'. If the image plane is moved closer to the lens, to a distance z_1', then the amount of blur is given by

$$b = \frac{d\left(z' - z_1'\right)}{z'} \tag{8.4}$$

since the ratio of $b/2$ to $(z' - z_1')$ must be equal to the ratio of $d/2$ to z' by similar triangles. We can solve the lens Equation 8.1 for z' and z_1', corresponding to z and z_1, respectively, and substitute these expressions in Equation 8.4 to obtain an expression that relates the amount of blur to distances in the scene:

$$b = \frac{d\,f\left(z - z_1\right)}{z\left(f + z_1\right)}. \tag{8.5}$$

Suppose that b is the maximum diameter of the blur circle for acceptable defocusing. Solve Equation 8.5 for z_1 to obtain an expression for the distance to the near plane that bounds the view volume:

$$z_1 = \frac{f\,z\left(d - b\right)}{d\,f + b\,z}. \tag{8.6}$$

To calculate the distance z_2 to the far plane that bounds the view volume, let

$$b = \frac{d\left(z_2' - z'\right)}{z'}, \tag{8.7}$$

where z_2' is the setting of the image plane distance (beyond the correct setting z') that corresponds to the maximum amount of blur. Solve the lens equation for z' and z_2' and substitute into the equation for the blur diameter:

$$b = \frac{d\,f\,(z_2 - z)}{z\,(f + z_2)}. \tag{8.8}$$

Solve this equation for the far plane distance:

$$z_2 = \frac{f\,z\,(d + b)}{d\,f - b\,z}. \tag{8.9}$$

These equations provide the positions of the near and far planes for a particular setting of the nominal plane of focus z and the aperture diameter d, focal length f, and maximum acceptable blur diameter b. There is a distance $z = d\,f/b$, called the hyperfocal distance, at which the far plane distance and the depth of field become infinite.

The depth of field D is the difference between the near and far plane distances, $z_2 - z_1$, given by

$$D = \frac{2\,b\,d\,f\,z\,(f + z)}{d^2\,f^2 - b^2\,z^2}. \tag{8.10}$$

8.4 View Volume

Consider the perspective model for image formation, introduced in Chapter 1, with the image plane in front of the center of projection. Let the center of projection be located at point (x_0, y_0, z_0) and the view vector from the center of projection to the origin of the image plane (the principal point) be (v_x, v_y, v_z). Assume that the image plane is perpendicular to the view vector. Note that the view vector defines both the direction in which the camera is pointed and the distance from the center of projection to the image plane (the camera constant). The view vector defines the orientation of the camera in space, except for the twist about the optical axis. Let the vector (u_x, u_y, u_z) from the origin in the image plane define the direction that corresponds to up. The projection of the up vector onto the image plane, normalized to unit length, defines the y' axis in the image plane. The x' axis is the unit vector perpendicular to y' and the view vector.

Assume that the viewable area on the image plane is a rectangle with height h and width w centered about the principal point. The corners of the rectangle and the location of the center of projection define four planes (top, bottom, left, and right) that bound the view volume. The near and far planes are perpendicular to the image plane at distances determined by the depth of field. By convention, the plane normals point to the outside of the view volume.

The six bounding planes can be calculated from the position and orientation of the camera, the dimensions of the imaging element, and the lens optics. The plane coordinates are in the same coordinate system used to define the position and orientation of the camera, typically the absolute coordinate system for the scene. The bounding planes can be used to determine if all scene points that should be within the view of the camera are inside the view volume. The position and orientation of the camera, the nominal focus, the aperture diameter, and the focal length can be adjusted to change the view volume until all scene points that should be viewable are within the view volume. The minumum aperture diameter is constrained by the ambient illumination and the shutter speed necessary to minimize motion blur.

8.5   Exposure

The amount of light collected by the camera depends on the intensity of light falling on the image plane (the image irradiance) and the duration of the exposure (shutter speed):

$$\mathcal{E} = Et. \tag{8.11}$$

Exposure is in units of joules (energy) per square meter while image irradiance is in units of watts (power) per square meter. Multiplying power by time gives energy, so the units work out correctly.

The F-number or F-stop is proportional to the ratio of the focal length f to the diameter d of the aperture:

$$\text{F-number} = \frac{f}{d}. \tag{8.12}$$

The lens aperture on cameras is designated in units of F-number because image intensity (or film exposure for constant shutter speed) is the same for different lenses at the same F-stop:

$$\text{F-number} \quad \sim \quad \frac{1}{E} \tag{8.13}$$

$$\sim \quad \frac{1}{\mathcal{E}} \quad \text{for constant shutter speed } t. \tag{8.14}$$

In other words, the F-stop is the aperture diameter provided in a system of units that factors out the different light gathering capabilities of lenses with different focal lengths.

The F-number on camera lenses is marked with numbers that are multiples of $\sqrt{2}$ because doubling the aperture area is equivalent to increasing the aperture diameter by $\sqrt{2}$,

$$2 \times \text{aperture area} \sim \sqrt{2}d \approx 1.4d. \tag{8.15}$$

The F-stops are numbered 2.8, 4, 5.6, 8, 11, and so on, so each F-stop changes the aperture diameter by a factor of 1.4 and increases the amount of light on the image plane by a factor of 2.

Further Reading

There are several good books on optics, but most texts give short coverage to radiometry. An excellent text on optics which includes some coverage of radiometry and the optics of the eye is a paperback book by Young [259]. The vision text by Horn [109] contains a good discussion of radiometry and imaging. Horn was the pioneer in the use of optics and radiometry in machine vision. The text by Beynon and Lamb [31] contains a chapter on CCD image sensors.

Exercises

8.1 The geometry of the view volume was formulated in Section 8.4. Assume that the camera position and orientation are given in camera coordinates, so that the position is $(0, 0, 0)$, the view vector is $(0, 0, 1)$, and the up vector is $(0, 1, 0)$.

a. Calculate the position of the corners of the view rectangle on the image plane in terms of the position and orientation of the camera and the height and width of the imaging element.

b. Calculate the formulas for the coefficients of the top, bottom, left, and right bounding planes in terms of the camera position and orientation and the height and width of the imaging element.

c. Calculate the formulas for the coefficients of the near and far planes in terms of the focal length, aperture diameter, and nominal distance to a scene point.

d. Generalize these formulas to the camera position and orientation in absolute coordinates.

e. Outline an iterative algorithm for adjusting the camera position and orientation, nominal focus, aperture diameter, and focal length to accommodate a given list of scene points that should be in the view volume.

8.2 Consider a moving object with position (p_x, p_y, p_z) and velocity vector (v_x, v_y, v_z) in camera coordinates. It is desirable to adjust the shutter speed so that the blur due to motion is reduced to an acceptable level, defined for this exercise to be the minimum distance between imaging elements.

a. Calculate the projected velocity vector on the image plane assuming perspective projection and camera constant f.

b. Calculate the minimum acceptable shutter speed.

c. Suppose that there are no moving objects in the scene, but the camera moves with velocity vector (v_x, v_y, v_z). Repeat the calculation for the minimum shutter speed. The result will depend on the distance to points in the scene.

d. Suppose that the camera motion is due to vibration, modeled as movement in a uniformly distributed direction with vector magnitude v. Repeat the calculation for minimum shutter speed in terms of the maximum magnitude of the velocity vector due to vibration.

Chapter 9

Shading

This chapter covers the physics of how light reflects from surfaces and describes a method called photometric stereo for estimating the shape of surfaces using the reflectance properties. Before covering photometric stereo and shape from shading, we must explain the physics of image formation. Imaging is the process that maps light intensities from points in a scene onto an image plane. The presentation follows the pioneering work of Horn [109].

9.1 Image Irradiance

The projections explained in Section 1.4 determine the location on the image plane of a scene point, but do not determine the image intensity of the point. Image intensity is determined by the physics of imaging, covered in this section. The proper term for image intensity is image irradiance, but terms such as intensity, brightness, or gray value are so common that they have been used as synonyms for image irradiance throughout this text.

The image irradiance of a point in the image plane is defined to be the power per unit area of radiant energy falling on the image plane. Radiance is outgoing energy; irradiance is incoming energy. The irradiance at a point in the image plane $E(x', y')$ is determined by the amount of energy radiated by the corresponding point in the scene $L(x, y, z)$ in the direction of the image point:

$$E(x', y') = L(x, y, z). \tag{9.1}$$

The scene point (x, y, z) lies on the ray from the center of projection through image point (x', y'). To find the source of image irradiance, we have to trace the ray back to the surface patch from which the ray was emitted and understand how the light from scene illumination is reflected by a surface patch.

Two factors determine the radiance reflected by a patch of scene surface:

- The illumination falling on the patch of scene surface

- The fraction of the incident illumination that is reflected by the surface patch.

The amount of illumination falling on a particular surface patch is determined by the position of the surface patch relative to the distribution of the light sources in the scene. The fraction of the incident illumination that is reflected by the surface patch in a particular direction depends on the optical properties of the surface material.

Consider an infinitesimal patch of surface in a scene that is illuminated with a single point light source. Establish a coordinate system on the surface patch as shown in Figure 9.1. The coordinate system represents the hemisphere of possible directions from which energy can arrive or depart from the surface. Let (θ_i, ϕ_i) denote the direction, in polar coordinates relative to the surface patch, of the point source of scene illumination and let (θ_e, ϕ_e) denote the direction in which energy from the surface patch is emitted. The energy arriving at the surface patch from a particular direction is $E(\theta_i, \phi_i)$ and the energy radiated in a particular direction from the surface patch is $L(\theta_e, \phi_e)$. The ratio of the amount of energy radiated from the surface patch in some direction to the amount of energy arriving at the surface patch from some direction is the *bidirectional reflectance distribution function*. The radiance is determined from the irradiance by

$$L(\theta_e, \phi_e) = f(\theta_i, \phi_i, \theta_e, \phi_e) E(\theta_i, \phi_i), \qquad (9.2)$$

where $f(\theta_i, \phi_i, \theta_e, \phi_e)$ is the bidirectional reflectance distribution function, called the BRDF for short. The BRDF depends on the optical properties of the surface material. This is the general formulation and can be very complicated, but in most cases of interest in machine vision, the effects are fairly simple. For most materials, the BRDF depends only on the difference between the incident and emitted angles:

$$f(\theta_i, \phi_i, \theta_e, \phi_e) = f(\theta_i - \theta_e, \phi_i - \phi_e). \qquad (9.3)$$

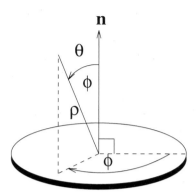

Figure 9.1: A polar coordinate system is established on an infinitesimal patch of surface to describe the directions of illumination and radiance in the hemisphere of directions visible from the surface patch.

9.1.1 Illumination

The radiance emitted from a surface patch can be calculated given the bidirectional reflectance distribution function (BRDF) for the surface material and the distribution of light sources. Two types of illumination will be covered:

- Point light source

- Uniform light source

First, the general formula for computing the total irradiance on a surface patch from a general distribution of light sources will be presented. The coordinate system is the polar coordinate system for the hemisphere of possible directions as diagrammed in Figure 9.1. The total irradiance on a surface patch is the sum of the contributions arriving at the surface patch from all of the directions in the hemisphere. Each small contribution of irradiance passing through a patch of the unit hemisphere must be counted in such a way that the total area of the hemisphere is used. A given section of the hemisphere with angular increments $\delta\theta_i$ and $\delta\phi_i$ covers an area of the hemisphere $\delta\omega$ called the solid angle:

$$\delta\omega = \sin\theta_i\,\delta\theta_i\,\delta\phi_i. \tag{9.4}$$

The $\sin \theta_i$ term accounts for the effect that the area of the portion of the hemisphere $\delta\theta_i \, \delta\phi_i$ is smaller near the top of the hemisphere. The area of a sphere of radius r is $4\pi r^2$, so the area of a hemisphere with unit radius is 2π. The area S of the hemisphere can be obtained by adding up the solid angles that comprise the hemisphere:

$$S = \int_0^{2\pi} d\omega \tag{9.5}$$

$$= \int_0^{\pi/2} \int_0^{2\pi} \sin\theta \, d\theta d\phi \tag{9.6}$$

$$= 2\pi \int_0^{\pi/2} \sin\theta \, d\theta \tag{9.7}$$

$$= 2\pi \left[-\cos\theta \right]_0^{\pi/2} \tag{9.8}$$

$$= 2\pi. \tag{9.9}$$

Without the $\sin\theta$ factor in Equation 9.4, the individual infinitesimal pieces of hemisphere would not add up to the correct total area. The total radiance passing through the sphere is the sum of the infinitesimal patches of sphere weighted by the amount of radiance per unit solid angle passing through each patch. Let $I(\theta_i, \phi_i)$ be the radiance per unit solid angle passing through the hemisphere from the direction (θ_i, ϕ_i). The total irradiance of the surface patch is

$$I_0 = \int_0^{2\pi} \int_0^{\pi/2} I(\theta_i, \phi_i) \sin\theta_i \cos\theta_i \, d\theta_i \, d\phi_i, \tag{9.10}$$

where the additional $\cos\theta_i$ term is needed because the surface patch appears smaller from the direction of illumination due to foreshortening. The amount of radiance reflected from the surface patch is

$$L(\theta_e, \phi_e) = \int_0^{2\pi} \int_0^{\pi/2} f(\theta_i, \phi_i, \theta_e, \phi_e) \, I(\theta_i, \phi_i) \sin\theta_i \cos\theta_i \, d\theta_i \, d\phi_i. \tag{9.11}$$

With the assumption that scene radiance is equal to image irradiance, the image irradiance at position (x', y') in the image plane is equal to the radiance from the corresponding surface patch in the scene:

$$E(x', y') = L(x, y, z) \tag{9.12}$$

$$= L(\theta_e, \phi_e), \tag{9.13}$$

where the angles of emittance of the scene radiance are determined from the geometry of the scene surfaces. Note that for each image position (x', y'), the corresponding scene position (x, y, z) can be calculated, as well as the surface normal $\hat{\mathbf{n}}$ for the surface patch and the angle (θ_e, ϕ_e) of the ray from the surface patch to the point (x', y') in the image plane, in polar coordinates relative to the surface normal or the surface patch.

To determine the irradiance of the entire image from the geometry of surfaces in the scene and the arrangement of light sources, it is necessary to know the BRDF for the scene surfaces. This is the subject of the next section.

9.1.2 Reflectance

Several different types of reflectance will be covered:

- Lambertian reflectance (also called diffuse reflectance)

- Specular reflectance

- Combinations of Lambertian and specular reflectance

- Reflectance in scanning electron microscopy

Lambertian Reflectance

A Lambertian surface appears equally bright from all viewing directions for a fixed distribution of illumination and a Lambertian surface does not absorb any incident illumination. Lambertian reflectance is also called diffuse reflectance since a Lambertian surface takes the incident illumination, whatever the distribution of illumination may be, and distributes all of the incident illumination in all surface directions such that the same amount of energy is seen from any direction. Note that this is not the same as saying that the surface emits energy equally in all directions, as will be explained in Section 9.3.2. Many matte surfaces are approximately Lambertian, and many surfaces, with the exceptions noted below, are qualitatively Lambertian.

The BRDF for a Lambertian surface is a constant:

$$f(\theta_i, \phi_i, \theta_e, \phi_e) = \frac{1}{\pi}. \tag{9.14}$$

The radiance is independent of the emitted direction and is obtained by summing the effect of the BRDF on the incident illumination coming from the hemisphere of possible directions:

$$L = \int_0^{2\pi} \int_0^{\pi/2} f(\theta_i, \phi_i, \theta_e, \phi_e) I(\theta_i, \phi_i) \sin\theta_i \cos\theta_i \, d\theta_i \, d\phi_i \qquad (9.15)$$

$$= \int_0^{2\pi} \int_0^{\pi/2} \frac{1}{\pi} I(\theta_i, \phi_i) \sin\theta_i \cos\theta_i \, d\theta_i \, d\phi_i \qquad (9.16)$$

$$= \frac{1}{\pi} I_0, \qquad (9.17)$$

where I_0 is the total incident illumination on the surface patch.

What is the perceived brightness of a Lambertian surface that is illuminated by a distant point source? The illumination from a point surface in a direction (θ_s, ϕ_s) relative to the normal of a surface patch is described by

$$I(\theta_i, \phi_i) = I_0 \frac{\delta(\theta_i - \theta_s)\,\delta(\phi_i - \phi_s)}{\sin\theta_i}, \qquad (9.18)$$

where I_0 is the total illumination. Essentially, the δ-functions merely restrict the directions from which illumination arrives at the surface patch to the single direction (θ_s, ϕ_s). Equation 9.18 has a sine term in the denominator so that when it is plugged into Equation 9.10, the total illumination comes out to be I_0.

Now plug the illumination function from Equation 9.18 and the BRDF from Equation 9.14 into Equation 9.11 for the radiance from the surface patch to get the equation for the perceived brightness:

$$L(\theta_e, \phi_e) = \int_0^{2\pi} \int_0^{\pi/2} f(\theta_i, \phi_i, \theta_e, \phi_e)\, I(\theta_i, \phi_i) \sin\theta_i \cos\theta_i \, d\theta_i \, d\phi_i$$

$$= \int_0^{2\pi} \int_0^{\pi/2} \frac{I_0}{\pi} \frac{\delta(\theta_i - \theta_s)\,\delta(\phi_i - \phi_s)}{\sin\theta_i} \sin\theta_i \cos\theta_i \, d\theta_i \, d\phi_i$$

$$= \frac{I_0}{\pi} \cos\theta_s. \qquad (9.19)$$

This is the Lambert cosine law, which says that the perceived brightness of a surface patch illuminated by a point source varies with the incident angle relative to the surface normal of the patch. The variation with incident angle is due to the foreshortening of the surface patch relative to the direction of illumination. In other words, a surface patch of a given area captures the most illumination if it is oriented so that the surface normal of the patch points in the direction of illumination. As the surface normal is pointed away from the direction of illumination, the area of the patch as seen from the direction of illumination, and hence the brightness of the patch, decreases. To see a demonstration of this effect for yourself, take a spherical object such as a white ball and turn out all of the lights in the room except for a single bulb. You will see that the brightest part of the sphere is the portion with the surface normal pointing toward the direction of illumination, regardless of where you stand relative to the ball, and the brightness decreases at the same rate in all directions from the point on the sphere that corresponds to the light source.

Suppose that instead of a point source, the illumination is uniform from all directions with total intensity I_0. Then the brightness is given by

$$
\begin{aligned}
L(\theta_e, \phi_e) &= \int_0^{2\pi} \int_0^{\pi/2} f(\theta_i, \phi_i, \theta_e, \phi_e)\, I(\theta_i, \phi_i) \sin\theta_i \cos\theta_i \, d\theta_i \, d\phi_i \\
&= \int_0^{2\pi} \int_0^{\pi/2} \frac{I_0}{\pi} \sin\theta_i \cos\theta_i \, d\theta_i \, d\phi_i \\
&= I_0.
\end{aligned}
\tag{9.20}
$$

Now the perceived brightness of the Lambertian surface patch is the same from all directions because no matter how the surface patch is oriented, it receives the same amount of illumination.

Specular Reflectance

A specular surface reflects all incident illumination in a direction that has the same angle with respect to the surface normal but is on the opposite side of the surface normal. In other words, a ray of light coming from the direction (θ_i, ϕ_i) is reflected in the direction $(\theta_e, \phi_e) = (\theta_i, \phi_i + \pi)$. The BRDF for a specular surface is

$$
f(\theta_i, \phi_i, \theta_e, \phi_e) = \frac{\delta(\theta_e - \theta_i)\,\delta(\phi_e - \phi_i - \pi)}{\sin\theta_i \cos\theta_i}.
\tag{9.21}
$$

The $\sin\theta_i$ and $\cos\theta_i$ factors are needed in the BRDF to cancel the corresponding factors due to foreshortening and solid angle in Equation 9.11. Plugging

Equation 9.21 into Equation 9.11 yields

$$L(\theta_e, \phi_e) = I(\theta_e, \phi_e - \pi), \tag{9.22}$$

which shows that the incoming rays of light are reflected from the surface patch as one would expect for a perfect mirror.

Combinations of Lambertian and Specular Reflectance

In computer graphics, it is common to use a combination of specular and diffuse reflectance to model the reflectance properties of objects:

$$f(\theta_i, \phi_i, \theta_e, \phi_e) = \frac{\eta}{\pi} + (1 - \eta)\frac{\delta(\theta_e - \theta_i)\,\delta(\phi_e - \phi_i - \pi)}{\sin \phi_i \cos \phi_i}, \tag{9.23}$$

where the constant η controls the mixture of the two reflectance functions. The relative proportion of specular and diffuse reflectance varies with the surface material of the objects. Objects that are glossy, in other words shiny objects, have a greater degree of specular reflectance than matte objects.

Scanning Electron Microscopy

In scanning electron microscopy, a surface emits the same amount of energy in all directions. This is not the same as a Lambertian surface, which appears equally bright from all directions. The difference is most easily seen by comparing the reflectance maps for the two surfaces, as will be done in Section 9.3.

9.2   Surface Orientation

The material in Section 9.1.2 discussed the relationship between illumination and perceived brightness in a coordinate system erected on a hypothetical surface patch. In order for this to be useful in vision, the discussion of surface reflectance and scene illumination must be reworked in the coordinate system of the image plane from Figure 9.2. Surface orientation must be formulated in camera coordinates.

Consider a sphere aligned with the optical axis as shown in Figure 9.3. Imagine a point on the sphere and suppose that a plane is tangent to the

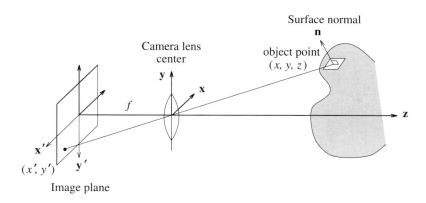

Figure 9.2: Projection of a point in the scene onto the image plane. The origin of the coordinate system is at the lens center.

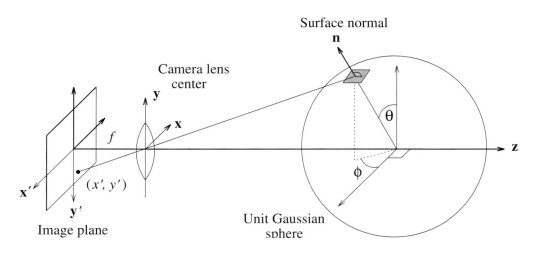

Figure 9.3: The Gaussian sphere illustrating the relationship between surface orientation and image coordinates.

sphere at that point. The surface normal of the plane is also the surface normal of the corresponding point on the sphere.

Suppose that the point is a distance z from the image plane and suppose that parallel projection is used to map the point onto the image plane. In camera coordinates, the point is at position (x, y, z). The following steps establish a nomenclature for the orientation of a surface patch in the scene in image plane coordinates. Consider a point nearby in the image plane at position $(x + \delta x, y + \delta y)$. The depth of the point will be $z + \delta z$. Let the depth be a function of image plane coordinates:

$$z = z(x, y). \tag{9.24}$$

How does the change in depth δz of the point relate to the change in image plane coordinates δx and δy? The answer is found in considering the Taylor series expansion of the function $z(x, y)$ about the point (x, y):

$$\delta z \approx \frac{\partial z}{\partial x} \delta x + \frac{\partial z}{\partial y} \delta y. \tag{9.25}$$

The size of the partial derivatives of z with respect to x and y are related to the amount of tilt in the tangent plane on the scene surface at the point corresponding to (x, y, z).

The gradient of the surface at (x, y, z) is the vector (p, q) given by

$$p = \frac{\partial z}{\partial x} \qquad \text{and} \qquad q = \frac{\partial z}{\partial y}. \tag{9.26}$$

The surface normal of a surface patch is related to the gradient by

$$\mathbf{n} = (p, q, 1), \tag{9.27}$$

which simply says that the amount of displacement in x and y corresponding to a unit change in depth z is p and q, respectively. The unit surface normal is obtained by dividing the surface normal by its length:

$$\hat{\mathbf{n}} = \frac{\mathbf{n}}{|\mathbf{n}|} = \frac{(p, q, 1)}{\sqrt{1 + p^2 + q^2}}. \tag{9.28}$$

Any point (x, y, z) in the scene can be specified by just the x and y coordinates, assuming that scene surfaces are opaque and the image is formed by parallel projection, so the coordinates of a point on a scene surface can just be denoted by image plane coordinates x and y with the primes omitted. Any function or property of the scene surface can be specified in terms of image plane coordinates; in particular, the orientation of a surface patch can be specified as functions of x and y, and since p and q have been developed to specify surface orientation, the orientation of any surface is specified by the two functions

$$p = p(x, y) \qquad \text{and} \qquad q = q(x, y). \tag{9.29}$$

The next step is to combine the ideas presented in this section for describing surface orientation in viewer-centered coordinates with the concepts of surface reflectance and scene illumination presented in Section 9.1.

9.3 The Reflectance Map

The combination of scene illumination, surface reflectance, and the representation of surface orientation in viewer-centered coordinates is called the *reflectance map*. It specifies the brightness of a patch of surface at a particular orientation for a given distribution of illumination and surface material. The presentation in this section assumes parallel projection, so the image plane coordinates will be denoted by (x, y) with the primes omitted.

9.3.1 Diffuse Reflectance

Consider a surface patch in the scene corresponding to image plane coordinates x and y with surface orientation p and q. Suppose that the surface patch has Lambertian reflectance and is illuminated by a point light source. In Section 9.1.2, the radiance of the patch was calculated to be

$$L(\theta_e, \phi_e) = \frac{I_0}{\pi} \cos \theta_s, \tag{9.30}$$

where θ_s is the angle between the surface normal of the patch and the direction vector to the light source. What is the corresponding representation in viewer-centered coordinates? In the viewer-centered coordinate system presented in Section 9.2, the surface normal is just $(-p, -q, 1)$ and the direction to the light source is $(-p_s, -q_s, 1)$. The cosine of the angle between any two vectors is the dot product of the two vectors divided by the length of each vector, so the cosine of the angle between the surface normal and the direction to the point light source is:

$$\cos\theta_s = \frac{(-p,-q,1)}{\sqrt{1+p^2+q^2}} \cdot \frac{(-p_s,-q_s,1)}{\sqrt{1+p_s^2+q_s^2}} \tag{9.31}$$

$$= \frac{1+p_sp+q_sq}{\sqrt{1+p^2+q^2}\sqrt{1+p_s^2+q_s^2}}. \tag{9.32}$$

For a given light source distribution and a given surface material, the reflectance for all surface orientations p and q can be cataloged or computed to yield the reflectance map $R(p,q)$. Since the precise value for the image irradiance depends on a variety of factors such as the strength of the light source, the light-gathering ability of the optical system, and many other factors that do not affect reflectance qualitatively, the reflectance map is normalized so that its maximum value is 1. Combining this normalization with the assumption that scene radiance equals image irradiance yields the *image irradiance equation*:

$$E(x,y) = R(p,q), \tag{9.33}$$

which says that the irradiance (brightness) at point (x,y) in the image plane is equal to the reflectance map value for the surface orientation p and q of the corresponding point on the scene surface. For a Lambertian reflector and point light source, the reflectance map $R(p,q)$ is given by Equation 9.32 and is shown in Figure 9.4.

9.3.2   Scanning Electron Microscopy

In scanning electron microscopy, the reflectance map is given by

$$R(p,q) = \sqrt{p^2+q^2+1} \tag{9.34}$$

$$= \sec\theta_i. \tag{9.35}$$

The same energy is emitted in all directions; hence a surface appears to be brighter if it is slanted away from the viewer because the viewer is seeing more surface within a given angle of viewing direction.

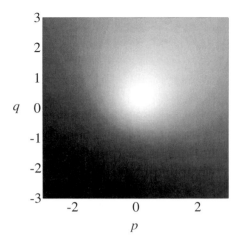 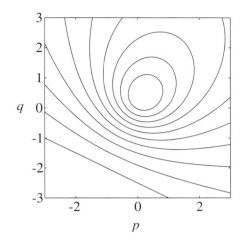

Figure 9.4: Typical reflectance map, $R(p, q)$, for a Lambertian surface illuminated by a point light source with $p_s = 0.2$ and $q_s = 0.4$. *Left:* Gray level representation. *Right:* Contour plot.

9.4 Shape from Shading

Image intensity at a pixel as a function of the surface orientation of the corresponding scene point is captured in a reflectance map. Thus, for fixed illumination and imaging conditions, and for a surface with known reflectance properties, changes in surface orientation translate into corresponding changes in image intensity. The inverse problem of recovering surface shape from changes in image intensity is known as the shape from shading problem. We now summarize the procedure for solving this problem using a surface smoothness constraint.

From the previous section, the relationship between image irradiance $E(x, y)$ and the orientation (p, q) of the corresponding point on the surface is given by

$$E(x, y) = R(p, q), \tag{9.36}$$

where $R(p, q)$ is the reflectance map of the surface. The goal is to recover surface shape by calculating the orientation (p, q) on the surface for each point (x, y) in the image. Note that we have only one equation, but there are two unknowns, p and q. Thus, this problem cannot be solved unless additional constraints are imposed. A commonly imposed constraint is that of surface smoothness. We assume that objects are made up of piecewise smooth surfaces which depart from the smoothness constraint only along their edges.

A smooth surface is characterized by slowly varying gradients, p and q. Thus, if p_x, p_y, q_x, and q_y represent the partial derivatives of p and q along x and y directions, we can specify the smoothness constraint as minimizing the integral of the sum of the squares of these partial derivatives as follows.

$$e_s = \int \int \left(\left(p_x^2 + p_y^2 \right) + \left(q_x^2 + q_y^2 \right) \right) dx \, dy. \tag{9.37}$$

Strictly speaking, we must minimize this integral subject to the constraint given in Equation 9.36. However, to account for noise which causes departure from the ideal, the problem is posed as that of minimizing total error e given by

$$e = e_s + \lambda e_i, \tag{9.38}$$

where λ is a parameter which weighs the error in smoothness constraint relative to the error in the image irradiance equation given by

$$e_i = \int \int \left(E(x, y) - R(p, q) \right)^2 dx \, dy. \tag{9.39}$$

This is a problem in the calculus of variations. An iterative solution for updating the value of (p, q) during the $(n + 1)$th iteration is given by

$$p_{ij}^{n+1} = p_{ij}^{*n} + \lambda \left[E_{ij} - R(p_{ij}^{*n}, q_{ij}^{*n}) \right] \frac{\partial R}{\partial p}, \tag{9.40}$$

$$q_{ij}^{n+1} = q_{ij}^{*n} + \lambda \left[E_{ij} - R(p_{ij}^{*n}, q_{ij}^{*n}) \right] \frac{\partial R}{\partial q}, \tag{9.41}$$

where * denotes the average values computed in a 2×2 neighborhood, and the subscripts i, j denote discrete coordinates in the image plane. Note that although the computations for a given iteration are local, global consistency is achieved by the propagation of constraints over many iterations.

The basic procedure described above has been improved in many ways. Details may be found in references at the end of the chapter. Although the basic principles of shape from shading are simple, there are many practical difficulties which limit its usefulness. In particular, the reflectance properties of the surfaces are not always known accurately, and it is difficult to control the illumination of the scene.

9.5 Photometric Stereo

Assume that all surfaces in the scene have Lambertian (diffuse) reflectance. For a point light source in a particular direction, the lines of constant reflectance (constant brightness) are defined by second-order polynomials. Each point (x, y) in the image will have brightness $E(x, y)$, and the possible surface orientations (p, q) will be restricted to lie along a curve in the reflectance map defined by some second-order polynomial.

Suppose that the same surface is illuminated by a point light source in a different location. Although the surface reflectance will be the same, the reflectance map will be different since it depends on both reflectance and illumination. The surface orientation corresponding to a given point in the image will be constrained to a different second-order curve in gradient space.

If a surface is illuminated with one light source and the image irradiance is measured, and the same surface is illuminated with a different light source and the image irradiance is measured again, then the pairs of image irradiance measurements will provide two equations for every point in the image that constrain the surface orientation. The basic idea is to solve the two equations at each point (x, y) in the image for the surface orientation $p(x, y)$ and $q(x, y)$. Since the equations of constraint are second-order polynomials, three equations will be needed since a pair of second-order equations in two unknowns does not have a unique solution. This is the method of photometric stereo, illustrated in Figure 9.5.

There is a radiometric effect that has been omitted in the discussion so far: all incident light is not radiated from a surface. This effect can be easily incorporated into the image irradiance equation with an albedo factor:

$$E(x, y) = \rho R(p, q), \tag{9.42}$$

with the albedo factor ρ such that $0 < \rho < 1$. The term *albedo* comes from Latin, meaning whiteness.

For a Lambertian surface with varying albedo, the surface orientation and albedo can be recovered simultaneously. Let the surface normal be denoted by

$$\hat{\mathbf{n}} = \frac{(p, q, -1)}{\sqrt{1 + p^2 + q^2}}. \tag{9.43}$$

Assume that there are three point sources of illumination. Let the direction

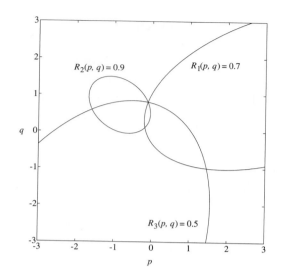

Figure 9.5: A diagram illustrating the principle of photometric stereo. The image irradiance measurements are normalized to the unit interval.

of point source i be denoted by the unit vector

$$\hat{\mathbf{s}}_i = \frac{(p_i, q_i, -1)}{\sqrt{1 + p_i^2 + q_i^2}}. \tag{9.44}$$

Recall that the brightness of a diffuse surface illuminated by a point source depends on the cosine of the angle between the surface normal and the direction of the illumination; hence the brightness is related to the dot product of these two vectors. For each point source of illumination, there will be a different image irradiance equation since there will be a different reflectance map. For point source i, the image irradiance equation will be

$$E_i = \rho \hat{\mathbf{s}}_i \cdot \hat{\mathbf{n}}. \tag{9.45}$$

Form a 3×3 matrix from the direction vectors for the point sources:

$$S = \begin{pmatrix} \hat{\mathbf{s}}_1 \\ \hat{\mathbf{s}}_2 \\ \hat{\mathbf{s}}_3 \end{pmatrix} = \begin{pmatrix} s_{1,x} & s_{1,y} & s_{1,z} \\ s_{2,x} & s_{2,y} & s_{2,z} \\ s_{3,x} & s_{3,y} & s_{3,z} \end{pmatrix}. \tag{9.46}$$

A vector of three image irradiance measurements is obtained at each point in the image:

$$\mathbf{E} = \begin{pmatrix} E_1 \\ E_2 \\ E_3 \end{pmatrix}. \tag{9.47}$$

The system of image irradiance equations for each point in the image can be written as

$$\mathbf{E} = \rho S \hat{\mathbf{n}}. \tag{9.48}$$

Note that $\mathbf{E}$, ρ, and $\hat{\mathbf{n}}$ depend on the position in the image, but S does not depend on position in the image plane and is constant for a given arrangement of light sources. For each point in the image, solve for the vector that represents albedo and surface orientation:

$$\rho \hat{\mathbf{n}} = S^{-1} \mathbf{E}. \tag{9.49}$$

The albedo $\rho(x, y)$ is the magnitude of this vector. The unit normal for surface orientation is obtained by dividing out the albedo.

For a given distribution of light sources, the S matrix can be inverted once using LU decomposition The inverse matrix does not have to be recomputed for every point in the image. In fact, the inverse of the S matrix can be determined once for the application and the stored inverse reused for subsequent surface orientation measurements. The values of $\rho \hat{\mathbf{n}}$ can be determined from the vector $\mathbf{E}$ of image irradiance values using back-substitution [197, pp. 39–45]. The set of image irradiance equations can be solved quickly using a lookup table that maps the image irradiance triple (E_1, E_2, E_3) into the albedo and surface orientation (ρ, p, q).

Further Reading

The classic paper on the properties of reflectance maps and shading in images was written by Horn [108]. Radiometric concepts are also described in detail in the book *Robot Vision* [109]. The first algorithm for shape from shading using the method of characteristics was developed by Horn [107]. There are several papers on shape from shading algorithms which are compiled in book form [112]. Photometric stereo was originally developed by Horn and Woodham. A recent reference is the paper by Woodham [256]. Nayar

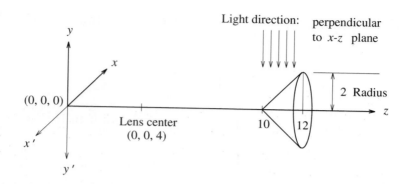

Figure 9.6: Diagram of the Lambertian cone for Exercise 9.1.

and Ikeuchi [181] present a photometric stereo algorithm that estimates the mixture of Lambertian to specular components in a scene surface.

Exercises

9.1 A Lambertian cone with its axis located along the z axis is illuminated by a distant light source as shown in Figure 9.6.

    a. Sketch the reflectance map of this illumination condition—in particular, the contour with $R(p, q) = 0$ and $R(p, q) = 0.5$ (i.e., find and sketch the equation of the corresponding contours in the (p, q) plane).

    b. Neglecting off-axis effects, sketch the image projected on the x–y plane and identify the contour of maximum brightness in the image.

9.2 Equation 9.11 relates the energy emitted in a particular direction to the amount of incident energy. Prove that this relationship is a linear system. A linear system has the properties of homogeneity and superposition. Assume that the incident illumination and the bidirectional reflectance distribution function are arbitrary.

    a. Show that if the illumination is $I = \alpha I_1$ for some constant α, then the radiance is αL_1, where L_1 is the radiance corresponding to illumination I_1. This property is called homogeneity.

b. Show that if the illumination is $I = I_1 + I_2$, then the radiance is $L_1 + L_2$, where L_1 and L_2 are the radiances for I_1 and I_2, respectively.

These properties show that the radiance for any linear combination of light sources is the linear combination of the radiances for the individual light sources.

Computer Projects

9.1 Reflectance map:

    a. Consider a camera with the origin of its image plane located at $(0, 0, 0)$. The lens center is located at $(0, 0, 4)$. Assume that a distant Lambertian object is centered around the positive z axis. The image plane is limited to $(-1, 1)$ in both x and y directions. A distant point source located in the direction (p_0, q_0, r_0) illuminates the object.

    b. Write a program to read the value of p_0, q_0, and r_0 from the keyboard, calculate the reflectance map, and display it as an image. Normalize the values of $R(p, q)$ such that maximum corresponds to 255 (for display as white). Locate $R(0, 0)$ at the center of the display. Note that both p and q are infinite in extent whereas the image is 256×256. Choose the appropriate truncation and scaling.

    c. Assume the object is spherical and has a radius of 3. The center of the sphere is at $(0, 0, 20)$. Calculate and display the image of the object captured by the camera. Normalize intensity ranges so that the image intensity ranges from 0 to 255. To distinguish the object from the background, set all background pixels to an intensity value of 64.

Chapter 10

Color

As discussed elsewhere in this book, light has intensity and images have gray value; but light consists of a spectrum of wavelengths and images can include samples from multiple wavelengths called channels. Perceived color depends on three factors:

- Spectral reflectance of scene surfaces, which determines how surfaces reflect color

- Spectral content of ambient illumination, which is the color content of the light shining on surfaces

- Spectral response of the sensors in the imaging system

In this chapter, it is assumed that surfaces are opaque so that the position of a point (x, y, z) on a scene surface can be represented by the image plane coordinates (x', y') for the line of sight to the scene point. This viewer-centered coordinate system will be used consistently throughout this chapter, so the primes on image plane coordinates will be omitted.

10.1 Color Physics

As with shading, presented in Chapter 9, ambient illumination is reflected from surfaces onto the image plane; but with color the illumination has some distribution of wavelengths and the surface reflectance is different for each wavelength. For any wavelength λ, the scene radiance is

$$S(x, y, \lambda)E(\lambda), \tag{10.1}$$

where S is the spectral reflectance of the scene point and E is the spectral distribution of the irradiance on the scene point. In shading, it is assumed that image irradiance is equal to scene radiance since the effect of the imaging system on the image irradiance is a constant that is included in the albedo; but with color the imaging system will filter the scene radiance through one or more sensors, and each sensor has a different spectral response. The image irradiance for sensor k is

$$\rho_k(x, y) = \int_0^\infty R_k(\lambda)S(x, y, \lambda)E(\lambda)\, d\lambda. \tag{10.2}$$

It is not necessary for the wavelengths to be restricted to visible light. The sensors could respond to energy in the infrared, ultraviolet, or X-ray portions of the electromagnetic spectrum, for example. This is the reason why the radiometric quantities used in the equations presented in Chapter 9 on shading were based on energy rather than visible light.

The imaging system will create multiple images, called channels, one for each sensor in the imaging system. In color imagery, there are usually three sensors with spectral responses that cover the red, green, and blue regions of the visible spectrum (RGB space). The range of all possible colors that can be captured by an imaging system or reproduced on a display is in the positive octant of the RGB space. It is common to work with normalized RGB values, so the set of all possible colors is a unit cube with corners at $(0, 0, 0)$ and $(1, 1, 1)$ in the RGB space.

10.2   Color Terminology

The physics of how colored light is reflected from surfaces is based on spectral distributions, but a practical imaging system, including the human vision system, works with a small number of samples from the distribution of wavelengths. The infinite-dimensional vector space of spectral distributions is reduced to a finite-dimensional vector space of samples (Equation 10.2). For any finite set of spectral response functions $\{R_k(\lambda), k = 1, \ldots, p\}$, there is an infinite number of spectral distributions that are filtered to the same set of response values $\{\rho_k, k = 1, \ldots, p\}$. For example, the spectral distribution at wavelengths where $R_k(\lambda) = 0$ can be varied arbitrarily without affecting the value of the response. For color imagery, the perceptually meaningful differences in spectral distributions are captured by the quantities of hue, saturation, and brightness (luminance).

Hue is determined by the dominant wavelength in the spectral distribution of light wavelengths. The spectral hues are single wavelengths in the visible spectrum. For example, the primary colors (red, green, and blue) are located at specific positions in the visible spectrum. Nonspectral hues are mixtures of wavelengths; for example; purple is a mixture of red and blue.

Saturation is the magnitude of the hue relative to the other wavelengths:

$$S = \frac{s_1}{s_2}, \tag{10.3}$$

where s_1 is the amount of light at the dominant wavelength and s_2 is the amount of light at all wavelengths. For example, a deep red color has saturation close to 1; but as other wavelengths are added, the color approaches the distribution of white light, the proportion of red and hence the saturation is reduced, and the color is desaturated to a shade of pink.

The brightness is a measure of the overall amount of light that passes through all of the spectral response functions. You can think of the brightness as a scale factor that is applied to the entire spectral distribution. The hue is the location of the peak in the spectral distribution (or the location and relative magnitudes of two peaks in the case of nonspectral hues such as purple). The saturation is the height of the peak relative to the entire spectral distribution. The location and shape of the peak in the spectral distribution (hue and saturation) determine the characteristics of light that are normally thought of as color.

10.3   Color Perception

The CIE (Commission Internationale de l'Eclairage—the International Commission on Illumination) color system is based on three spectral curves for the CIE primaries. Colors are specified by the relative amounts of the CIE primaries X, Y, and Z that match a given color. The Y value is luminance, a measure of the amount of light at all wavelengths that corresponds to perceived brightness. The chromaticity values depend on the dominant wavelength and saturation, independent of the luminance:[1]

$$x = \frac{X}{X + Y + Z} \qquad (10.4)$$

$$y = \frac{Y}{X + Y + Z} \qquad (10.5)$$

$$z = \frac{Z}{X + Y + Z}. \qquad (10.6)$$

Since $x + y + z = 1$, only two chromaticity values are needed. Colors are conveniently represented by the x and y chromaticities and luminance Y.

The x and y chromaticities represent the components of color independent of luminance. Two colors, such as dark green and light green, may appear different but may actually have the same relative distribution of wavelengths. If the spectral distribution is scaled by a constant, the color may appear lighter or darker, but the shape of the spectral distribution is not changed and the hue (dominant wavelength) and saturation (relative amount of the dominant wavelength) are not changed.

The perceptually significant chromaticities lie inside the arch-shaped region in Figure 10.1. White light is at the center of the chromaticity diagram. Draw a line from white at position W, through a particular color P, to a position H along the outer boundary of the chromaticity diagram. The hue is H, and the saturation S is the length of WP relative to WH. You can think of a color as an additive mixture of white light and pure spectral hue,

$$P = SH + (1 - S)W, \qquad (10.7)$$

where the saturation S controls the relative proportions of white tint and hue.

Hue, saturation, and luminance are encoded in the RGB color values in a way that makes it hard to use hue and saturation in vision algorithms. For example, it may be easy to identify objects with different hues by setting thresholds on the range of hues (spectral wavelengths) that bracket the objects. But where are these thresholds in the RGB cube, what is the shape of

---

[1]Technically, the chromaticities are calculated from luminance and the other CIE values, but the chromaticities are normalized and used in a way that describes color independent of, or at least conceptually separate from, luminance [81].

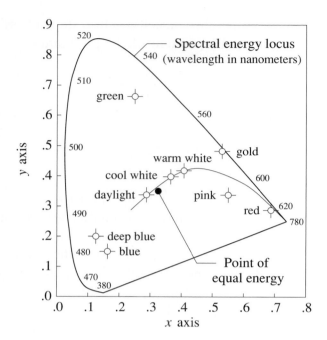

Figure 10.1: The diagram of CIE chromaticities. White light is at the center of the arch-shaped region. Fully saturated colors are along the outer edge of the diagram. The hue for a specific color is obtained by extending a line from white to the edge of the diagram passing through the color.

the surfaces that divide the color regions corresponding to different objects, and what is the formula for applying the thresholds to the RGB values in an image? These questions are hard to answer in the RGB color space but become simple when the RGB values are converted to hue, saturation, and luminance.

10.4   Color Processing

The HSI color model represents a color in terms of hue, saturation, and intensity. The intensity is the gray level of the pixels in a monochrome (black and white) image, such as the images used as examples for the machine vision algorithms presented in other chapters of this book.

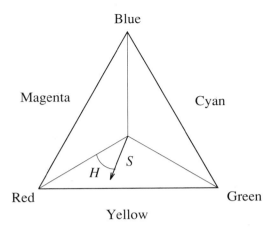

Figure 10.2: The HSI triangle represents the combinations of colors that can be represented by a linear combination of the primary colors at the corners of the triangle.

The HSI color triangle, shown in Figure 10.2, represents the combinations of hue and saturation that can be represented by combinations of the RGB primaries. The corners of the triangle correspond to the maximum amounts of each primary color (red, green, and blue) available from the imaging system. Achromatic (colorless) pixels are shades of gray, corresponding to equal amounts of the primary colors, and lie at the center of the HSI triangle.

The HSI solid adds the dimension of image intensity (Figure 10.3), with black at the bottom of the solid and white at the top. Shades of gray run along the axis of the solid. Each cross section of the solid is an HSI triangle with the proportions of the primary colors constrained to produce a particular intensity value. The HSI solid narrows to a point at the top and bottom because white and black can only be represented by unique combinations of the RGB primaries.

The RGB components of an image can be converted to the HSI color representation. Assume that the RGB components have been normalized to 1. This allows the derivation to be done in device-independent units. The intensity is the sum of the RGB values normalized to 1:

$$I = \frac{1}{3}(R + G + B). \tag{10.8}$$

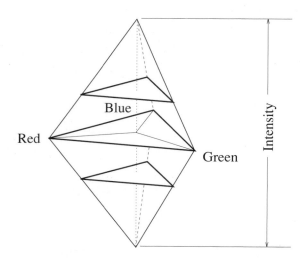

Figure 10.3: The HSI solid bounds the range of colors that can be represented by combinations of the primary colors.

The derivation of the formulas for hue and saturation begins by removing intensity from the RGB values:

$$r = \frac{R}{R + G + B} \tag{10.9}$$

$$g = \frac{G}{R + G + B} \tag{10.10}$$

$$b = \frac{B}{R + G + B}. \tag{10.11}$$

The locus of all possible values for r, g, and b is the triangle in the positive octant of rgb-space with corners $(1, 0, 0)$, $(0, 1, 0)$, and $(0, 0, 1)$. Let point P on the triangle denote the position in rgb-space of some color. Let $p = (r, g, b)$ be the vector to point P on the triangle, w be the vector to the point at the center of the triangle that represents white, and p_r be the vector to the corner of the triangle that corresponds to fully saturated red. The hue is the angle from vector $p_r - w$ to vector $p - w$, counterclockwise around the triangle as seen from the side of the triangle away from the origin. This is the right-hand rule with the thumb aligned along the normal to the triangle pointing away from the origin. The cosine of the hue is

$$\cos H = \frac{(p - w) \cdot (p_r - w)}{\|p - w\|\|p_r - w\|}.$$

(10.12)

Vector w to the center of the triangle is $(1/3, 1/3, 1/3)$. The magnitude of vector $p - w$ is

$$\|p - w\| = \sqrt{(r - 1/3)^2 + (g - 1/3)^2 + (b - 1/3)^2}.$$

(10.13)

Since $p_r = (1, 0, 0)$ and $w = (1/3, 1/3, 1/3)$, the magnitude of $p_r - w$ is

$$\|p_r - w\| = \sqrt{2/3}.$$

(10.14)

The dot product between $p - w$ and $p_r - w$ is

$$(p - w) \cdot (p_r - w) = \frac{2(r - 1/3) - (g - 1/3) - (b - 1/3)}{3}.$$

(10.15)

Divide this dot product by $\|p - w\|$ and $\|p_r - w\|$, substitute Equations 10.9 through 10.11, and simplify to get the formula for computing the hue from the R, G, and B values:

$$\cos H = \frac{2R - G - B}{2\sqrt{(R - G)^2 + (R - B)(G - B)}}.$$

(10.16)

In order to have the value for hue in the range from 0 to 360 degrees, it is necessary to subtract H from 360 when $B/I > G/I$. Even though this derivation began with normalized RGB values, the formula in Equation 10.16 will work with RGB values on any scale since the scale factors in the numerator and denominator will cancel.

The saturation is the distance on the triangle in the rgb-space from white relative to the distance from white to the fully saturated color with the same hue. Fully saturated colors are on the edges of the triangle. Let d_p be the distance from point W at the center of the triangle to point P for some color, and let d_q be the distance to point Q at the edge of the triangle along the line from W to Q passing through P. The saturation is the ratio of the distances, d_p/d_q. The formula for the saturation is

$$S = 1 - \frac{3}{R + G + B} \min(R, G, B).$$

(10.17)

The derivation is provided by Gonzalez and Woods [91].

Equations 10.16, 10.17, and 10.18 can be used to convert the RGB image from a color image acquisition system to the HSI representation for further processing. The hue is not defined when the saturation is zero, that is, for any colors along the axis of the HSI solid. Saturation is not defined when the intensity $I = 0$.

The transformation from RGB to HSI is used to convert color images into a form appropriate for machine vision. Gray-level algorithms can be performed on the I component of the HSI representation. Segmentation can be performed on the H component to discriminate between objects with different hues. However, hue is not reliable for discrimination when the saturation is low, so the segmentation algorithms must be modified to leave pixels with low saturation unassigned. Region growing algorithms can use thresholds on hue to form core regions, leaving unassigned any pixels that have low saturation or are outside the threshold boundaries. The algorithms for growing core regions by assigning neighboring pixels are unchanged.

More general algorithms can divide the HSI solid into regions using thresholds on hue, saturation, and intensity. These thresholds are easier to formulate and apply in the HSI representation than in the RGB representation provided by the imaging system.

10.5   Color Constancy

The color content of outdoor light varies considerably, yet people are able to correctly perceive the colors of objects in the scene independent, for the most part, from the color of the ambient illumination. This phenomenon is called *color constancy*.

Ambient light has the spectral distribution $E(\lambda)$, which describes the power at each wavelength. Assume that scene surfaces are opaque, so that scene coordinates can be specified using the coordinates of the corresponding point (x, y) on the image plane. The fraction of light at wavelength λ reflected from the surface point at location (x, y) is $S(x, y, \lambda)$. The light arriving at each location in the image is determined by the spectral distribution of the ambient light that falls on the scene surfaces and the fraction of light reflected at various wavelengths:

$$S(x, y, \lambda)E(\lambda). \tag{10.18}$$

Assume that there are p sensors at each image location (x, y) and each sensor has a different spectral response function. The spectral response of sensor k is $R_k(\lambda)$. Each sensor at location (x, y) samples a different distribution of light:

$$\rho_k(x, y) = \int_0^\infty R_k(\lambda)S(x, y, \lambda)E(\lambda)\, d\lambda. \tag{10.19}$$

The information about the color (spectral reflectance) of a point on the scene surface corresponding to location (x, y) in the image plane is encoded in the values $\rho_1(x, y), \rho_2(x, y), \ldots, \rho_p(x, y)$ obtained from the p sensors at that location. Color constancy is the problem of recovering the spectral reflectance $S(x, y, \lambda)$ of scene surfaces from the sensor responses $\{\rho_k(x, y), k = 1, \ldots, p\}$ independent of the spectral distribution of the ambient light $E(\lambda)$.

In principle, the color constancy problem can be formulated as an inverse problem in a finite-dimensional linear space and solved using matrix methods. Assume that the surface reflectance is a linear combination of basis functions,

$$S(x, y, \lambda) = \sum_{i=1}^n \sigma_i(x, y)S_i(\lambda). \tag{10.20}$$

The number n of basis functions is the number of degrees of freedom in the surface reflectance. Assume that the basis functions $S_i(\lambda)$ are known. Linear models with as few as three basis functions may be sufficient to account for typical surface reflectances.

Represent the ambient light by a linear model with m degrees of freedom:

$$E(\lambda) = \sum_{j=1}^m \epsilon_j E_j(\lambda). \tag{10.21}$$

Assume that the spectral distributions $E_j(\lambda)$ of the basis lights are known. Only three or four basis lights are needed to model natural daylight under a wide variety of weather conditions and times of day.

The color determination problem can be represented in matrix notation. Combine the m values of ϵ_j into a column vector ϵ, and combine the n values of σ_i into a column vector σ. Substitute the column vectors into Equation 10.19 to yield a matrix model for each pixel in the image:

$$\rho = \Lambda_\epsilon \sigma. \tag{10.22}$$

The lighting matrix Λ_ϵ is a p by n matrix, and the ij entry is

$$\int_0^\infty R_i(\lambda) S_j(\lambda) E(\lambda)\, d\lambda. \tag{10.23}$$

If the ambient illumination is known, then the lighting matrix Λ_ϵ is known. If the number of sensors equals the number of degrees of freedom in the surface reflectivity, $p = n$, then the lighting matrix can be inverted to obtain the coefficients of the surface spectral reflectivity σ at each point in the image which characterizes the color of the corresponding points on the scene surfaces.

If the ambient illumination is not known, then solving the problem will require more sensors than the number of degrees of freedom in surface reflectance. Since it was assumed that the ambient illumination is the same for all points in the scene, the information at multiple scene points can be used to estimate the ambient illumination. Suppose $p = n+1$. From s different spatial locations, $sp = s(n+1)$ different measurements can be obtained. There are sn unknowns for the surface reflectance and m unknowns for the ambient light spectrum. It is necessary to sample at $s > m$ locations to have more data than unknowns. This analysis suggests that the problem of color constancy can be solved without knowing the ambient illumination if there are $n+1$ sensors.

The matrix Λ_ϵ maps an n-dimensional surface space into an $(n+1)$-dimensional sensor space. For example, if $p = 3$ and $n = 2$, then the subspace is a plane in a three-dimensional space. This suggests the following two-step algorithm for determining surface reflectivity independent of scene illumination:

1. Determine the plane (subspace) spanning the data points in the sensor space to recover the ambient light vector ϵ.

2. Determine the lighting matrix Λ_ϵ from the ambient light vector ϵ and invert it to obtain the surface reflectivity vector σ.

10.6   Discussion

This chapter has barely scratched the surface of the topic of color vision. The intention has been to introduce the minimum number of formulas and concepts required to generalize the material on shading (Chapter 9) to cover colored light, reflectance, and imaging. The brief presentation on color terminology and the CIE color model provides an introduction to color theory sufficient to allow the reader to pursue the topic in other sources. The presentation of the HSI color model gives the reader some glimpse of the transformations that can be applied to multispectral imagery to allow the segmentation and edge detection algorithms covered elsewhere in this text to be used. The key concept in segmenting multispectral imagery is to find a transformation that reduces the dimensionality (number of channels) and makes it easy to use thresholds to determine core regions.

Image processing algorithms have been developed for correcting differences in the color capabilities of various displays and printers, and realistic graphics renderings require approximations to the spectral distributions for ambient illumination and spectral reflectivity. There are many image enhancement techniques that use color to highlight interesting features in images. Since this text is not concerned with the generation, reproduction, or enhancement of images, many interesting topics in color imagery have been omitted; the reader is encouraged to pursue the topic in the many excellent texts on image processing and computer graphics that are widely available.

Further Reading

There are many sources in computer graphics that provide very readable accounts of color theory [81]. Image processing texts provide extensive coverage of color representations, including the CMY color model used in color printing and the YIQ color model used in broadcast television. The HSI color model and the derivation of the formulas for converting RGB values to HSI values was adapted from Gonzalez and Woods [91]. The material on color constancy was adapted from the work of Maloney and Wandell (see Wandell [246]).

Exercises

10.1 What determines color at a point in an image? What is the role of the illumination in color images?

10.2 How is a color image represented in an image? Why are the particular frequencies of the electromagnetic spectrum selected for use in color vision? Name the basic colors used in defining characteristics at a point.

10.3 Define Hue, Saturation, and Brightness. Which of these is important in characterizing color at a point? Why?

10.4 How can you compute HSI characteristics from RGB characteristics? Why would you want to convert data in one representation to the other?

10.5 In machine vision, which color representation is more suitable? Justify your answer.

10.6 Design an edge detector that will detect prominent edges in a color image. Apply this to edge images. How will you combine the outputs of the different channels to provide you with edges in color images?

10.7 What are subtractive and additive models of color? Which one will be more suitable for displays, photos, and printers? Why?

10.8 What is the HSI solid? Where and how is it used in color processing?

10.9 Define and explain color constancy. Can machine vision systems display color constancey?

10.10 Why has color not been used much in machine vision? Do you think its application will increase? If so, what will be the leading application?

Chapter 11

Depth

Calculating the distance of various points in the scene relative to the position of the camera is one of the important tasks for a computer vision system. A common method for extracting such depth information from intensity images is to acquire a pair of images using two cameras displaced from each other by a known distance. As an alternative, two or more images taken from a moving camera can also be used to compute depth information. In contrast to intensity images, images in which the value at each pixel is a function of the distance of the corresponding point in the scene from the sensor are called range images. Such images are acquired directly using range imaging systems. Two of the most commonly used principles for obtaining such range images are radar and triangulation. In addition to these methods in which the depth information is computed directly, 3-D information can also be estimated indirectly from 2-D intensity images using image cues such as shading and texture. These methods are described briefly in this chapter.

11.1 Stereo Imaging

The geometry of binocular stereo is shown in Figure 11.1. The simplest model is two identical cameras separated only in the x direction by a *baseline distance b*. The image planes are coplanar in this model. A feature in the scene is viewed by the two cameras at different positions in the image plane. The displacement between the locations of the two features in the image plane is called the *disparity*. The plane passing through the camera centers and

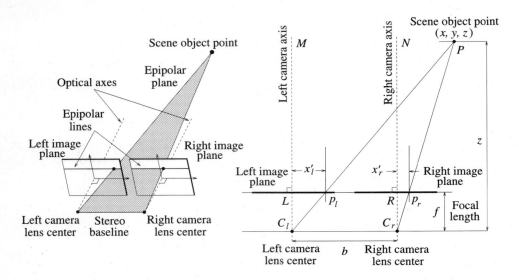

Figure 11.1: Any point in the scene that is visible in both cameras will be projected to a pair of image points in the two images, called a *conjugate pair*. The displacement between the positions of the two points is called the *disparity*.

the feature point in the scene is called the *epipolar plane*. The intersection of the epipolar plane with the image plane defines the *epipolar line*. For the model shown in the figure, every feature in one image will lie on the same row in the second image. In practice, there may be a vertical disparity due to misregistration of the epipolar lines. Many formulations of binocular stereo algorithms assume zero vertical disparity. Chapter 12 describes the relative orientation problem for calibrating stereo cameras.

Definition 11.1 *A conjugate pair is two points in different images that are the projections of the same point in the scene.*

Definition 11.2 *Disparity is the distance between points of a conjugate pair when the two images are superimposed.*

In Figure 11.1 the scene point P is observed at points p_l and p_r in the left and right image planes, respectively. Without loss of generality, let us assume that the origin of the coordinate system coincides with the left lens center. Comparing the similar triangles PMC_l and p_lLC_l, we get

$$\frac{x}{z} = \frac{x'_l}{f} \qquad (11.1)$$

Similarly, from the similar triangles PNC_r and p_rRC_r, we get

$$\frac{x - b}{z} = \frac{x'_r}{f} \qquad (11.2)$$

Combining these two equations, we get

$$z = \frac{bf}{(x'_l - x'_r)} \qquad (11.3)$$

Thus, the depth at various scene points may be recovered by knowing the disparities of corresponding image points.

Note that due to the discrete nature of the digital images, the disparity values are integers unless special algorithms are used to compute disparities to subpixel accuracy. Thus, for a given set of camera parameters, the accuracy of depth computation for a given scene point is enhanced by increasing the baseline distance b so that the corresponding disparity is large. Such wide-angle stereopsis methods introduce other problems, however. For instance, when the baseline distance is increased, the fraction of all scene points that are seen by both cameras decreases. Furthermore, even those regions that are seen by both cameras are likely to appear different in one image compared to the corresponding regions in the other image due to distortions introduced by perspective projection, making it difficult to identify conjugate pairs.

Before we discuss the problem of detecting and matching features in image pairs to facilitate stereopsis, we now briefly consider imaging systems in which the cameras are in any general position and orientation.

11.1.1 Cameras in Arbitrary Position and Orientation

Even when the two cameras are in any general position and orientation, the image points corresponding to a scene point lie along the lines of intersection between the image planes and the epipolar plane containing the scene point and the two lens centers as shown in Figure 11.2. It is clear from this figure that the epipolar lines are no longer required to correspond to image rows.

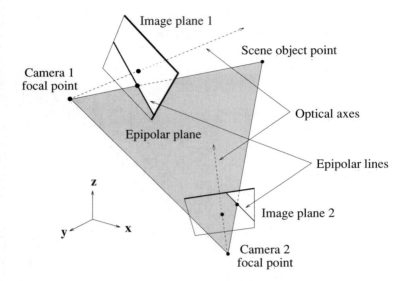

Figure 11.2: Two cameras in arbitrary position and orientation. The image points corresponding to a scene point must still lie on the epipolar lines.

In certain systems, the cameras are oriented such that their optical axes intersect at a point in space. In this case, the disparity is relative to the vergence angle. For any angle there is a surface in space corresponding to zero disparity as shown in Figure 11.3. Objects that are farther than this surface have disparity greater than zero, and objects that are closer have disparity less than zero. Within a region the disparities are grouped into three pools:

$$
\begin{array}{cl}
+ & d > 0 \\
- & d < 0 \\
0 & d = 0
\end{array}
$$

These pools are then used to resolve ambiguous matches.

More recent research work has addressed the issue of dynamically controlling the position, orientation, and other camera parameters to facilitate better image analysis. In systems known as *active vision systems*, the image analysis process dynamically controls camera parameters and movements. Computing the depths of various points in a scene is a common task in such systems.

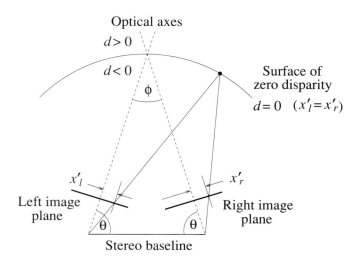

Figure 11.3: Stereo cameras focused at a point in space. The angle of the cameras defines a surface in space for which the disparity is zero.

11.2 Stereo Matching

Implicit in the stereopsis technique is the assumption that we can identify conjugate pairs in the stereo images. Detecting conjugate pairs in stereo images, however, has been an extremely challenging research problem known as the *correspondence problem.*

The correspondence problem can be stated as follows: *for each point in the left image, find the corresponding point in the right image.* To determine that two points, one in each image, form a conjugate pair, it is necessary to measure the similarity of the points. Clearly, the point to be matched should be distinctly different from its surrounding pixels; otherwise (e.g., for pixels in homogeneous intensity regions), every point would be a good match. Thus, before stereo matching, it is necessary to locate matchable features. Both edge features and region features have been used in stereo matching.

The implication of selecting a subset of all image pixels for matching is that depth is only computed at these feature points. Depth at other points is then estimated using interpolation techniques.

Note that the epipolar constraint significantly limits the search space for finding conjugate pairs. However, due to measurement errors and other uncertainties in camera position and orientation, matching points may not occur exactly on the estimated epipolar lines in the image plane; in this case, a search in a small neighborhood is necessary.

11.2.1   Edge Matching

We first present an algorithm for binocular stereo. The basic idea behind this and similar algorithms is that features are derived from the left and right images by filtering the images, and the features are matched along the epipolar lines. In this discussion, the epipolar lines are along the image rows. This algorithm uses edges detected by the first derivative of Gaussian. Edges computed using the gradient of Gaussian are more stable with respect to noise. The steps in the stereo algorithm are:

1. Filter each image in the stereo pair with Gaussian filters at four filter widths such that each filter is twice as wide as the next smaller filter. This can be done efficiently by repeated convolution with the smallest filter.

2. Compute the edge positions within the row.

3. Match edges at coarse resolutions by comparing their orientations and strengths; clearly, horizontal edges cannot be matched.

4. Refine the disparity estimates by matching at finer scales.

Note that computing the edge pixels to subpixel resolution would improve the precision of depth computation. In order to simplify the matching processes, the search for a match for each feature in one image takes place along the corresponding epipolar line in the second image for a limited distance centered around its expected position. In addition, the orientation of edges is recorded in 30° increments and the coarse orientations are used in matching. The orientation can be efficiently computed by coarsely quantizing the x and y partial derivatives and using a lookup table. One could also evaluate potential matches by using a composite norm that includes terms that penalize differences in edge orientation, edge contrast, and other measures of lack of similarity between potential features.

With active convergent cameras, the edges are matched at a coarse scale, and then the angle between the cameras is adjusted so that the region has a disparity of around zero, and the matching is redone at a finer scale. This adjustment limits the value of the maximum disparity and hence reduces the number of false matches and speeds up the matching process even when a small-scale filter is used to compute the disparity accurately. The matching process must begin at a coarse scale to determine an approximate value for the disparity. There are fewer edges at a coarse filter scale, so there is little chance of false matches.

11.2.2 Region Correlation

An important limitation of edge-based methods for stereo matching is that the value of the computed depth is not meaningful along occluding edges where the depth is not well defined. Along these edges the value of the depth is anywhere from the distance of the foreground object's occluding edge to the distance of the background scene point. In particular, for curved objects occluding edges are silhouette edges, and the observed image plane curves in the two images do not correspond to the same physical edge. Unfortunately, strong edges in the image plane are detected only along such occluding edges unless the object has other high-contrast nonoccluding edges or other features. Thus one of the primary problems in recovering depth is the problem of identifying more features distributed throughout the image as candidates for correspondence. One of the many methods developed for finding potential features for correspondence is to identify *interesting points* in both images and match these points in the two images using *region correlation* methods.

Detection of Interesting Points in Regions

In matching points from two images, we need points that can be easily identified and matched in the images. Obviously, the points in a uniform region are not good candidates for matching. The interest operator finds areas of image with high variance. It is expected that there will be enough of such isolated areas in images for matching.

The variances along different directions computed using all pixels in a window centered about a point are good measures of the distinctness of the point along different directions. The directional variances are given by

$$I_1 = \sum_{(x,y) \in S} [f(x,y) - f(x, y+1)]^2 \tag{11.4}$$

$$I_2 = \sum_{(x,y) \in S} [f(x,y) - f(x+1, y)]^2 \tag{11.5}$$

$$I_3 = \sum_{(x,y) \in S} [f(x,y) - f(x+1, y+1)]^2 \tag{11.6}$$

$$I_4 = \sum_{(x,y) \in S} [f(x,y) - f(x+1, y-1)]^2, \tag{11.7}$$

where S represents the pixels in the window. Typical window sizes range from 5×5 to 11×11 pixels. Since simple edge points have no variance in the direction of the edge, the minimum value of the above directional variances is taken as the *interest* value at the central pixel, (x_c, y_c). This eliminates edge pixels from consideration since an edge pixel in one image would match all pixels along the same edge in the second image, making it difficult to determine exact disparity (especially when the edge is along the epipolar line). Thus, we have

$$I(x_c, y_c) = \min(I_1, I_2, I_3, I_4). \tag{11.8}$$

Finally, to prevent multiple neighboring points from being selected as *interesting* for the same feature, feature points are chosen where the interest measure has a local maxima. A point is considered a "good" interesting point if, in addition, this local maxima is greater than a preset threshold.

Once features are identified in both images, they can be matched using a number of different methods. A simple technique is to compute the correlation between a small window of pixels centered around a feature in the first image and a similar window centered around every potential matching feature in the second image. The feature with the highest correlation is considered as the match. Clearly, only those features which satisfy the epipolar constraint are considered for matching. To allow for some vertical disparity, features which are near the epipolar line are also included in the potential matching feature set.

Consider two images f_1 and f_2. Let the pair of candidate feature points to be matched have a disparity of (d_x, d_y). Then a measure of similarity between the two regions centered around the features is given by the *correlation coefficient* $r(d_x, d_y)$ defined as

$$r(d_x, d_y) = \frac{\sum_{(x,y)\in S}\left[f_1(x,y) - \bar{f}_1\right]\left[f_2(x+d_x, y+d_y) - \bar{f}_2\right]}{\left\{\sum_{(x,y)\in S}\left[f_1(x,y) - \bar{f}_1\right]^2 \sum_{(x,y)\in S}\left[f_2(x+d_x, y+d_y) - \bar{f}_2\right]^2\right\}^{1/2}}.$$

(11.9)

Here $\bar{f}_1$ and $\bar{f}_2$ are the average intensities of the pixels in the two regions being compared, and the summations are carried out over all pixels within small windows centered around the feature points.

Instead of using the image intensity values in the above equation, the accuracy of correlation is improved by using thresholded signed gradient magnitudes at each pixel. This is done by computing the gradient at each pixel in the two images without smoothing and then mapping these into three values, -1, 0, and 1, using two thresholds, one above zero and the other below zero. This transforms the images into square waves that produce more sensitive correlations. If this is done, it is not necessary to include the normalization terms shown in the equation for correlation, and $r(d_x, d_y)$ simplifies to the sum of the products of corresponding pixel values.

In most situations, the depths of scene points corresponding to nearby feature points are likely to be close to one another. This heuristic is exploited in the iterative relaxation method described in Section 14.3.

As noted earlier, stereo matching based on features produces a sparse depth map at scene points which correspond to the image feature points. Surface interpolation or approximation must be performed on the sparse depth map to reconstruct the surface as described in Chapter 13.

One of the principal difficulties in stereo reconstruction is in the selection of *interesting points*. Such points are typically selected based on high local variance in intensity. Unfortunately, such points occur more frequently at corners and other surface discontinuities where the smoothness constraint does not hold. In some machine vision applications, this problem is solved by using structured light. Patterns of light are projected onto the surface, creating interesting points even in regions which would be otherwise smooth as shown in Figure 11.4.

Finding and matching such points are further simplified by knowing the geometry of the projected patterns. Since these patterns create artificial texture on the surfaces, the *shape from texture* techniques described in Chapter 7 may also be used.

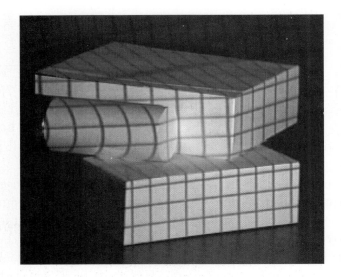

Figure 11.4: Patterns of light are projected onto a surface to create interesting points on an otherwise smooth surface.

11.3   Shape from X

In addition to the stereo imaging method described above, numerous other methods known as *shape from X* techniques have been developed for extracting shape information from intensity images. Many of these methods estimate local surface orientation rather than absolute depth at each point. If the actual depth to at least one point on each object is known, then the depth at other points on the same object can be computed by integrating the local surface orientation. Hence these methods are called indirect methods for depth computation. We briefly describe some of these methods here and provide pointers to other chapters where they are described in more detail.

Photometric Stereo

In the photometric stereo method, three images of the same scene are obtained using light sources from three different directions. Both camera and objects in the scene are required to be stationary during the acquisition of the three images. By knowing the surface reflectance properties of the objects in the scene, the local surface orientation at points illuminated by all three light sources can be computed. This method is described in detail in Chapter 9. One of the important advantages of the photometric stereo method is that the points in all three images are perfectly registered with one another since both camera and scene are stationary. Thus, this method does not suffer from the correspondence problem. The primary disadvantages of this method are that it is an indirect method and it may not be practical to employ an imaging system in which the illumination is so carefully controlled.

Shape from Shading

Shape from shading methods exploit the changes in the image intensity (shading) to recover surface shape information. This is done by calculating the orientation of the scene surface corresponding to each point (x', y') in the image. In addition to the constraint imposed by the radiometric principles, shape from shading methods assume that the surfaces are smooth in order to calculate surface orientation parameters. This method is described in detail in Chapter 9. Clearly, shape from shading is an indirect method for depth computation. Furthermore, the smoothness constraint is not satisfied at all points and the surface reflectance properties are not always known accurately, resulting in inaccurate reconstructions.

Shape from Texture

Image plane variations in the texture properties such as density, size, and orientation are the cues exploited by shape from texture algorithms. For example, the *texture gradient*, defined as the magnitude and direction of maximum change in the primitive size of the texture elements, determines the orientation of the surface. Quantifying the changes in the shape of texture elements (e.g., circles appearing as ellipses) is also useful to determine surface orientation. From images of surfaces with textures made up of regular grids of lines, possibly due to structured lighting (described in the following section), orientation may be uniquely determined by finding the *vanishing points*. Besides being indirect methods for depth computation, shape from texture methods also suffer from difficulties in accurately locating and quantifying texture primitives and their properties. Shape from texture techniques are described in Chapter 7.

Shape from Focus

Due to the finite depth of field of optical systems (see Chapter 8), only objects which are at a proper distance appear focused in the image whereas those at other depths are blurred in proportion to their distances. Algorithms to exploit this blurring effect have been developed. The image is modeled as a convolution of focused images with a point spread function determined by the camera parameters and the distance of the object from the camera. The depth is recovered by estimating the amount of blur in the image and using the known or estimated line spread function. Such reconstruction problems are mathematically ill posed. However, in some applications, especially those requiring qualitative depth information, depth from focus methods are useful.

Shape from Motion

When images of a stationary scene are acquired using a moving camera, the displacement of the image plane coordinate of a scene point from one frame to another depends on the distance of the scene point from the camera. This is thus similar to the stereo imaging described in earlier sections. Alternatively, a moving object also produces motion disparity in image sequences captured by a stationary camera. Such a disparity also depends upon the position and velocity of the object point. Methods for recovering structure and motion of objects are described in detail in Chapter 14.

11.4   Range Imaging

Cameras which measure the distance to every scene point within the viewing angle and record it as a two-dimensional function are called *range imaging systems*, and the resulting images are called *range images*. Range images are also known as *depth maps*. A range image is shown in Figure 11.5.

Two of the most commonly used principles for range imaging are triangulation and radar. *Structured lighting* systems, which are used extensively in machine vision, make use of the principle of triangulation to compute depth. *Imaging radar* systems use either acoustic or laser range finders to compute the depth map.

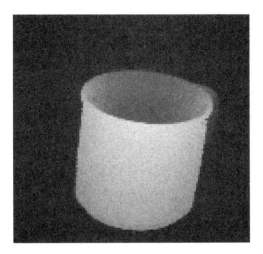

Figure 11.5: A range image of a coffee mug.

11.4.1 Structured Lighting

Imaging using structured lighting refers to systems in which the scene is illuminated by a known geometrical pattern of light. In a simple point projection system, a light projector and a camera are separated by a baseline distance b as shown in Figure 11.6. The object coordinates (x, y, z) are related to the measured image coordinates (x', y') and projection angle θ by

$$[x \ y \ z] = \frac{b}{f \cot \theta - x'}[x' \ y' \ f]. \tag{11.10}$$

The range resolution of such a triangulation system is determined by the accuracy with which the angle θ and the horizontal position of the image point x' can be measured.

To compute the depth at all points, the scene is illuminated one point at a time in a two-dimensional grid pattern. The depth at each point is then calculated using the above equation to obtain a two-dimensional range image. Because of its sequential nature, this method is slow and is not suitable for use with dynamically changing scenes. In a typical structured lighting system either planes of light or two-dimensional patterns of light are projected on the scene. The projected pattern of light on the surfaces of the objects in the scene is imaged by a camera which is spatially displaced with respect to the source of illumination. The observed image of the light patterns contain

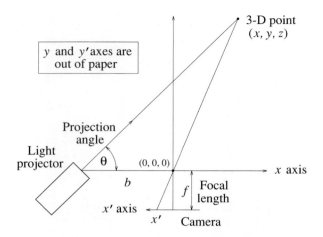

Figure 11.6: Camera-centered triangulation geometry [26].

distortions as determined by the shape and orientation of object surfaces on which the patterns are projected. This is illustrated in Figure 11.7 (also see Figure 11.4). Note that the light pattern as seen by the camera contains discontinuities and changes in orientation and curvature. The 3-D object coordinate corresponding to any point in the image plane may be calculated by computing the intersection of the camera's line of sight with the light plane. To obtain the complete object description, either the light source is panned as shown in the figure or the object is moved on a conveyor belt to acquire multiple images. Different surfaces in the object are detected by clustering stripes of light having similar spatial attributes.

In dynamically changing situations it may not be practical to project light stripes in sequence to acquire the complete image set covering the entire scene. Note that if multiple stripes of light are simultaneously projected to acquire the entire depth map, there would be potential ambiguities in matching stripe segments resulting from object surfaces at different depths. In such a case, patterns of multiple stripes of light in which each stripe is uniquely encoded are projected. For example, using a binary encoding scheme, it is possible to acquire a complete set of data by projecting only $\log_2 N$ patterns where $(N - 1)$ is the total number of stripes. This method is illustrated in Figure 11.8 for $N = 8$.

Each of the seven stripes has a unique binary code from (001) to (111).

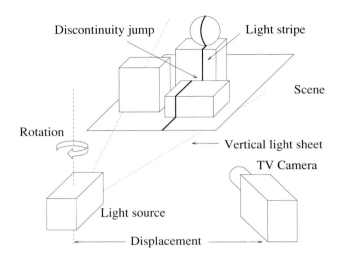

Figure 11.7: Illustration of striped lighting technique [131].

Since $\log_2 8$ is 3, only three images are acquired. Each image is identified by the bit position 1, 2, or 3 of the 3-bit binary code. A particular stripe of light is turned ON in an image only if its corresponding bit position is 1. For example, stripe 2 (010) is ON only in the second image, whereas stripe 7 (111) is ON in all three images. The stripes in the set of three images are now uniquely identified and hence there would be no ambiguities in matching stripe segments. In rapidly changing scenes, a single color-coded image is used instead of several binary-encoded images.

Structured lighting techniques have been used extensively in industrial vision applications in which it is possible to easily control the illumination of the scene. In a typical application, objects on a conveyor belt pass through a plane of light creating a distortion in the image of the light stripe. The profile of the object at the plane of the light beam is then calculated. This process is repeated at regular intervals to recover the shape of the object.

The primary drawback of structured lighting systems is that it is not possible to obtain data for object points which are not visible to either the light source or the imaging camera.

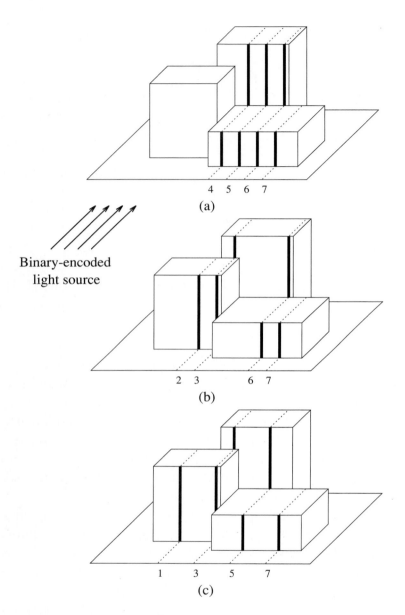

Figure 11.8: Illustration of binary-coded structured lighting where the sequence of projections determines the binary code for each stripe.

11.4.2 Imaging Radar

A second method for range imaging is *imaging radar*. In a time-of-flight pulsed radar, the distance to the object is computed by observing the time difference between the transmitted and received electromagnetic pulse. The depth information can also be obtained by detecting the phase difference between the transmitted and received waves of an amplitude-modulated beam or the beat frequency in a coherently mixed transmitted and received signal in a frequency-modulated beam. Several commercial laser beam imaging systems are built using these principles.

Range images are useful due to their explicit specification of depth values. At one time it was believed that if depth information is explicitly available, later processing would be easy. It became clear that though the depth information helps, the basic task of image interpretation retains all its difficulties.

11.5 Active Vision

Most computer vision systems rely on data captured by systems with fixed characteristics. These include both passive sensing systems such as video cameras and active sensing systems such as laser range finders. In contrast to these modes of data capture, it is argued that an active vision system in which the parameters and characteristics of data capture are dynamically controlled by the scene interpretation system is crucial for perception. The concept of active vision is not new. Biological systems routinely acquire data in an active fashion. Active vision systems may employ either passive or active sensors. However, in an active vision system, the state parameters of the sensors such as focus, aperture, vergence, and illumination are controlled to acquire data that would facilitate the scene interpretation task. Active vision is essentially an intelligent data acquisition process controlled by the measured and calculated parameters and errors from the scene. Precise definitions of these scene- and context-dependent parameters require a thorough understanding of not only the properties of the imaging and processing system, but also their interdependence. Active vision is a very active area of research.

Further Reading

The Marr-Poggio-Grimson algorithm for binocular stereo is described in the book on stereo by Grimson [92] and in the book on computational vision by Marr [163]. Barnard and Fischler [20] published a survey paper on binocular stereo. Stereo vision is also described in detail in the book by Faugeras [78]. An iterative algorithm for disparity analysis of images is described by Barnard and Thompson [21]. Many different forms of stereo have been investigated, including trinocular stereo [12, 98] and motion stereo [127, 185]. For solving difficulties in correspondence problems, different features such as interesting points [172], edge and line segments [11, 16, 167, 255], regions [60, 161, 167], and multiple primitives [162] have been tried to make stereo practical and useful.

One of the most active research topics in computer vision has been shape from X techniques. Shape from shading methods are described in the book with the same title [112]. The problem of concave objects resulting in secondary illuminations has been studied by Nayar et al. [179]. A recent approach to shape from focus is given in [180]. Shape from texture methods have been investigated by several groups [5, 36, 136].

Range image sensing, processing, interpretation, and applications are described in detail in the book [128]. Various methods for acquiring range images and a relative comparison of their merits are given by Besl [26]. An earlier survey by Jarvis includes not only direct range measuring techniques but also methods in which the range is calculated from 2-D image cues [131]. Boyer and Kak [44] describe a method in which color coding is used to acquire the range information using a single image. The classical paper by Will and Pennington [250] describes grid coding and Fourier domain processing techniques to locate different planar surfaces in the scene.

Arguments in favor of active perception systems and a control strategy for an active vision system were presented by Bajcsy [14]. Krotkov [150] has described a stereo image capture system in which the focus, zoom, aperture, vergence, and illumination are actively controlled to obtain depth maps. Aloimonos et al. [8] have described the advantages of active vision systems for performing many computer vision tasks such as shape recovery from image cues. Ahuja and Abbot [3] have integrated disparity, camera vergence, and lens focus for surface estimation in an active stereo system.

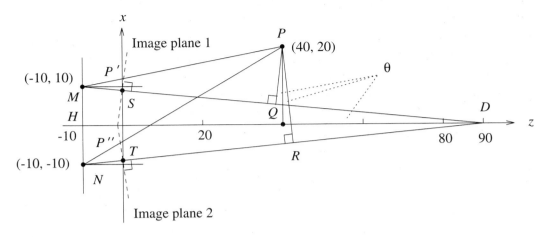

Figure 11.9: Convergent binocular imaging system for Exercise 11.3.

Exercises

11.1 A cube with 10 cm long edges is being imaged using a range camera system. The camera axis passes through two diagonally opposite vertices of the cube, and the nearest vertex along the axis is 10 cm away from the camera center. The intensity recorded by the camera is equal to $1000/d$ where d is the distance from the camera plane measured along the camera axis (not Euclidean distance). Sketch the range image captured by the camera. Calculate the intensities of the vertices seen by the camera.

11.2 Derive an equation for the surface of zero disparity shown in Figure 11.3.

11.3 Consider the convergent binocular imaging system shown in Figure 11.9. The cameras and all the points are in the $y = 0$ plane. The image planes are perpendicular to their respective camera axes. Find the disparity corresponding to the point P. *Hint*: The perpendicular distance between any point (x_0, z_0) and a line given by $Ax + Bz + C = 0$ is $(Ax_0 + Bz_0 + C)/\sqrt{A^2 + B^2}$.

11.4 Consider the binocular stereo imaging system shown in Figure 11.10. Find the disparity, $x_d = |x_1 - x_2|$, for the point P located at $(10, 20, 10)$.

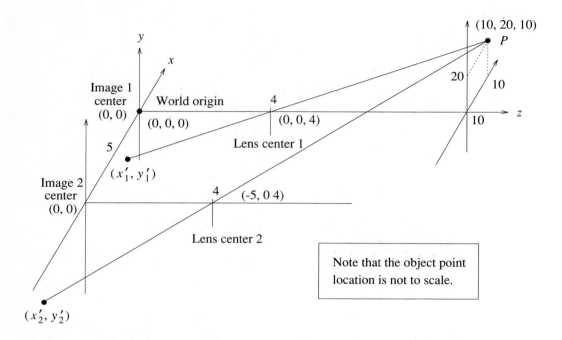

Figure 11.10: Binocular stereo imaging system for Exercise 11.4.

Chapter 12

Calibration

This chapter will present methods for calibrating cameras and depth measurement systems such as binocular stereo and range cameras. The machine vision algorithms developed in previous chapters involved the extraction of measurements within the image plane. Measurements, such as the location of an edge or the centroid of a region, were made in the coordinate system of the image array. We assumed that image plane coordinates could be determined from image locations using simple formulas based on the premise that the location of the central axis of projection in the image plane (the principal point) was at the center of the image. We also assumed that the distance of the image plane from the center of projection (the camera constant) was known. This chapter will cover the calibration problems associated with these assumptions. Measurements based on the simple image geometry assumed in previous discussions can be corrected to true image plane coordinates using the solution to the interior orientation problem, covered in Section 12.9. The position and orientation of a camera in the scene can be determined by solving the exterior orientation problem, covered in Section 12.8. The solution to the interior and exterior orientation problems provides an overall solution to the camera calibration problem that relates the location of pixels in the image array to points in the scene. Camera calibration is covered in Section 12.10. The relative orientation problem for calibrating the relationship between two cameras for binocular stereo is covered in Section 12.4, the method for computing depth measurements from binocular stereo disparities is explained in Section 12.6, and the absolute orientation problem for converting depth measurements in viewer-centered coordinates to an absolute coordinate system for the scene is covered in Section 12.3. Several different coordinate systems encountered in camera calibration are described in Section 12.1, and the mathematics of rigid body transformations is covered in Section 12.2.

The mathematics behind solutions to the problems of calibrating cameras and range sensors has been developed in the field of photogrammetry. Photogrammetry provides a collection of methods for determining the position and orientation of cameras and range sensors in the scene and relating camera positions and range measurements to scene coordinates. There are four calibration problems in photogrammetry:

Absolute orientation determines the transformation between two coordinate systems or the position and orientation of a range sensor in an absolute coordinate system from the coordinates of calibration points.

Relative orientation determines the relative position and orientation between two cameras from projections of calibration points in the scene.

Exterior orientation determines the position and orientation of a camera in an absolute coordinate system from the projections of calibration points in the scene.

Interior orientation determines the internal geometry of a camera, including the camera constant, the location of the principal point, and corrections for lens distortions.

These calibration problems are the classic problems in photogrammetry and originated with the techniques used to create topographic maps from aerial images. In addition to the four basic calibration problems, photogrammetry also deals with the problem of determining the position of points in the scene from binocular stereo disparities and provides methods for resampling stereo images so that the epipolar lines correspond to image rows.

All of the photogrammetric problems for determining the transformation between coordinate systems assume that a set of conjugate pairs is available. The conjugate pairs are obtained by matching feature points between views. These matches must be correct since the classic methods of photogrammetry use least-squares criteria and are very sensitive to outliers due to mismatched features. In fact, calibration algorithms can be ill conditioned, and one should not tempt fate by adding outliers to the normally distributed errors. Section 12.13 includes some discussion on using robust regression for handling mismatched feature points. In some applications, it is necessary to match two sets of points that are not conjugate pairs or to match two curves or surfaces that do not have discrete features. This is the registration problem, discussed in Section 13.9.

12.1 Coordinate Systems

The image processing operations in machine vision are usually done in the coordinate system of the image array, with the origin at the upper left pixel. The rows and columns correspond to integer coordinates for the pixel grid. Subpixel measurements add a fractional part to the image array (pixel) coordinate system, leading to pixel coordinates being represented as floating-point numbers; however, subpixel resolution does not change the geometry of the image array coordinate system. We can convert from pixel coordinates to image plane coordinates using some assumptions about the camera geometry. In Section 12.9, we will show how the camera parameters can be calibrated so that the mapping from image array (pixel) coordinates to image plane coordinates uses the actual geometry of the camera, including accounting for the effects of lens distortions.

The approximate transformation from pixel coordinates to image coordinates assumes that the principal axis intersects the image plane in the center of the image array. If the image array has n rows and m columns, then the center of the image array is

$$\hat{c}_x = \frac{m-1}{2} \tag{12.1}$$

$$\hat{c}_y = \frac{n-1}{2}. \tag{12.2}$$

We use the hat notation to stress that these are estimates for the location of the principal point. The x axis is in the direction of increasing column index, but the direction of increasing row index and the y axis point in opposite directions. The transformation from pixel coordinates $[i, j]$ to image coordinates (x', y') is

$$x' = j - \frac{m-1}{2} \tag{12.3}$$

$$y' = -\left(i - \frac{n-1}{2}\right). \tag{12.4}$$

This transformation assumes that the spacing between rows and columns in the image array is the same and that image plane coordinates should be expressed in this system of units. Let the spacing between columns be s_x and the spacing between rows be s_y. We can add these conversion factors to the transformation from pixel to image coordinates:

$$x' = s_x\left(j - \frac{m-1}{2}\right) \tag{12.5}$$

$$y' = -s_y\left(i - \frac{n-1}{2}\right). \tag{12.6}$$

If the image sensor has square pixels, then the conversion factors are identical and can be omitted. This simplifies the image processing algorithms. The image measurements (such as distances) can be converted to real units later, if necessary. If the image sensor has nonsquare pixels, then $s_x \neq s_y$ and it may be necessary to convert pixel coordinates to image coordinates, using the formulas above, before performing measurements. However, some measurements are not affected by nonsquare pixels. For example, the centroid can be computed using pixel coordinates and converted to image plane coordinates later, but distances and angles are not invariant to unequal scale factors and should be computed from point locations in image coordinates. For example, the centroids of two regions must be converted to image plane coordinates before calculating the distance between the centroids. Because of these problems, it is very common to require cameras with square pixels in machine vision applications. If an application uses a camera with nonsquare pixels, then you must carefully consider how the measurements will be affected by nonsquare pixels. When in doubt, convert to image plane coordinates before making any measurements.

The image plane coordinate system is part of the camera coordinate system, a viewer-centered coordinate system located at the center of projection with x and y axes parallel to the x' and y' axes in the image plane and a z axis for depth. The camera coordinate system is positioned and oriented relative to the coordinate system for the scene, and this relationship is determined through camera calibration, discussed in Section 12.10.

In summary, there are several coordinate systems in camera calibration:

Scene coordinates for points in the scene

Camera coordinates for the viewer-centered representation of scene points

Image coordinates for scene points projected onto the image plane

Pixel coordinates for the grid of image samples in the image array

Image coordinates can be true image coordinates, corrected for camera errors such as lens distortions, or uncorrected image coordinates. Pixel coordinates are also called image array coordinates or grid coordinates.

There may be multiple cameras in the scene, each with its own coordinate system. For example, in binocular stereo there is the left camera coordinate system, the right camera coordinate system, and the stereo coordinate system in which stereo depth measurements are represented. Determining the relationships between these coordinate systems is the purpose behind the various calibration problems discussed in this chapter.

12.2 Rigid Body Transformations

Any change in the position or orientation of an object is a rigid body transformation, since the object moves (changes position or orientation) but does not change size or shape. Suppose that a point $\mathbf{p}$ is visible from two viewpoints. The position of point $\mathbf{p}$ in the coordinate system of the first viewpoint is

$$\mathbf{p}_1 = (x_1, y_1, z_1)^T \tag{12.7}$$

and the position of point $\mathbf{p}$ in the coordinate system of the second viewpoint is

$$\mathbf{p}_2 = (x_2, y_2, z_2)^T. \tag{12.8}$$

The transformation between the two camera positions is rigid body motion, so each point at position $\mathbf{p}_1$ in the first view is transformed to its coordinates in the second view by rotation and translation:

$$\mathbf{p}_2 = R\,\mathbf{p}_1 + \mathbf{p}_0, \tag{12.9}$$

where matrix R is a 3×3 orthonormal matrix for rotation,

$$R = \begin{bmatrix} r_{xx} & r_{xy} & r_{xz} \\ r_{yx} & r_{yy} & r_{yz} \\ r_{zx} & r_{zy} & r_{zz} \end{bmatrix},$$ (12.10)

and vector $\mathbf{p}_0$ is the vector for the amount and direction of translation. Point $\mathbf{p}_0$ is the position of the origin of coordinate system one in coordinate system two.

Equation 12.9 can be viewed as a formula for computing the new coordinates of a point that has been rotated and translated or as a formula for computing the coordinates of the same point in space in different coordinate systems. The first interpretation is used in rigid body mechanics: the new coordinates of a point on an object must be computed after the object has been moved to a new position and orientation. The second interpretation is used for calibration problems: the same point has different coordinates when seen from different viewing positions and orientations. The change in coordinates is determined by the rigid body transformation between the two viewpoints, and the calibration problem is to determine the transformation from a set of calibration points (conjugate pairs). For example, consider the same point seen by two identical range cameras at different positions and orientations in space. Since the viewpoints are different, the coordinates are different even though the coordinates represent the same point. Imagine that the first range camera is rotated so that it has the same orientation as the second range camera. Now the coordinate systems of the range cameras have the same orientation but different positions in space. Now imagine that the first range camera is translated to the same position in space as the second range camera. Now the point has the same coordinates in both cameras. This process of aligning the coordinate systems of the two cameras so that identical points have the same coordinates is modeled by the rigid body transformation in Equation 12.9, which says that a point $\mathbf{p}_1$ in the viewer-centered coordinate system of the first camera is first rotated and then translated to change its coordinates to point $\mathbf{p}_2$ in the second camera. This process is illustrated in Figure 12.1. In practice, there may be one or more range cameras, or depth measuring systems such as binocular stereo, that generate point coordinates in their own camera-centered coordinate system, and these

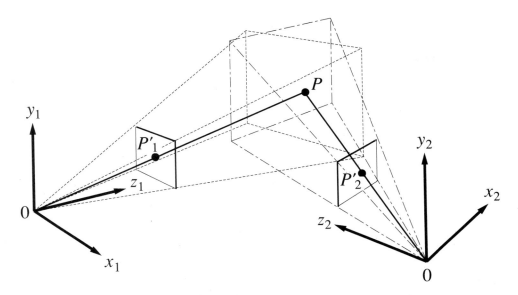

Figure 12.1: Point P is visible from both viewpoints but has different coordinates in each system. The drawing illustrates the rigid body transformation: the coordinate system on the left is first rotated and then translated until it is aligned with the coordinate system on the right.

measurements must be transformed into a common coordinate system that has been predefined for the scene. This common coordinate system is called the world or absolute coordinate system, sometimes called scene coordinates. The term *absolute coordinates* is used in this book to make it clear that we are talking about a global coordinate system for points in the scene, rather than the viewer-centered coordinate system for a particular camera or depth measuring system. The terms *viewer-centered* and *camera-centered coordinates* are synonymous. When we speak about camera coordinates, we are referring to the three-dimensional coordinate system for a camera, that is, the coordinate system for a particular viewpoint. As explained in Section 1.4.1, we generally use a left-handed coordinate system with the x and y axes corresponding to coordinates in the image plane in the usual way and the z axis pointing out into the scene. A point in the scene with coordinates

(x, y, z) in a camera-centered coordinate system is projected to a point (x', y') in the image plane through perspective projection:

$$x' = \frac{xf}{z} \tag{12.11}$$

$$y' = \frac{yf}{z}. \tag{12.12}$$

The coordinates (x', y') of the projected point are called image plane coordinates, or just image coordinates, for short. The origin of the image plane coordinate system is located at the intersection of the image plane and the z axis of the camera coordinate system. The z axis is called the principal axis or optical axis for the camera system, and the origin of the image plane coordinate system is called the principal point.

An affine transformation is an arbitrary linear transformation plus a translation. In other words, the vector of point coordinates is multiplied by an arbitrary matrix and a translation vector is added to the result. Affine transformations include changes in position and orientation (rigid body transformations) as a special case, as well as other transformations such as scaling, which changes the size of the object. A transformation that changes the shape of an object in some general way is a nonlinear transformation, also called warping, morphing, or deformation. There are several ways to represent rotation, including Euler angles and quaternions, covered in the following sections.

12.2.1   Rotation Matrices

Angular orientation can be specified by three angles: rotation ω about the x axis, rotation ϕ about the new y axis, and rotation κ about the new z axis. Angle ω is the pitch (vertical angle) of the optical axis, angle ϕ is the yaw (horizontal angle) of the optical axis, and angle κ is the roll or twist about the optical axis. These angles are called the Euler angles. No rotation (zero values for all three angles) means that two coordinate systems are perfectly aligned. Positive ω raises the optical axis above the x–z plane in the direction of positive y, positive ϕ turns the optical axis to the left of the y–z plane in the direction of negative x, and positive κ twists the coordinate system clockwise about the optical axis as seen from the origin. The entries of the rotation matrix R defined in Equation 12.10 in terms of these angles are

$$
\begin{aligned}
r_{xx} &= \cos\phi\cos\kappa \\
r_{xy} &= \sin\omega\sin\phi\cos\kappa + \cos\omega\sin\kappa \\
r_{xz} &= -\cos\omega\sin\phi\cos\kappa + \sin\omega\sin\kappa \\
r_{yx} &= -\cos\phi\sin\kappa \\
r_{yy} &= -\sin\omega\sin\phi\sin\kappa + \cos\omega\cos\kappa \\
r_{yz} &= \cos\omega\sin\phi\sin\kappa + \sin\omega\cos\kappa \\
r_{zx} &= \sin\phi \\
r_{zy} &= -\sin\omega\cos\phi \\
r_{zz} &= \cos\omega\cos\phi.
\end{aligned}
\tag{12.13}
$$

Although this is a common representation for rotation, determining the rotation by solving for the Euler angles leads to algorithms that are not numerically well conditioned since small changes in the Euler angles may correspond to large changes in rotation. Calibration algorithms either solve for the entries of the rotation matrix or use other representations for the rotation angles such as quaternions.

The rotation matrix is an orthonormal matrix,

$$
R^T R = I,
\tag{12.14}
$$

where I is the identity matrix. This means that the matrix inverse is just the transpose of the rotation matrix. A calibration algorithm will produce a rigid body transformation between coordinate systems in one direction; for example, from coordinate system 1 to coordinate system 2,

$$
\mathbf{p}_2 = R\mathbf{p}_1 + \mathbf{p}_{2,0}.
\tag{12.15}
$$

The inverse rigid body transform that converts coordinates in system 2 to coordinates in system 1 is

$$
\mathbf{p}_1 = R^T(\mathbf{p}_2 - \mathbf{p}_{2,0}) = R^T\mathbf{p}_2 + \mathbf{p}_{1,0},
\tag{12.16}
$$

where the notation $\mathbf{p}_{i,0}$ means the point in coordinate system i that is the origin of the other coordinate system. Note that the inverse translation is not just $-\mathbf{p}_{2,0}$ but must be multiplied by the inverse rotation matrix, because the translation $\mathbf{p}_{2,0}$ is in coordinate system 2 and the inverse translation must be expressed in the same orientation as coordinate system 1.

12.2.2   Axis of Rotation

Rotation can also be specified as a counterclockwise (right-handed) rotation about the axis specified by the unit vector $(\omega_x, \omega_y, \omega_z)$. This is a very intuitive way of viewing rotation, but it has the same problems as Euler angles in numerical algorithms. The angle and axis representation can be converted into a rotation matrix for use in the formula for rigid body transformation (Equation 12.9), but it would be nice to have a scheme for working directly with the angle and axis representation that produced good numerical algorithms. This is part of the motivation for the quaternion representation for rotation, discussed in the next section.

12.2.3   Unit Quaternions

The quaternion is a representation for rotation that has been shown through experience to yield well-conditioned numerical solutions to orientation problems. A quaternion is a four-element vector,

$$\mathbf{q} = (q_0, q_1, q_2, q_3). \tag{12.17}$$

To understand how quaternions encode rotation, consider the unit circle in the x–y plane with the implicit equation

$$x^2 + y^2 = 1. \tag{12.18}$$

Positions on the unit circle correspond to rotation angles. In three dimensions, the unit sphere is defined by the equation

$$x^2 + y^2 + z^2 = 1. \tag{12.19}$$

Positions on the unit sphere in three dimensions encode the rotation angles of ω and ϕ about the x and y axes but cannot represent the twist κ about the z axis. One more degree of freedom is required to represent all three rotation angles. The unit sphere in four dimensions is defined by the implicit equation

$$x^2 + y^2 + z^2 + w^2 = 1. \tag{12.20}$$

All three rotation angles in three-dimensional space can be represented by points on the unit sphere in four dimensions.

Rotation is represented by unit quaternions with

$$q_0^2 + q_1^2 + q_2^2 + q_3^2 = 1. \tag{12.21}$$

Each unit quaternion and its antipole $-\mathbf{q} = (-q_0, -q_1, -q_2, -q_3)$ represent a rotation in three dimensions.

The rotation matrix for rigid body transformation can be obtained from the elements of the unit quaternion:

$$R(\mathbf{q}) = \begin{bmatrix} q_0^2 + q_1^2 - q_2^2 - q_3^2 & 2(q_1 q_2 - q_0 q_3) & 2(q_1 q_3 + q_0 q_2) \\ 2(q_1 q_2 + q_0 q_3) & q_0^2 + q_2^2 - q_1^2 - q_3^2 & 2(q_2 q_3 - q_0 q_1) \\ 2(q_1 q_3 - q_0 q_2) & 2(q_2 q_3 + q_0 q_1) & q_0^2 + q_3^2 - q_1^2 - q_2^2 \end{bmatrix}. \tag{12.22}$$

After the unit quaternion is computed, Equation 12.22 can be used to compute the rotation matrix so that the rotation can be applied to each point using matrix multiplication.

The unit quaternion is closely related to the angle and axis representation for rotation, described in Section 12.2.2. A rotation can be represented as a scalar θ for the amount of rotation and a vector $(\omega_x, \omega_y, \omega_z)$ for the axis of rotation. A quaternion has a scalar part, which is related to the amount of rotation, and a vector part, which is the axis of rotation.

Let the axis of rotation be represented by the unit vector $(\omega_x, \omega_y, \omega_z)$ and use $\mathbf{i}$, $\mathbf{j}$, and $\mathbf{k}$ to represent the coordinate axes so that the unit vector for the rotation axis can be represented as

$$\omega_x \mathbf{i} + \omega_y \mathbf{j} + \omega_z \mathbf{k}. \tag{12.23}$$

The unit quaternion for a counterclockwise rotation by θ about this axis is

$$\mathbf{q} = \cos\frac{\theta}{2} + \sin\frac{\theta}{2}(\omega_x \mathbf{i} + \omega_y \mathbf{j} + \omega_z \mathbf{k}) \tag{12.24}$$

$$= q_0 + q_x \mathbf{i} + q_y \mathbf{j} + q_z \mathbf{k}. \tag{12.25}$$

The first term is called the scalar (real) part of the quaternion, and the other terms are called the vector (imaginary) part. A point $\mathbf{p} = (x, y, z)$ in space has a quaternion representation $\mathbf{r}$ which is the purely imaginary quaternion with vector part equal to $\mathbf{p}$,

$$\mathbf{r} = x\mathbf{i} + y\mathbf{j} + z\mathbf{k}. \tag{12.26}$$

Let $\mathbf{p}'$ be point $\mathbf{p}$ rotated by matrix $R(\mathbf{q})$,

$$\mathbf{p}' = R(\mathbf{q})\mathbf{p}. \tag{12.27}$$

If $\mathbf{r}$ is the quaternion representation for point $\mathbf{p}$, then the quaternion representation $\mathbf{r}'$ for the rotated point can be computed directly from the elements of quaternion $\mathbf{q}$,

$$\mathbf{r}' = \mathbf{qrq}^*, \tag{12.28}$$

where $\mathbf{q}^* = (q_0, -q_x, -q_y, -q_z)$ is the conjugate of quaternion $\mathbf{q}$ and quaternion multiplication is defined as

$$\begin{aligned}
\mathbf{rq} = \ & (r_0 q_0 - r_x q_x - r_y q_y - r_z q_z, \\
& r_0 q_x + r_x q_0 + r_y q_z - r_z q_y, \\
& r_0 q_y - r_x q_z + r_y q_0 + r_z q_x, \\
& r_0 q_z + r_x q_y - r_y q_x + r_z q_0).
\end{aligned} \tag{12.29}$$

Rigid body transformations can be conveniently represented using a seven-element vector, $(q_0, q_1, q_2, q_3, q_4, q_5, q_6)$, in which the first four elements are a unit quaternion and the last three elements are the translation. If we let $R(\mathbf{q})$ denote the rotation matrix corresponding to the unit quaternion in this representation, then the rigid body transformation is

$$\mathbf{p}_2 = R(\mathbf{q})\,\mathbf{p}_1 + (q_4, q_5, q_6)^T. \tag{12.30}$$

We will use quaternions in the next section to present an algorithm for solving the absolute orientation problem.

12.3   Absolute Orientation

The absolute orientation problem is the recovery of the rigid body transformation between two coordinate systems. One application for the absolute orientation problem is to determine the relationship between a depth measuring device, such as a range camera or binocular stereo system, and the absolute coordinate system defined for a scene so that all measurement points may be expressed in a common coordinate system. Let $\mathbf{p}_c = (x_c, y_c, z_c)$ denote the coordinates of a point in camera coordinates and $\mathbf{p}_a = (x_a, y_a, z_a)$ denote the coordinates of a point in absolute coordinates. The input to the absolute orientation problem is a set of conjugate pairs: $\{(\mathbf{p}_{c,1}, \mathbf{p}_{a,1}), (\mathbf{p}_{c,2}, \mathbf{p}_{a,2}), \ldots,$ $(\mathbf{p}_{c,n}, \mathbf{p}_{a,n})\}.$

To develop a solution to the absolute orientation problem, expand the equation for the rigid body transformation from a point $\mathbf{p}_c$ in camera coordinates to a point $\mathbf{p}_a$ in absolute coordinates to expose the components of the rotation matrix:

$$
\begin{aligned}
x_a &= r_{xx}x_c + r_{xy}y_c + r_{xz}z_c + p_x \\
y_a &= r_{yx}x_c + r_{yy}y_c + r_{yz}z_c + p_y \\
z_a &= r_{zx}x_c + r_{zy}y_c + r_{zz}z_c + p_z.
\end{aligned}
\tag{12.31}
$$

The unknowns are the 12 parameters of the transformation: the 9 elements of the rotation matrix and the 3 components of translation. Each conjugate pair yields three equations. At least four conjugate pairs are needed to get 12 equations for the 12 unknowns; in practice, a much larger number of calibration points is used to improve the accuracy of the result.

If the system of linear equations is solved without constraining the rotation matrix R to be orthonormal, the result may not be a valid rotation matrix. Using a nonorthonormal matrix as a rotation matrix can produce unexpected results: the matrix transpose is not necessarily the inverse of the matrix, and measurement errors in the conjugate pairs may influence the solution in ways that do not yield the best rigid body approximation. Some approaches orthogonalize the rotation matrix after each iteration, but there is no guarantee that the orthogonalized matrix will be the best approximation to the rotation. In Section 12.7, we present a method for solving the absolute orientation problem that guarantees that the solution matrix will be a rotation matrix. Another approach is to solve for the rotation angles rather than the entries in the rotation matrix; however, using the Euler angles leads to nonlinear algorithms with numerical difficulties. In photogrammetry, the nonlinear equations are linearized and solved to get corrections to the nominal values [103], but this approach assumes that good initial estimates are available.

There are many other representations for rotation that may yield good numerical methods, such as unit quaternions. Let $R(\mathbf{q})$ be the rotation matrix corresponding to a unit quaternion $\mathbf{q}$. The rigid body transformation that converts the coordinates of each point in camera coordinates to absolute coordinates is

$$
\mathbf{p}_{a,i} = R(\mathbf{q})\,\mathbf{p}_{c,i} + \mathbf{p}_c,
\tag{12.32}
$$

where $\mathbf{p}_c$ is the location of the origin of the camera in the absolute coordinate system. Now the regression problem has seven parameters: the four elements in the unit quaternion for rotation plus the three elements in the translation vector.

As stated earlier, the input to the absolute orientation problem is a set of conjugate pairs: $\{(\mathbf{p}_{c,1}, \mathbf{p}_{a,1}), (\mathbf{p}_{c,2}, \mathbf{p}_{a,2}), \ldots, (\mathbf{p}_{c,n}, \mathbf{p}_{a,n})\}$. Consider the set of points in camera coordinates and the set of points in absolute coordinates as two sets of points: $P_a = \{\mathbf{p}_{a,1}, \mathbf{p}_{a,2}, \ldots, \mathbf{p}_{a,n}\}$ and $P_c = \{\mathbf{p}_{c,1}, \mathbf{p}_{c,2}, \ldots, \mathbf{p}_{c,n}\}$. The absolute orientation problem is to align these two clouds of points in space. Compute the centroid of each point cloud,

$$\bar{\mathbf{p}}_a = \frac{1}{n} \sum_{i=1}^{n} \mathbf{p}_{a,i} \tag{12.33}$$

$$\bar{\mathbf{p}}_c = \frac{1}{n} \sum_{i=1}^{n} \mathbf{p}_{c,i}, \tag{12.34}$$

and subtract the centroid from each point:

$$\mathbf{r}_{a,i} = \mathbf{p}_{a,i} - \bar{\mathbf{p}}_a \tag{12.35}$$

$$\mathbf{r}_{c,i} = \mathbf{p}_{c,i} - \bar{\mathbf{p}}_c. \tag{12.36}$$

After the rotation has been determined, the translation is given by

$$\mathbf{p}_c = \bar{\mathbf{p}}_a - R(\mathbf{q})\,\bar{\mathbf{p}}_c. \tag{12.37}$$

Now we are left with the problem of determining the rotation that will align the two bundles of rays.

In the rest of this derivation of the rotation, the points will be expressed as rays about the centroids and all coordinates will be ray coordinates. Since the bundles of rays were derived from the set of conjugate pairs, we know which ray in the camera bundle corresponds to each ray in the bundle for absolute coordinates. When the two bundles of rays are aligned, each corresponding pair of rays will be coincident, as illustrated in Figure 12.2. In other words, each pair of rays will lie along the same line and point in the same direction. Neglecting the effects of measurement errors, the angle between each pair of rays will be 0 and the cosine of this angle will be 1. Measurement errors will prevent the bundles from being perfectly aligned, but we can achieve

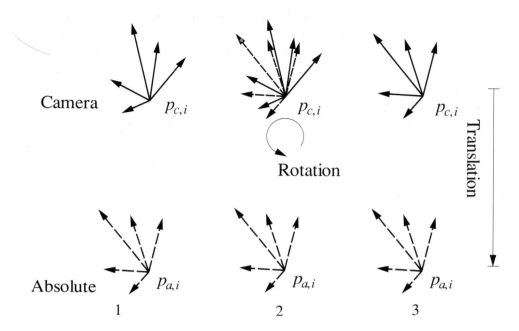

Figure 12.2: After the two bundles of rays (vectors of points about the centroids) are aligned and scaled, the centroid of one bundle can be translated to bring the two coordinate systems into alignment.

the best alignment in a least-squares sense by finding the rotation $R(\mathbf{q})$ that maximizes the scalar product of each ray pair:

$$\chi^2 = \sum_{i=1}^{n} \mathbf{r}_{a,i} \cdot R(\mathbf{q})\mathbf{r}_{c,i}. \tag{12.38}$$

In quaternion notation, this sum is

$$\sum_{i=1}^{n} \mathbf{r}_{a,i} \cdot \mathbf{q}\mathbf{r}_{c,i}\mathbf{q}^* = \sum_{i=1}^{n}(\mathbf{q}\mathbf{r}_{c,i}) \cdot (\mathbf{q}\mathbf{r}_{a,i}). \tag{12.39}$$

The sum can be successively changed into the notation of a quadratic form,

$$\sum_{i=1}^{n}(\mathbf{q}\mathbf{r}_{c,i}) \cdot (\mathbf{r}_{a,i}\mathbf{q}) = \sum_{i=1}^{n}(N_{c,i}\mathbf{q})^T(N_{a,i}\mathbf{q}) \tag{12.40}$$

$$= \sum_{i=1}^{n}\mathbf{q}^T N_{c,i}^T N_{a,i}\mathbf{q} \tag{12.41}$$

$$= \mathbf{q}^T \left(\sum_{i=1}^{n} N_{c,i}^T N_{a,i} \right) \mathbf{q} \qquad (12.42)$$

$$= \mathbf{q}^T \left(\sum_{i=1}^{n} N_i \right) \mathbf{q} \qquad (12.43)$$

$$= \mathbf{q}^T N \mathbf{q}, \qquad (12.44)$$

assuming that $\mathbf{q}$ corresponds to a column vector. The unit quaternion that maximizes this quadratic form is the eigenvector corresponding to the most positive eigenvalue. The eigenvalues can be determined by solving a fourth-order polynomial using the formulas published by Horn [110], or the eigenvalues and eigenvectors can be calculated using standard numerical methods [197].

The matrices $N_{c,i}$ and $N_{a,i}$ are formed from the elements of each ray. Let $\mathbf{r}_{c,i} = (x_{c,i}, y_{c,i}, z_{c,i})$ and $\mathbf{r}_{a,i} = (x_{a,i}, y_{a,i}, z_{a,i})$; then

$$N_{c,i} = \begin{bmatrix} 0 & -x_{c,i} & -y_{c,i} & -z_{c,i} \\ x_{c,i} & 0 & z_{c,i} & -y_{c,i} \\ y_{c,i} & -z_{c,i} & 0 & x_{c,i} \\ z_{c,i} & y_{c,i} & -x_{c,i} & 0 \end{bmatrix} \qquad (12.45)$$

$$N_{a,i} = \begin{bmatrix} 0 & -x_{a,i} & -y_{a,i} & -z_{a,i} \\ x_{a,i} & 0 & -z_{a,i} & y_{a,i} \\ y_{a,i} & z_{a,i} & 0 & -x_{a,i} \\ z_{a,i} & -y_{a,i} & x_{a,i} & 0 \end{bmatrix} \qquad (12.46)$$

and the matrix N is

$$N = \begin{bmatrix} (S_{xx}+S_{yy}+S_{zz}) & S_{yz}-S_{zy} & S_{zx}-S_{xz} & S_{xy}-S_{yx} \\ S_{yz}-S_{zy} & (S_{xx}-S_{yy}-S_{zz}) & S_{xy}+S_{yx} & S_{zx}+S_{xz} \\ S_{zx}-S_{xz} & S_{xy}+S_{yx} & (-S_{xx}+S_{yy}-S_{zz}) & S_{yz}+S_{zy} \\ S_{xy}-S_{yx} & S_{zx}+S_{xz} & S_{yz}+S_{zy} & (-S_{xx}-S_{yy}+S_{zz}) \end{bmatrix},$$

$$(12.47)$$

where the sums are taken over the elements of the ray coordinates in the camera and absolute coordinate systems:

$$S_{xx} = \sum_{i=1}^{n} x_{c,i} x_{a,i} \qquad (12.48)$$

$$S_{xy} = \sum_{i=1}^{n} x_{c,i} y_{a,i} \tag{12.49}$$

$$S_{xz} = \sum_{i=1}^{n} x_{c,i} z_{a,i} \tag{12.50}$$

$$\vdots$$

In general, S_{kl} is the sum over all conjugate pairs of the product of coordinate k in the camera point and coordinate l in the absolute point:

$$S_{kl} = \sum_{i=1}^{n} k_{c,i} l_{a,i}. \tag{12.51}$$

The result of these calculations is the unit quaternion that represents the rotation that aligns the ray bundles. A rotation matrix can be obtained from the quaternion using Equation 12.22, and the translation part of the rigid body transformation can be obtained using Equation 12.37. The rigid body transformation can be applied to any point measurements generated by the depth measurement system, whether from a range camera, binocular stereo, or any other scheme, to transform the points into the absolute coordinate system.

12.4 Relative Orientation

The problem of relative orientation is to determine the relationship between two camera coordinate systems from the projections of corresponding points in the two cameras. The relative orientation problem is the first step in calibrating a pair of cameras for use in binocular stereo. We covered binocular stereo algorithms for matching features along epipolar lines in Section 11.2. To simplify the presentation, we assumed that the corresponding epipolar lines in the left and right image planes corresponded to the same rows in the left and right image arrays. This section will cover the solution to the relative orientation problem and show how the location of the epipolar lines in the two image planes can be determined. Section 12.5 will show how the left and right images can be resampled so that the epipolar lines correspond to the image rows, as assumed by the algorithms presented in Section 11.2. The disparities

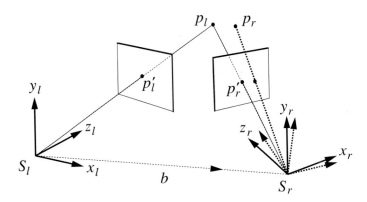

Figure 12.3: Illustration of the relative orientation problem for calibrating stereo cameras using the corresponding projections of scene points.

found by stereo matching are actually conjugate pairs. Section 12.6 will show how these conjugate pairs can be converted to point measurements in the coordinate system of the stereo device. The relative orientation problem is illustrated in Figure 12.3.

Suppose that a point $\mathbf{p}$ in the scene is within the view volume of two cameras, designated as the left and right cameras. Point $\mathbf{p}$ is denoted $\mathbf{p}_l$ in the coordinate system of the left camera and $\mathbf{p}_r$ in the coordinate system of the right camera. The projection of point $\mathbf{p}$ onto the image plane of the left camera is $\mathbf{p}_l' = (x_l', y_l')$ and the projection of the point onto the image plane of the right camera is $\mathbf{p}_r' = (x_r', y_r')$. From the equations for perspective projection:

$$\frac{x_l'}{f_l} = \frac{x_l}{z_l} \qquad\qquad \frac{y_l'}{f_l} = \frac{y_l}{z_l} \tag{12.52}$$

$$\frac{x_r'}{f_r} = \frac{x_r}{z_r} \qquad\qquad \frac{y_r'}{f_r} = \frac{y_r}{z_r}. \tag{12.53}$$

The rigid body transformation that transforms coordinates in the left camera system to coordinates in the right camera system is

$$x_r = r_{xx}x_l + r_{xy}y_l + r_{xz}z_l + p_x \tag{12.54}$$

$$y_r = r_{yx}x_l + r_{yy}y_l + r_{yz}z_l + p_y \tag{12.55}$$

$$z_r = r_{zx}x_l + r_{zy}y_l + r_{zz}z_l + p_z. \tag{12.56}$$

Solve the equations for perspective projection for x_l, y_l, x_r, and y_r and plug into the equations for a rigid body transformation to obtain a set of equations for the relationship between the projections of the conjugate pairs:

$$r_{xx}x_l' + r_{xy}y_l' + r_{xz}f_l + p_x\frac{f_l}{z_l} = x_r'\frac{z_r}{z_l}\frac{f_l}{f_r} \tag{12.57}$$

$$r_{yx}x_l' + r_{yy}y_l' + r_{yz}f_l + p_y\frac{f_l}{z_l} = x_r'\frac{z_r}{z_l}\frac{f_l}{f_r} \tag{12.58}$$

$$r_{zx}x_l' + r_{zy}y_l' + r_{zz}f_l + p_z\frac{f_l}{z_l} = x_r'\frac{z_r}{z_l}\frac{f_l}{f_r}. \tag{12.59}$$

The rotation part of the transformation changes the orientation of the left camera so that it coincides with the orientation of the right camera. The translation is the baseline between the two cameras. The variables for translation and depth appear as ratios in the equations, which means that the length of the baseline and depth can be scaled arbitrarily. For example, you can separate the cameras by twice as much and move the points in the scene twice as far away without changing the perspective geometry.

It is not possible to determine the baseline distance from the projections of the calibration points. This is not a serious problem, as the scale factor can be determined later by other means. For now, assume that the translation between cameras is a unit vector. Solving the relative orientation problem provides the three parameters for rotation and the two parameters for a unit vector that represents the direction of baseline. The binocular stereo depth measurements scale with the baseline distance. Assuming a unit baseline distance means that the binocular stereo measurements will be in an arbitrary system of units. The measurements obtained under the assumption of unit baseline distance will be correct except for the unknown scale factor. Relative distances between points will be correct. These arbitrary units can be converted to real units by multiplying by the baseline distance after it is obtained. Section 12.7 shows how the baseline distance can be determined as part of the solution to the absolute orientation problem. The conversion of stereo measurements from arbitrary units to real units and the transformation of point coordinates from viewer-centered coordinates to absolute coordinates can be done simultaneously by applying the transformation obtained from this augmented absolute orientation problem.

The rotation matrix is orthonormal, and this provides six additional constraints in addition to the artificial constraint of unit baseline distance. Given

n calibration points, there are $12 + 2n$ unknowns and $7 + 3n$ constraints. At least five conjugate pairs are needed for a solution, but in practice many more calibration points would be used to provide more accuracy.

The relative orientation problem starts with a set of calibration points and determines the rigid body transformation between the left and right cameras using the projections of these calibration points in the left and right image planes. Each calibration point $\mathbf{p}$ in the scene projects to point $\mathbf{p}'_l$ in the left camera and point $\mathbf{p}'_r$ in the right camera. Each projected point corresponds to a ray from the center of projection of its camera, through the projected point, and into the scene. The rays corresponding to $\mathbf{p}'_l$ and $\mathbf{p}'_r$ should intersect at point $\mathbf{p}$ in the scene, but may not intersect due to errors in measuring the projected locations in the image planes. We want to find the relative position and orientation of the two cameras in space, subject to the constraint of unit baseline distance, so that the errors in the locations of the rays in the image planes are minimized.

Let $\mathbf{r}_l$ be the ray (vector) from the center of projection of the left camera through point $\mathbf{p}'_l$ in the left image plane, let $\mathbf{r}_r$ be the ray from the center of projection of the right camera through point $\mathbf{p}'_r$ in the right image plane, and let $\mathbf{b}$ be the vector from the center of projection of the left camera to the center of projection of the right camera. We need to work with each ray in the same coordinate system, so rotate $\mathbf{r}_l$ so that it is in the same coordinate system as ray $\mathbf{r}_r$ and let $\mathbf{r}'_l$ denote this rotated ray. If the two rays intersect, then they lie in the plane normal to $\mathbf{r}'_l \times \mathbf{r}_r$. The baseline lies in this same plane, so the baseline is normal to $\mathbf{r}'_l \times \mathbf{r}_r$. This relationship is expressed mathematically by saying that the dot product of the baseline with the normal to the plane is zero:

$$\mathbf{b} \cdot (\mathbf{r}'_l \times \mathbf{r}_r) = 0. \tag{12.60}$$

This relationship is called the coplanarity condition.

Due to measurement errors, the rays will not intersect and the coplanarity condition will be violated. We can formulate a least-squares solution to the relative orientation problem by minimizing the sum of the squared errors of deviations from the triple products that represent the coplanarity condition:

$$\chi^2 = \sum_{i=1}^{n} w_i (\mathbf{b} \cdot (\mathbf{r}'_{l,i} \times \mathbf{r}_{r,i})^2), \tag{12.61}$$

where the weight counteracts the effect that when the triple product is near zero, changes in the baseline and rotation make a large change in the triple product. It is good practice in any regression problem to scale the relationship between the parameters and the data values so that small changes in the parameter space correspond to small changes in the data space. The weight on each triple product is

$$
w_i = \frac{\|\mathbf{r}'_{l,i} \times \mathbf{r}_{r,i}\|^2 \sigma_0^2}{\left[(\mathbf{b} \times \mathbf{r}_{r,i}) \cdot (\mathbf{r}'_{l,i} \times \mathbf{r}_{r,i})\right]^2 \|\mathbf{r}'_{l,i}\|^2 \sigma_l^2 + \left[(\mathbf{b} \times \mathbf{r}'_{l,i}) \cdot (\mathbf{r}'_{l,i} \times \mathbf{r}_{r,i})\right]^2 \|\mathbf{r}'_{r,i}\|^2 \sigma_r^2}.
$$
(12.62)

We need to switch to matrix notation, so assume that all vectors are column vectors. The error to be minimized is

$$
\chi^2 = \sum_{i=1}^{n} w_i (\mathbf{b} \cdot \mathbf{c}_i)^2 \tag{12.63}
$$

$$
= \mathbf{b}^T \left(\sum_{i=1}^{n} w_i \mathbf{c}_i \mathbf{c}_i^T \right) \mathbf{b} \tag{12.64}
$$

$$
= \mathbf{b}^T C \mathbf{b} \tag{12.65}
$$

subject to the constraint that $\mathbf{b}^T \mathbf{b} = 1$. The term $\mathbf{c}_i \mathbf{c}_i^T$ is an outer product, a 3×3 matrix formed by multiplying a 3×1 matrix with a 1×3 matrix, and is a real, symmetric matrix. The constrained minimization problem is

$$
\chi^2 = \mathbf{b}^T C \mathbf{b} + \lambda(1 - \mathbf{b}^T \mathbf{b}), \tag{12.66}
$$

where λ is the Lagrange multiplier. Differentiating with respect to vector $\mathbf{b}$,

$$
C \mathbf{b} = \lambda \mathbf{b}. \tag{12.67}
$$

The solution for the baseline is the unit eigenvector corresponding to the smallest eigenvalue of the C matrix.

The results above say that given an initial estimate for the rotation between the two cameras, we can determine the unit vector for the baseline. There is no closed-form solution for determining the rotation, but the estimates for the rotation and baseline can be refined using iterative methods. We will develop a system of linear equations for computing incremental improvements for the baseline and rotation.

The improvement $\delta\mathbf{b}$ to the baseline must be perpendicular to the baseline, since the unit vector representing the baseline cannot change length,

$$\mathbf{b} \cdot \delta\mathbf{b} = 0, \tag{12.68}$$

and the improvement to the rotation of the left ray into the right coordinate system is the infinitesimal rotation vector $\delta\boldsymbol{\omega}$. The corrections to the baseline and rotation will change each triple product for each ray from $t_i = \mathbf{b} \cdot (\mathbf{r}'_{l,i} \times \mathbf{r}_{r,i})$ to $t_i + \delta t_i$, with the increment in the triple product given by

$$\mathbf{c}_i \cdot \delta\mathbf{b} + \mathbf{d}_i \cdot \delta\boldsymbol{\omega}, \tag{12.69}$$

where

$$\mathbf{c}_i = \mathbf{r}'_{l,i} \times \mathbf{r}_{r,i} \tag{12.70}$$

$$\mathbf{d}_i = \mathbf{r}'_{l,i} \times (\mathbf{r}_{r,i} \times \mathbf{b}). \tag{12.71}$$

The corrections are obtained by minimizing

$$\chi^2 = \sum_{i=1}^{n} w_i (t_i + \mathbf{c}_i \cdot \delta\mathbf{b} + \mathbf{d}_i \cdot \delta\boldsymbol{\omega})^2, \tag{12.72}$$

subject to the constraint that $\delta\mathbf{b} \cdot \mathbf{b} = 0$. The constraint can be added onto the minimization problem using the Lagrange multiplier λ to get a system of linear equations for the baseline and rotation increments and the Lagrange multiplier:

$$\begin{bmatrix} C & F & \mathbf{b} \\ F^T & D & 0 \\ \mathbf{b}^T & 0 & 0 \end{bmatrix} \begin{bmatrix} \delta\mathbf{b} \\ \delta\boldsymbol{\omega} \\ \lambda \end{bmatrix} = - \begin{bmatrix} \bar{\mathbf{c}} \\ \bar{\mathbf{d}} \\ 0 \end{bmatrix}, \tag{12.73}$$

where

$$C = \sum_{i=1}^{n} w_i \mathbf{c}_i \mathbf{c}_i^T \tag{12.74}$$

$$F = \sum_{i=1}^{n} w_i \mathbf{c}_i \mathbf{d}_i^T \tag{12.75}$$

$$D = \sum_{i=1}^{n} w_i \mathbf{d}_i \mathbf{d}_i^T \tag{12.76}$$

$$\bar{c} = \sum_{i=1}^{n} w_i t_i \mathbf{c}_i^T \tag{12.77}$$

$$\bar{d} = \sum_{i=1}^{n} w_i t_i \mathbf{d}_i^T. \tag{12.78}$$

Once we have the corrections to the baseline and rotation, we have to apply the corrections in a way that preserves the constraints that the baseline is a unit vector and the rotation is represented correctly. For example, if rotation is represented as an orthonormal matrix, the corrected matrix must be orthonormal. It is difficult to update a rotation matrix without violating orthonormality, so the rotation will be represented as a unit quaternion.

The baseline is updated using the formula

$$\mathbf{b}^{n+1} = \mathbf{b}^n + \delta\mathbf{b}^n. \tag{12.79}$$

The updated baseline should be explicitly normalized to guarantee that numerical errors do not lead to violation of the unit vector constraint.

The infinitesimal rotation can be represented by the unit quaternion

$$\delta\mathbf{q} = \sqrt{1 - \frac{1}{4}\|\delta\boldsymbol{\omega}\|^2} + \frac{1}{2}\delta\boldsymbol{\omega}. \tag{12.80}$$

This formula guarantees that the quaternion will be a unit quaternion even if the rotation is large. If $\mathbf{r}$ is the quaternion representing the rotation, then the updated rotation is the quaternion $\mathbf{r}'$ given by

$$\mathbf{r}' = \mathbf{q}\mathbf{r}\mathbf{q}^*, \tag{12.81}$$

where multiplication is performed according to the rules for multiplying quaternions (Section 12.2.3), and $\mathbf{q}^*$ is the conjugate of quaternion $\mathbf{q}$.

12.5 Rectification

Rectification is the process of resampling stereo images so that the epipolar lines correspond to image rows. The basic idea is simple: if the left and right image planes are coplanar and the horizontal axes are colinear (no rotation about the optical axes), then the image rows are epipolar lines and stereo correspondences can be found by searching for matches along corresponding rows.

In practice, this condition can be difficult to achieve and some vergence (inward rotation about the vertical camera axes) may be desirable, but if the pixels in the left and right images are projected onto a common plane, then the ideal epipolar geometry is achieved. Each pixel in the left (right) camera corresponds to a ray in the left (right) camera coordinate system. Let T_l and T_r be the rigid body transformations that bring rays from the left and right cameras, respectively, into the coordinate system of the common plane. Determine the locations in the common plane of the corners of each image, create new left and right image grids, and transform each grid point back into its original image. Bilinear interpolation, discussed in Section 13.6.2, can be used to interpolate pixel values to determine the pixel values for the new left and right images in the common plane.

12.6   Depth from Binocular Stereo

Binocular stereo matches feature points in the left and right images to create a set of conjugate pairs, $\{(\mathbf{p}_{l,i}, \mathbf{p}_{r,i})\}, i = 1, \ldots, n$. Each conjugate pair defines two rays that (ideally) intersect in space at a scene point. The space intersection problem is to find the three-dimensional coordinates of the point of intersection. Due to errors in measuring the image plane coordinates and errors in the cameras, the rays will not intersect, so the problem of computing depth from stereo pairs is to find the coordinates of the scene point that is closest to both rays.

We will assume that stereo measurements will be made in a coordinate system that is different from the coordinate systems of either camera. For example, the stereo coordinate system might be attached to the frame that holds the two cameras. There will be two rigid body transformations: one aligns the left camera with stereo coordinates and the other aligns the right camera with stereo coordinates. The left transformation has rotation matrix R_l and translation $\mathbf{p}_l = (x_l, y_l, z_l)$, and the right transformation has rotation matrix R_r and translation $\mathbf{p}_r = (x_r, y_r, z_r)$. To represent point measurements in the coordinate system of the right (left) camera, use the rigid body transformation obtained from solving the relative orientation problem (or its inverse) for the left camera transformation and use the identity transformation for the right (left) camera.

The coordinates of the conjugate pair in three dimensions are $(x'_{l,i}, y'_{l,i}, f_l)$ and $(x'_{r,i}, y'_{r,i}, f_r)$. Rotate and translate the left camera coordinates into stereo coordinates,

$$\begin{pmatrix} x \\ y \\ z \end{pmatrix} = \begin{pmatrix} x_l \\ y_l \\ z_l \end{pmatrix} + t_l R_l \begin{pmatrix} x'_{l,i} \\ y'_{l,i} \\ f_l \end{pmatrix}, \tag{12.82}$$

and rotate and translate the right camera coordinates into stereo coordinates,

$$\begin{pmatrix} x \\ y \\ z \end{pmatrix} = \begin{pmatrix} x_r \\ y_r \\ z_r \end{pmatrix} + t_r R_r \begin{pmatrix} x'_{r,i} \\ y'_{r,i} \\ f_r \end{pmatrix}. \tag{12.83}$$

In order to find the point that is close to both rays, find the values for t_l and t_r that correspond to the minimum distance between the rays by minimizing the norm

$$\chi^2 = \left[\begin{pmatrix} x_l \\ y_l \\ z_l \end{pmatrix} + t_l R_l \begin{pmatrix} x'_{l,i} \\ y'_{l,i} \\ f_l \end{pmatrix} - \begin{pmatrix} x_r \\ y_r \\ z_r \end{pmatrix} - t_r R_r \begin{pmatrix} x'_{r,i} \\ y'_{r,i} \\ f_r \end{pmatrix} \right]^2 \tag{12.84}$$

$$= [\mathbf{b} + t_l \mathbf{r}_l - t_r \mathbf{r}_r]^2, \tag{12.85}$$

where $\mathbf{b}$ is the baseline in stereo coordinates, and $\mathbf{r}_l$ and $\mathbf{r}_r$ are the left and right rays rotated into stereo coordinates. To solve the ray intersection problem, differentiate with respect to t_l and t_r and set the result equal to zero. Solve the equations for t_l and t_r and plug the solution values for the parameters into the ray equations (Equation 12.82 and 12.83) to obtain the point on each ray that is closest to the other ray. Average the two point locations to obtain the depth estimate.

The stereo point measurements are in the coordinate system of the stereo system, either the left or right camera, or a neutral coordinate system. If the algorithm for relative orientation presented in Section 12.4 was used to determine the baseline, then the measurements are in a unitless system of measurement. If the rigid body transformations between the left and right cameras and another coordinate system for the stereo hardware were obtained by solving the exterior orientation problem (Section 12.8) or by other means, then the stereo point measurements are in the units that were used for the calibration points. Regardless of how the stereo system was calibrated, we have to transform the point measurements into an absolute coordinate system for the scene. We can convert the measurements into a system of units appropriate to the scene at the same time by solving the absolute orientation problem with a scale factor.

12.7   Absolute Orientation with Scale

The formulation of the absolute orientation problem in Section 12.3 does not allow transformations that include a scale change; the transformation between coordinate systems is a rigid body transformation which includes only rotation and translation. Scale changes occur, for example, in binocular stereo when the baseline between the stereo cameras is unknown or incorrect or between range cameras with different measurement units.

The absolute orientation problem, presented in Section 12.3, can be extended to include scale changes. The solution to this extended problem will be a transformation that includes rotation and translation to align the viewpoint to an absolute coordinate system and includes a scale factor to convert camera-specific measurement units to the common system of units. Consider a point $\mathbf{p}$ with coordinates $\mathbf{p}_1 = (x_1, y_1, z_1)$ in one coordinate system and coordinates $\mathbf{p}_2 = (x_2, y_2, z_2)$ in another coordinate system. The transformation between coordinates is

$$\mathbf{p}_2 = sR\mathbf{p}_1 + \mathbf{p}_0, \qquad (12.86)$$

where s is the scale change. This increases the number of parameters in the absolute orientation problem to seven: three parameters for rotation, three parameters for translation, and the scale factor. The scaling transformation is uniform scaling: the coordinates of each axis are scaled by the same amount.

The input to the absolute orientation problem is a set of n conjugate pairs for the first and second views: $\{(\mathbf{p}_{1,i}, \mathbf{p}_{2,i})\}$. The regression problem is to find the rotation R, translation $\mathbf{p}_0$, and scale s that minimize

$$\sum_{i=1}^{n} \left(\mathbf{p}_{2,i} - sR\mathbf{p}_{1,i} - \mathbf{p}_0\right)^2. \qquad (12.87)$$

The solution requires at least three points to get nine equations for the seven unknowns. In practice, more points are used to provide better accuracy.

Ignore for the moment that the correspondence between points is known and imagine that the two sets of points (the set of points in the first coordinate system and the set in the second system) are two clouds of points in absolute coordinate space. Compute the centroid of each cloud of points:

$$\bar{\mathbf{p}}_1 = \frac{1}{n}\sum_{i=1}^{n}\mathbf{p}_{1,i} \qquad (12.88)$$

$$\bar{\mathbf{p}}_2 = \frac{1}{n} \sum_{i=1}^{n} \mathbf{p}_{2,i} \tag{12.89}$$

and transform each cloud of points to a bundle of vectors about the centroid:

$$\mathbf{r}_{1,i} = \mathbf{p}_{1,i} - \bar{\mathbf{p}}_1, \qquad \mathbf{r}_{2,i} = \mathbf{p}_{2,i} - \bar{\mathbf{p}}_2. \tag{12.90}$$

The scale parameter can be determined by computing the mean length of the vectors in each bundle:

$$s^2 = \frac{\sum_{i=1}^{n} \|\mathbf{r}_{2,i}\|^2}{\sum_{i=1}^{n} \|\mathbf{r}_{1,i}\|^2}. \tag{12.91}$$

The scale factor can be computed without knowing either the rotation or the translation. This is a very useful formula for calibrating the baseline distance in binocular stereo and is more accurate than using only a few points.

After the rotation and scale factor are determined, the translation can be easily computed from the centroids:

$$\mathbf{p}_0 = \bar{\mathbf{p}}_2 - sR\bar{\mathbf{p}}_1. \tag{12.92}$$

Computing the rotation is essentially the problem of determining how to align the bundles of rays about the centroids. Form the matrix M from the sum of scalar products of the coodinates of the rays in the first and second views:

$$M = \sum_{i=1}^{n} \mathbf{r}_{2,i} (\mathbf{r}_{1,i})^T. \tag{12.93}$$

Let matrix $Q = M^T M$. The rotation matrix is

$$R = MS^{-1}, \tag{12.94}$$

where matrix S is

$$S = Q^{1/2}. \tag{12.95}$$

The eigenvalue-eigenvector decomposition of matrix Q is

$$Q = \lambda_1 \mathbf{v}_1 \mathbf{v}_1^T + \lambda_2 \mathbf{v}_2 \mathbf{v}_2^T + \lambda_3 \mathbf{v}_3 \mathbf{v}_3^T \tag{12.96}$$

The eigenvalues of $M^T M$ are obtained by solving a cubic equation. The roots of the cubic equation can be computed from direct formulas [197]. Use the eigenvalues to solve the linear equations

$$\left(M^T M - \lambda_i I \right) \mathbf{v}_i = 0, \tag{12.97}$$

for the orthogonal eigenvectors $\mathbf{v}_1$, $\mathbf{v}_2$, and $\mathbf{v}_3$. Matrix S is the square root of matrix Q. Fortunately, matrix square roots and their inverses are easy to compute in the eigensystem representation (Equation 12.96). The inverse of matrix S is

$$S^{-1} = \left(M^T M\right)^{-1/2} = \frac{1}{\sqrt{\lambda_1}}\mathbf{v}_1\mathbf{v}_1^T + \frac{1}{\sqrt{\lambda_2}}\mathbf{v}_2\mathbf{v}_2^T + \frac{1}{\sqrt{\lambda_3}}\mathbf{v}_3\mathbf{v}_3^T. \qquad (12.98)$$

Compute the outer products of the eigenvectors, divided by the square root of the eigenvectors, and multiply this matrix by M to get the rotation matrix R. This method of construction guarantees that matrix R will be an orthonormal matrix.

This algorithm provides a closed form (noniterative) solution for the rotation matrix. The scale can be determined without determining either the translation or rotation using the formula in Equation 12.91 and, finally, the translation can be determined using the formula in Equation 12.92. The transformation in Equation 12.86 can be applied to the point measurements from any depth measurement system, including binocular stereo or a range camera, to align the point measurements to an absolute coordinate system and convert the measurements into the units of the absolute coordinate system. The measurement units will be whatever system of units was used for the coordinates of the calibration points used for the absolute orientation problem. For example, if the calibration points in the absolute coordinate system are in millimeters and the depth measurements are from binocular stereo with a unit baseline, then the rigid body transformation obtained by solving the absolute orientation problem with these calibration points will transform stereo measurements to millimeters. The system of measurement must be the same along each coordinate axis since the same scale factor s is applied to each coordinate.

12.8   Exterior Orientation

The problem of exterior orientation is to determine the relationship between image plane coordinates (x', y') and the coordinates (x, y, z) of scene points in an absolute coordinate system. The exterior orientation problem is called the hand–eye problem in robotics and machine vision. Recall from Section 1.4.1 on perspective projection that a point (x, y, z) in the viewer-centered coordinate system of the camera is projected to a point (x', y') in the image plane. Until this section, we have been content to represent the coordinates of points in the scene in the coordinate system of the camera; in many applications, though, it is necessary to relate the coordinates of measurements computed in the image plane coordinate system to the absolute coordinate system defined for the scene. Each point (x', y') in the image plane defines a ray from the center of projection, passing through (x', y') in the image plane, and continuing on into the scene. The position of the camera in the scene is the location of the center of projection, and the orientation of the camera determines the orientation of the bundle of rays from the center of projection, passing through image plane points. An image plane point does not correspond to a unique point in the scene, but we may be able to use the equation for the ray passing through the image point, along with other information about the scene geometry, to determine a unique point in absolute coordinates. For example, if we know that an image plane point corresponds to a point on a wall, and if we know the equation for the plane that models the wall, then the exact location of the point on the wall can be obtained by solving the system of ray and plane equations for the intersection.

The exterior orientation problem is to determine the position and orientation of the bundle of rays corresponding to image plane points so that the coordinates of each image plane point may be transformed to its ray in the absolute coordinate system of the scene. The problem is illustrated in Figure 12.4. The position and orientation of the bundle of rays is represented as the rigid body transformation from camera coordinates to absolute coordinates. An image plane point (x', y') has coordinates (x', y', f) in the three-dimensional coordinate system of the camera, with the image plane at a distance f in front of the center of projection. The center of projection corresponds to the origin of the camera coordinate system. The position of the camera in the scene is the location of the center of projection in absolute coordinates. In camera coordinates, the parametric equation of the ray passing through point (x', y') in the image plane is

$$(x, y, z) = t(x', y', f), \qquad (12.99)$$

with parameter t going from zero (at the center of projection) to infinity. At $t = 1$, the point (x, y, z) in the camera coordinate system is the image plane point (x', y', f). Given a measured location (x', y') in the image and

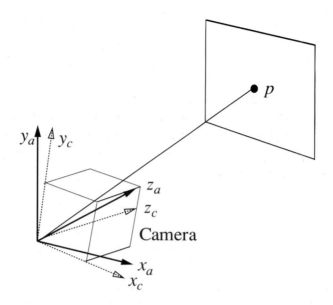

Figure 12.4: Illustration of how the position and orientation of the camera are determined by aligning the bundle of rays with corresponding scene points.

an estimate for the camera constant f, we have the equation for the ray in camera-centered coordinates.

Let $\mathbf{p} = (x, y, z)^T$ and $\mathbf{p}' = (x', y', f)^T$. The equation for the ray in the absolute coordinate system for the scene is obtained by applying the rigid body transformation from camera coordinates to absolute coordinates to the parametric equation for the ray (Equation 12.99),

$$\mathbf{p} = tRp' + \mathbf{p}_0. \tag{12.100}$$

The rigid body transformation can be determined by measuring the position (x_i', y_i') of the projection of calibration points in the scene with known position (x_i, y_i, z_i), relating the image plane points to the scene points using the equations for perspective projection, and solving for the rigid body transformation. The parameters for the exterior orientation of the camera (rotation angles and translation vector to the camera origin) are called the extrinsic parameters, as opposed to the intrinsic parameters of the internal geometry of the camera such as the camera constant.

The exterior orientation problem can be succinctly stated: what is the rigid body transformation from absolute coordinates to camera coordinates that positions and orients the bundle of rays in space so that each ray passes through its corresponding calibration point? To make the coordinate systems clear, we will use subscripts to distinguish absolute coordinates from camera coordinates. The position of a point in absolute coordinates is

$$\mathbf{p}_a = (x_a, y_a, z_a)^T \tag{12.101}$$

and the position of a point in camera coordinates is

$$\mathbf{p}_c = (x_c, y_c, z_c)^T. \tag{12.102}$$

We will develop the exterior orientation problem as a transformation from absolute (scene) coordinates to camera coordinates. The inverse of this transformation, needed for practical applications, is given by Equation 12.16. The rigid body transformation from absolute coordinates to camera coordinates is

$$x_c = r_{xx}x_a + r_{xy}y_a + r_{xz}z_a + p_x \tag{12.103}$$
$$y_c = r_{yx}x_a + r_{yy}y_a + r_{yz}z_a + p_y \tag{12.104}$$
$$z_c = r_{zx}x_a + r_{zy}y_a + r_{zz}z_a + p_z. \tag{12.105}$$

The positions of the points in the coordinate system of the camera are unknown, but the projection of the points onto the image plane is determined by the equations for perspective projection:

$$\frac{x'}{f} = \frac{x_c}{z_c} \tag{12.106}$$

$$\frac{y'}{f} = \frac{y_c}{z_c}. \tag{12.107}$$

Solve the perspective equations for x_c and y_c, and combine the results for the first two equations for the transformation from absolute to camera coordinates (Equations 12.103 and 12.104) with Equation 12.105 as the denominator to obtain two equations relating image plane coordinates (x', y') to the absolute coordinates for a calibration point:

$$\frac{x'}{f} = \frac{r_{xx}x_a + r_{xy}y_a + r_{xz}z_a + p_x}{r_{zx}x_a + r_{zy}y_a + r_{zz}z_a + p_z} \tag{12.108}$$

$$\frac{y'}{f} = \frac{r_{yx}x_a + r_{yy}y_a + r_{yz}z_a + p_y}{r_{zx}x_a + r_{zy}y_a + r_{zz}z_a + p_z}. \tag{12.109}$$

Each calibration point yields two equations that constrain the transformation:

$$x'(r_{zx}x_a+r_{zy}y_a+r_{zz}z_a+p_z) - f(r_{xx}x_a+r_{xy}y_a+r_{xz}z_a+p_x) = 0 \quad (12.110)$$

$$y'(r_{zx}x_a+r_{zy}y_a+r_{zz}z_a+p_z) - f(r_{yx}x_a+r_{yy}y_a+r_{yz}z_a+p_y) = 0. \quad (12.111)$$

Six calibration points would yield 12 equations for the 12 transformation parameters, but using the orthonormality constraints for the rotation matrix reduces the minimum number of calibration points to four. Many more points would be used in practice to provide an accurate solution. Replace the elements of the rotation matrix with the formulas using Euler angles and solve the nonlinear regression problem.

12.8.1   Calibration Example

A robot is equipped with a suction pickup tool at the end of the arm. The tool is good for picking up small, flat objects if the suction tool is positioned near the center of the object. There are flat objects on a table within reach of the arm. The absolute coordinate system is at one corner of the table. A camera is positioned above the table, with the table within the field of view. The position of a point in the image plane is (x', y'). If the object has good contrast against the background of the table top, then the image plane position can be estimated using first moments. The position of a point in the absolute coordinate system of the table is (x, y, z). The position and orientation of the camera relative to the absolute coordinate system can be determined by solving the exterior orientation problem.

Given the position (x', y') of the center of the object in the image plane, the location (x, y, z) in the absolute coordinate system of the center of the part on the table is given by intersecting the ray from the camera origin through (x', y') with the plane of the table top. Both the equation for the table top

$$ax + by + cz + d = 0 \qquad (12.112)$$

and the equation for the ray from the camera origin

$$\begin{pmatrix} x \\ y \\ z \end{pmatrix} = t \begin{pmatrix} x' \\ y' \\ f \end{pmatrix} \qquad (12.113)$$

must be in the absolute coordinate system. The ray from the camera can be transformed to the absolute coordinate system using the transformation obtained by solving the exterior orientation problem. If the origin of the absolute coordinate system is in the plane of the table with the z axis normal to the table, then Equation 12.112 reduces to $z = 0$ and the intersection is easy to compute.

12.9 Interior Orientation

The problem of interior orientation is to determine the internal geometry of the camera. The geometry is represented by a set of camera parameters:

Camera constant for the distance of the image plane from the center of projection

Principal point for the location of the origin of the image plane coordinate system

Lens distortion coefficients for the changes in image plane coordinates caused by optical imperfections in the camera

Scale factors for the distances between the rows and columns

The interior orientation problem is the problem of compensating for errors in the construction of the camera so that the bundle of rays inside the camera obeys the assumptions of perspective projection. The camera parameters are called the intrinsic parameters, as opposed to the extrinsic parameters for the exterior orientation of the camera. The interior orientation problem is the regression problem for determining the intrinsic parameters of a camera.

The camera constant is not the same as the focal length of the lens. When the lens is focused at infinity, then the camera constant is equal to the focal length; otherwise, the camera constant is less than the focal length. The principal point is where the optical axis intersects the image plane. It establishes the origin of the image plane coordinate system, which up to this point has been assumed to be the center of the image array (see Equations 12.1 and 12.2). Although the camera constant is close to the focal length and the principal point is close to the center of the image, these approximations may not be good enough for many applications. The spacing between the rows and columns of pixels in the image sensor can be determined from the camera specifications, but frame grabbers may introduce errors that must be calibrated.

Some calibration algorithms solve the interior orientation problems and exterior orientation problems at the same time. The motivation for this is that the true location of the calibration points on the image plane cannot be known until the exterior orientation of the camera has been determined. However, the interior orientation problem can be solved by itself, and there are several methods for determining the camera constant, location of the principal point, and lens distortions without knowing the exterior orientation of the camera in absolute coordinates. Methods for determining both the intrinsic and extrinsic parameters are covered in Section 12.10.

The fundamental idea in determining the intrinsic camera parameters independent of the extrinsic parameters is to use a calibration image with some regular pattern, such as a grid of lines. Distortions in the pattern are used to estimate the lens distortions and calculate corrections to the nominal values for the other intrinsic parameters.

Lens distortions include two components: radial distortion that bends the rays of light by more or less than the correct amount, and decentering caused by a displacement of the center of the lens from the optical axis. The radial distortion and decentering effects are modeled as polynomials; the interior orientation algorithm estimates the coefficients of these polynomials. Figure 12.5 illustrates the radially symmetric nature of most lens distortions, in the absence of lens decentering errors. Light rays are bent toward the optical axis by more or less than the correct amount, but this error is the same at all positions on the lens (or in the image plane) that are the same distance from the principal point.

The radial distortion can be modeled as a polynomial in even powers of the radius, since error in the amount of bending of the rays of light is rotationally symmetric. Let (x', y') denote the true image coordinates and $(\tilde{x}, \tilde{y})$ denote the uncorrected image coordinates obtained from the pixel coordinates i and j using an estimate for the location of the principal point:

$$\tilde{x} = j - \hat{c}_x \tag{12.114}$$
$$\tilde{y} = -(i - \hat{c}_y). \tag{12.115}$$

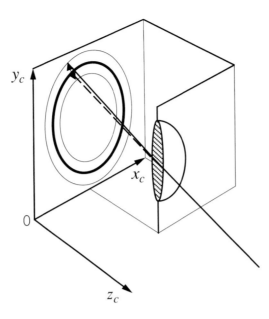

Figure 12.5: Most lens distortions are radially symmetric. The rays are bent toward the center of the image by more or less than the correct amount. The amount of radial distortion is the same at all points in the image plane that are the same distance from the true location of the principal point.

The corrections $(\delta x, \delta y)$ will be added to the uncorrected coordinates to get the true image plane coordinates:

$$x' = \tilde{x} + \delta x \tag{12.116}$$

$$y' = \tilde{y} + \delta y. \tag{12.117}$$

The corrections for radial lens distortions are modeled by a polynomial in even powers of the radial distance from the center of the image plane:

$$\delta x = (\tilde{x} - x_p)(\kappa_1 r^2 + \kappa_2 r^4 + \kappa_3 r^6) \tag{12.118}$$

$$\delta y = (\tilde{y} - y_p)(\kappa_1 r^2 + \kappa_2 r^4 + \kappa_3 r^6), \tag{12.119}$$

where (x_p, y_p) is the refinement to the location of the principal point and

$$r^2 = (\tilde{x} - x_p)^2 + (\tilde{y} - y_p)^2 \tag{12.120}$$

is the square of the radial distance from the center of the image. Note that x_p and y_p are not the same as $\hat{c}_x$ and $\hat{c}_y$ in Equations 12.114 and 12.115, which are used to compute the uncorrected image coordinates; x_p and y_p are corrections to $\hat{c}_x$ and $\hat{c}_y$. After calibration, the corrections can be applied to the initial estimates:

$$c_x = \hat{c}_x + x_p \tag{12.121}$$

$$c_y = \hat{c}_y - y_p. \tag{12.122}$$

The calibration problem for correcting radial distortion is to find the coefficients, κ_1, κ_2, and κ_3, of the polynomial. Lens distortion models beyond the sixth degree are rarely used; in fact, it may not be necessary to go beyond a second-degree polynomial. The location of the principal point is included in the calibration problem since an accurate estimate for the location of the principal point is required to model the lens distortions. More powerful models for lens distortions can include tangential distortions due to effects such as lens decentering:

$$\delta x = (\tilde{x} - x_p)(\kappa_1 r_2 + \kappa_2 r_4 + \kappa_3 r_6) \tag{12.123}$$
$$+ \left[p_1 \left(r^2 + 2(\tilde{x} - x_p)^2 \right) + 2 p_2 (\tilde{x} - x_p)(\tilde{y} - y_p) \right] (1 + p_3 r^2)$$

$$\delta y = (\tilde{y} - y_p)(\kappa_1 r_2 + \kappa_2 r_4 + \kappa_3 r_6) \tag{12.124}$$
$$+ \left[2 p_1 (\tilde{x} - x_p)(\tilde{y} - y_p) + 2 p_2 \left(r^2 + 2(\tilde{y} - y_p)^2 \right) \right] (1 + p_3 r^2).$$

Use a calibration target consisting of several straight lines at different positions and orientations in the field of view. The only requirement is that the lines be straight; the lines do not have to be perfectly horizontal or vertical. This method does not involve solving the exterior orientation problem simultaneously. It is easy to make a grid of horizontal and vertical lines using a laser printer. Diagonal lines are rendered less accurately, but since the exterior orientation does not matter, you can shift and rotate the grid to different positions in the field of view, acquire several images, and gather a large set of digitized lines for the calibration set. Mount the grid on a flat, rigid surface that is normal to the optical axis. Since the lines do not have to be parallel, any tilt in the target will not affect the calibration procedure. Determine the location of the edge points to subpixel resolution by computing the first moment over small windows throughout the image. The window size should be somewhat larger than the width of the lines, but smaller than the spacing between lines. The Hough transform can be used to group edges into lines and determine initial estimates for the line parameters.

The equation for each line l in true (corrected) image coordinates is

$$x' \cos \theta_l + y' \sin \theta_l - \rho_l = 0. \tag{12.125}$$

Since the precise position and orientation of each line is unknown, the estimates for the line parameters must be refined as part of the interior orientation problem. Let $(\tilde{x}_{kl}, \tilde{y}_{kl})$ denote the coordinates of edge point k along line l. Replace the true image coordinates (x', y') in Equation 12.125 with the uncorrected coordinates $\tilde{x}_{kl} + \delta x$ and $\tilde{y}_{kl} + \delta y$ using the model for the corrections given above. This yields an equation of the form

$$f(\tilde{x}_{kl}, \tilde{y}_{kl}; x_p, y_p, \kappa_1, \kappa_2, \kappa_3, p_1, p_2, p_3, \rho_l, \theta_l) = 0 \tag{12.126}$$

for each observation (edge point) and the intrinsic parameters. The set of equations for n edge points is a system of n nonlinear equations that must be solved using nonlinear regression. The initial values for corrections to the location of the principal point are zero, and the coefficients for radial lens distortion and decentering can also be initialized to zero. The overall minimization criterion in

$$\chi^2 = \sum_{k=1}^{n} (f(x'_{kl}, y'_{kl}; x_p, y_p, \kappa_1, \kappa_2, \kappa_3, p_1, p_2, p_3, \rho_l, \theta_l))^2. \tag{12.127}$$

This nonlinear regression problem can be solved for the location of the principal point (x_p, y_p); the parameters for the radial lens distortion κ_1, κ_2, and κ_3; and the parameters for the lens decentering p_1, p_2, and p_3. The parameters of each line are estimated as a byproduct of determining the intrinsic parameters and can be discarded.

Example 12.1 *Suppose that a calibration table has been prepared to compensate for lens and camera distortions. The table provides a correction $(\delta x, \delta y)$ for every row and column in the image array. How would this information be used in the overall system presented as an example in Section 12.8.1? After computing the position $(\tilde{x}, \tilde{y})$ of the centroid of the object in the image plane, interpolate the corrections between pixels and add the corrections to the centroid to get correct coordinates for the ray from the camera origin to the object.*

12.10   Camera Calibration

The camera calibration problem is to relate the locations of pixels in the image array to points in the scene. Since each pixel is imaged through perspective projection, it corresponds to a ray of points in the scene. The camera calibration problem is to determine the equation for this ray in the absolute coordinate system of the scene. The camera calibration problem includes both the exterior and interior orientation problems, since the position and orientation of the camera and the camera constant must be determined to relate image plane coordinates to absolute coordinates, and the location of the principal point, the aspect ratio, and lens distortions must be determined to relate image array locations (pixels coordinates) to positions in the image plane. The camera calibration problem involves determining two sets of parameters: the extrinsic parameters for rigid body transformation (exterior orientation) and the intrinsic parameters for the camera itself (interior orientation).

We can use an initial approximation for the intrinsic parameters to get a mapping from image array (pixel) coordinates to image plane coordinates. Suppose that there are n rows and m columns in the image array and assume that the principal point is located at the center of the image array:

$$c_x = \frac{m-1}{2} \tag{12.128}$$

$$c_y = \frac{n-1}{2}. \tag{12.129}$$

The image plane coordinates for the pixel at grid location $[i, j]$ are

$$\tilde{x} = \tau_x d_x (j - c_x) \tag{12.130}$$

$$\tilde{y} = -d_y (i - c_y), \tag{12.131}$$

where d_x and d_y are the center-to-center distances between pixels in the x and y directions, respectively, and τ_x is a scale factor that accounts for distortions in the aspect ratio caused by timing problems in the digitizer electronics. The row and column distances, d_x and d_y, are available from the specifications for the CCD camera and are very accurate, but the scale factor τ_x must be added to the list of intrinsic parameters for the camera and determined through calibration. Note that these are uncorrected image coordinates, marked with a tilde to emphasize that the effects of lens distortions have not been removed. The coordinates are also affected by errors in the estimates for the location of the principal point (c_x, c_y) and the scale factor τ_x.

We must solve the exterior orientation problem before attempting to solve the interior orientation problem, since we must know how the camera is positioned and oriented in order to know where the calibration points project into the image plane. Once we know where the projected points should be, we can use the projected locations p_i' and the measured locations $\tilde{p}_i$ to determine the lens distortions and correct the location of the principal point and the image aspect ratio. The solution to the exterior orientation problem must be based on constraints that are invariant to the lens distortions and camera constant, which will not be known at the time that the problem is solved.

12.10.1 Simple Method for Camera Calibration

This section explains the widely used camera calibration method published by Tsai [234]. Let $\mathbf{p}_0'$ be the location of the origin in the image plane, $\mathbf{r}_i'$ be the vector from $\mathbf{p}_0'$ to the image point $\mathbf{p}_i' = (x_i', y_i')$, $\mathbf{p}_i = (x_i, y_i, z_i)$ be a calibration point, and $\mathbf{r}_i$ be the vector from the point $(0, 0, z_i)$ on the optical axis to $\mathbf{p}_i$. If the difference between the uncorrected image coordinates $(\tilde{x}_i, \tilde{y}_i)$ and the true image coordinates (x_i', y_i') is due only to radial lens distortion, then $\mathbf{r}_i'$ is parallel to $\mathbf{r}_i$. The camera constant and translation in z do not affect the direction of $\mathbf{r}_i'$, since both image coordinates will be scaled by the same amount. These constraints are sufficient to solve the exterior orientation problem [234].

Assume that the calibration points lie in a plane with $z = 0$ and assume that the camera is placed relative to this plane to satisfy the following two crucial conditions:

1. The origin in absolute coordinates is not in the field of view.

2. The origin in absolute coordinates does not project to a point in the image that is close to the y axis of the image plane coordinate system.

Condition 1 decouples the effects of radial lens distortion from the camera constant and distance to the calibration plane. Condition 2 guarantees that the y component of the rigid body translation, which occurs in the denominator of many equations below, will not be close to zero. These conditions are easy to satisfy in many imaging situations. For example, suppose that the camera is placed above a table, looking down at the middle of the table. The absolute coordinate system can be defined with $z = 0$ corresponding to the plane of the table, with the x and y axes running along the edges of the table, and with the corner of the table that is the origin in absolute coordinates outside of the field of view.

Suppose that there are n calibration points. For each calibration point, we have the absolute coordinates of the point (x_i, y_i, z_i) and the uncorrected image coordinates $(\tilde{x}_i, \tilde{y}_i)$. Use these observations to form a matrix A with rows a_i,

$$a_i = (\tilde{y}_i x_i, \tilde{y}_i y_i, -\tilde{x}_i x_i, -\tilde{x}_i y_i, \tilde{y}_i). \qquad (12.132)$$

Let $u = (u_1, u_2, u_3, u_4, u_5)$ be a vector of unknown parameters that are related to the parameters of the rigid body transformation:

$$u_1 = \frac{r_{xx}}{p_y} \qquad (12.133)$$

$$u_2 = \frac{r_{xy}}{p_y} \qquad (12.134)$$

$$u_3 = \frac{r_{yx}}{p_y} \qquad (12.135)$$

$$u_4 = \frac{r_{yy}}{p_y} \qquad (12.136)$$

$$u_5 = \frac{p_x}{p_y}. \qquad (12.137)$$

Form a vector $\mathbf{b} = (\tilde{x}_1, \tilde{x}_2, \ldots, \tilde{x}_n)$ from the n observations of the calibration points. With more than five calibration points, we have an overdetermined system of linear equations,

$$A\mathbf{u} = \mathbf{b}, \qquad (12.138)$$

for the parameter vector $\mathbf{u}$. Solve this linear system using singular value decomposition, and use the solution parameters, u_1, u_2, u_3, u_4, and u_5, to compute the rigid body transformation, except for p_z, which scales with the camera constant f and will be determined later.

First, compute the magnitude of the y component of translation. If u_1 and u_2 are not both zero and u_3 and u_4 are not both zero, then

$$p_y^2 = \frac{U - [U^2 - 4(u_1 u_4 - u_2 u_3)^2]^{1/2}}{2(u_1 u_4 - u_2 u_3)^2}, \qquad (12.139)$$

where $U = u_1^2 + u_2^2 + u_3^2 + u_4^2$; otherwise, if u_1 and u_2 are both zero, then

$$p_y^2 = \frac{1}{u_3^2 + u_4^2}; \qquad (12.140)$$

otherwise, using u_1 and u_2,

$$p_y^2 = \frac{1}{u_1^2 + u_2^2}. \qquad (12.141)$$

Second, determine the sign of p_y. Pick the calibration point $\mathbf{p} = (x, y, z)$ that projects to an image point that is farthest from the center of the image (the scene point and corresponding image point that are farthest in the periphery of the field of view). Compute r_{xx}, r_{xy}, r_{yx}, r_{yy}, and p_x from the solution vector obtained above:

$$r_{xx} = u_1 p_y \qquad (12.142)$$
$$r_{xy} = u_2 p_y \qquad (12.143)$$
$$r_{yx} = u_3 p_y \qquad (12.144)$$
$$r_{yy} = u_4 p_y \qquad (12.145)$$
$$p_x = u_5 p_y. \qquad (12.146)$$

Let $\xi_x = r_{xx} x + r_{xy} y + p_x$ and $\xi_y = r_{yx} x + r_{yy} y + p_y$. If ξ_x and $\tilde{x}$ have the same sign and ξ_y and $\tilde{y}$ have the same sign, then p_y has the correct sign (positive); otherwise, negate p_y. Note that the parameters of the rigid body transformation computed above are correct, regardless of the sign of p_y, and do not need to be changed.

Third, compute the remaining parameters of the rigid body transformation:

$$r_{xz} = \sqrt{1 - r_{xx}^2 - r_{xy}^2} \qquad (12.147)$$
$$r_{yz} = \sqrt{1 - r_{yx}^2 - r_{yy}^2}. \qquad (12.148)$$

Since the rotation matrix must be orthonormal, it must be true that $R^T R = I$. Use this fact to compute the elements in the last row of the rotation matrix:

$$r_{zx} = \frac{1 - r_{xx}^2 - r_{xy}r_{yx}}{r_{xz}} \tag{12.149}$$

$$r_{zy} = \frac{1 - r_{yx}r_{xy} - r_{yy}^2}{r_{yz}} \tag{12.150}$$

$$r_{zz} = \sqrt{1 - r_{zx}r_{xz} - r_{zy}r_{yz}}. \tag{12.151}$$

If the sign of $r_{xx}r_{yx} + r_{xy}r_{yy}$ is positive, negate r_{yz}. The signs of r_{zx} and r_{zy} may need to be adjusted after computing the camera constant in the following step.

Fourth, compute the camera constant f and p_z, the z component of translation. Use all of the calibration points to form a system of linear equations,

$$A\mathbf{v} = \mathbf{b}, \tag{12.152}$$

for estimating f and p_z. Use each calibration point to compute the corresponding row of the matrix,

$$a_i = (r_{yx}x_i + r_{yy}y_i + p_y, -d_y\tilde{y}_i), \tag{12.153}$$

and the corresponding element of the vector on the right side of Equation 12.152,

$$b_i = (r_{zx}x_i + r_{zy}y_i)d_y\tilde{y}_i. \tag{12.154}$$

The vector $\mathbf{v}$ contains the parameters to be estimated:

$$\mathbf{v} = (f, p_z)^T. \tag{12.155}$$

Use singular value decomposition to solve this system of equations. If the camera constant $f < 0$, then negate r_{zx} and r_{zy} in the rotation matrix for the rigid body transformation.

Fifth, use the estimates for f and p_z obtained in the previous step as the initial conditions for nonlinear regression to compute the first-order lens distortion κ_1 and better estimates for f and p_z. The true (corrected) image plane coordinates (x', y') are related to the calibration points in camera coordinates (x_c, y_c, z_c) through perspective projection:

$$x' = f\frac{x_c}{z_c} \tag{12.156}$$

$$y' = f\frac{y_c}{z_c}. \tag{12.157}$$

Assume that the true (corrected) image plane coordinates are related to the measured (uncorrected) image plane coordinates using the first term in the model for radial lens distortion:

$$x' = \tilde{x}(1 + \kappa_1 r^2) \tag{12.158}$$
$$y' = \tilde{y}(1 + \kappa_1 r^2), \tag{12.159}$$

where the radius r is given by

$$r = \sqrt{\tilde{x}^2 + \tilde{y}^2}. \tag{12.160}$$

Note that the uncorrected (measured) image plane coordinates $(\tilde{x}, \tilde{y})$ are not the same as the pixel coordinates $[i, j]$ since the location of the image center (c_x, c_y), the row and column spacing d_x and d_y, and the estimated scale factor τ_x have already been applied.

Use the y components of the equations for perspective projection, lens distortion, and the rigid body transformation from absolute coordinates to camera coordinates to get a constraint on the camera constant f, z translation, and lens distortion:

$$\tilde{y}_i(1 + \kappa_1 r^2) = f\frac{r_{yx}x_{a,i} + r_{yy}y_{a,i} + r_{yz}z_{a,i} + p_y}{r_{zx}x_{a,i} + r_{zy}y_{a,i} + r_{zz}z_{a,i} + p_z}. \tag{12.161}$$

This leads to a nonlinear regression problem for the parameters p_z, f, and κ_1. We use the measurements for y, rather than x, because the x measurements are affected by the scale parameter τ_x. The spacing between image rows d_y is very accurate and readily available from the camera specifications and is not affected by problems in the digitizing electronics.

Since the calibration points were in a plane, the scale factor τ_x cannot be determined. Also, the location of the image center, c_x and c_y, has not been calibrated. The list of further readings provided at the end of this chapter provides references to these calibration problems.

12.10.2   Affine Method for Camera Calibration

The interior orientation problem can be combined with the exterior orientation problem to obtain an overall transformation that relates (uncalibrated) image coordinates to the position and orientation of rays in the absolute coordinate system. Assume that the transformation from uncorrected image coordinates to true image coordinates can be modeled by an affine transformation within the image plane. This transformation accounts for several sources of camera error:

Scale error due to an inaccurate value for the camera constant

Translation error due to an inaccurate estimate for the image origin (principal point)

Rotation of the image sensor about the optical axis

Skew error due to nonorthogonal camera axes

Differential scaling caused by unequal spacing between rows and columns in the image sensor (nonsquare pixels)

However, an affine transformation cannot model the errors due to lens distortions.

In the development of the exterior orientation problem (Section 12.8), we formulated equations for the transformation from absolute coordinates to image coordinates. Now we will add an affine transformation from true image coordinates to measured (uncorrected) image coordinates to get the overall transformation from absolute coordinates to measured image coordinates.

The affine transformation in the image plane that models the distortions due to errors and unknowns in the intrinsic parameters is

$$\tilde{x} = a_{xx}x' + a_{xy}y' + b_x \tag{12.162}$$
$$\tilde{y} = a_{yx}x' + a_{yy}y' + b_y, \tag{12.163}$$

where we are mapping from true image plane coordinates (x', y') to uncorrected (measured) image coordinates $(\tilde{x}, \tilde{y})$. Use the equations for perspective projection,

$$\frac{x'}{f} = \frac{x_c}{z_c} \tag{12.164}$$

$$\frac{y'}{f} = \frac{y_c}{z_c}, \tag{12.165}$$

to replace x' and y' with ratios of the camera coordinates:

$$\frac{\tilde{x}}{f} = a_{xx}\left(\frac{x_c}{z_c}\right) + a_{xy}\left(\frac{y_c}{z_c}\right) + \frac{b_x}{f} \tag{12.166}$$

$$\frac{\tilde{y}}{f} = a_{yx}\left(\frac{x_c}{z_c}\right) + a_{yy}\left(\frac{y_c}{z_c}\right) + \frac{b_y}{f}. \tag{12.167}$$

Camera coodinates are related to absolute coordinates by a rigid body transformation:

$$x_c = r_{xx}x_a + r_{xy}y_a + r_{xz}z_a + p_x \tag{12.168}$$

$$y_c = r_{yx}x_a + r_{yy}y_a + r_{yz}z_a + p_y \tag{12.169}$$

$$z_c = r_{zx}x_a + r_{zy}y_a + r_{zz}z_a + p_z. \tag{12.170}$$

We can use these equations to replace the ratios of camera coordinates in the affine transformation with expressions for the absolute coordinates,

$$\frac{\tilde{x} - b_x}{f} = \frac{s_{xx}x_a + s_{xy}y_a + s_{xz}z_a + t_x}{s_{zx}x_a + s_{zy}y_a + s_{zz}z_a + t_z} \tag{12.171}$$

$$\frac{\tilde{y} - b_y}{f} = \frac{s_{yx}x_a + s_{yy}y_a + s_{yz}z_a + t_y}{s_{zx}x_a + s_{zy}y_a + s_{zz}z_a + t_z}, \tag{12.172}$$

where the coefficients are sums of products of the coefficients in the affine transformation and the rigid body transformation. What we have is a pair of equations, similar to the equations for exterior orientation (Equations 12.108 and 12.109), that relate absolute coordinates to uncorrected image coordinates. The affine model for camera errors has been absorbed into the transformation from absolute to camera coordinates. Equations 12.171 and 12.172 can be written as

$$\frac{\tilde{x} - b_x}{f} = \frac{\tilde{x}_c}{\tilde{z}_c} = \frac{s_{xx}x_a + s_{xy}y_a + s_{xz}z_a + t_x}{s_{zx}x_a + s_{zy}y_a + s_{zz}z_a + t_z} \tag{12.173}$$

$$\frac{\tilde{y} - b_y}{f} = \frac{\tilde{y}_c}{\tilde{z}_c} = \frac{s_{yx}x_a + s_{yy}y_a + s_{yz}z_a + t_y}{s_{zx}x_a + s_{zy}y_a + s_{zz}z_a + t_z} \tag{12.174}$$

to show that the (uncorrected) image coordinates are related to the camera coordinates by perspective projection, but the space of camera coordinates has been warped to account for the camera errors.

Returning to Equations 12.171 and 12.172, we can absorb the corrections to the location of the principal point, b_x and b_y, into the affine transformation to get

$$\tilde{x}_i(s_{zx}x_{a,i}+s_{zy}y_{a,i}+s_{zz}z_{a,i}+t_z) - f(s_{xx}x_{a,i}+s_{xy}y_{a,i}+s_{xz}z_{a,i}+t_x) = 0 \quad (12.175)$$

$$\tilde{y}_i(s_{zx}x_{a,i}+s_{zy}y_{a,i}+s_{zz}z_{a,i}+t_z) - f(s_{yx}x_{a,i}+s_{yy}y_{a,i}+s_{yz}z_{a,i}+t_y) = 0, \quad (12.176)$$

which shows that each calibration point and its corresponding measured location in the image plane provides two linear equations for the parameters of the transformation. The nominal value f for the camera constant is not absorbed into the affine transformation since it is needed for constructing the ray in camera coordinates.

The set of calibration points yields a set of homogeneous linear equations that can be solved for the coefficients of the transformation. At least six points are needed to get 12 equations for the 12 unknowns, but more calibration points should be used to increase accuracy. To avoid the trivial solution with all coefficients equal to zero, fix the value of one of the parameters, such as t_x or t_y, and move it to the right side of the equation. Form a system of linear equations,

$$A\mathbf{u} = \mathbf{b}, \quad (12.177)$$

where $\mathbf{u}$ is the vector of transformation coefficients; row i of the A matrix is filled with absolute coordinates for calibration point i and products of the absolute coordinates and $\tilde{x}_i$, $\tilde{y}_i$, or f; and element i of the $\mathbf{b}$ vector is the constant chosen for t_x or t_y. Since the affine transformation within the image plane is combined with the rotation matrix for exterior orientation, the transformation matrix is no longer orthonormal. The system of linear equations can be solved, without the orthonormality constraints, using common numerical methods such as singular value decomposition.

The transformation maps absolute coordinates to measured image coordinates. Applications require the inverse transformation, given by

$$\begin{pmatrix} x \\ y \\ z \end{pmatrix} = S^{-1} \left[\begin{pmatrix} \tilde{x}_i \\ \tilde{y}_i \\ f \end{pmatrix} - \begin{pmatrix} t_x \\ t_y \\ t_z \end{pmatrix} \right], \quad (12.178)$$

which can be used to determine the equation of a ray in absolute coordinates from the measured coordinates in the image. Note that the camera constant f has been carried unchanged through the formulation of the calibration algorithm. Since corrections to the camera constant are included in the affine transformation (Equations 12.162 and 12.163), the focal length of the lens can be used for f. Finally, the transformation from pixel coordinates $[i, j]$ to image coordinates,

$$\tilde{x} = s_x(j - c_x) \tag{12.179}$$
$$\tilde{y} = -s_y(i - c_y), \tag{12.180}$$

is an affine transformation that can be combined with the model for camera errors (Equations 12.162 and 12.163) to develop a transformation between absolute coordinates and pixel coordinates.

12.10.3 Nonlinear Method for Camera Calibration

Given a set of calibration points, determine the projections of the calibration points in the image plane, calculate the errors in the projected positions, and use these errors to solve for the camera calibration parameters. Since it is necessary to know where the calibration points should project to in the image plane, the exterior orientation problem is solved simultaneously. The method presented in this section is different from the procedure explained in Section 12.10.2, where the interior and exterior orientation problems were combined into a single affine transformation, in that the actual camera calibration parameters are obtained and can be used regardless of where the camera is later located in scene.

The principle behind the solution to the camera calibration problem is to measure the locations (x_i', y_i') of the projections of the calibration points onto the image plane, calculate the deviations $(\delta x_i, \delta y_i)$ of the points from the correct positions, and plug these measurements into the equations that model the camera parameters. Each calibration point yields two equations. The solution requires at least enough equations to cover the unknowns, but for increased accuracy more equations than unknowns are used and the overdetermined set of equations is solved using nonlinear regression.

Assume that the approximate position and orientation of the camera in absolute coordinates is known. Since we have initial estimates for the rotation angles, we can formulate the exterior orientation problem in terms of the Euler angles in the rotation matrix. The parameters of the regression problem are the rotation angles ω, ϕ, and κ; the position of the camera in absolute coordinates p_x, p_y, and p_z; the camera constant f; the corrections to the location of the principal point (x_p, y_p); and the polynomial coefficients for radial lens distortion κ_1, κ_2, and κ_3. The equations for the exterior orientation problem are

$$\frac{x'}{f} = \frac{r_{xx}x_a + r_{xy}y_a + r_{xz}z_a + p_x}{r_{zx}x_a + r_{zy}y_a + r_{zz}z_a + p_z} \tag{12.181}$$

$$\frac{y'}{f} = \frac{r_{yx}x_a + r_{yy}y_a + r_{yz}z_a + p_y}{r_{zx}x_a + r_{zy}y_a + r_{zz}z_a + p_z}. \tag{12.182}$$

Replace x' and y' with the corrected positions from the camera model,

$$\frac{(\tilde{x}-x_p)(1+\kappa_1 r^2+\kappa_2 r^4+\kappa_3 r^6)}{f} = \frac{r_{xx}x_a + r_{xy}y_a + r_{xz}z_a + p_x}{r_{zx}x_a + r_{zy}y_a + r_{zz}z_a + p_z} \tag{12.183}$$

$$\frac{(\tilde{y}-y_p)(1+\kappa_1 r^2+\kappa_2 r^4+\kappa_3 r^6)}{f} = \frac{r_{yx}x_a + r_{yy}y_a + r_{yz}z_a + p_y}{r_{zx}x_a + r_{zy}y_a + r_{zz}z_a + p_z}, \tag{12.184}$$

and replace the elements of the rotation matrix with the formulas for the rotation matrix entries in terms of the Euler angles, provided in Equation 12.13. Solve for the camera parameters and exterior orientation using nonlinear regression. The regression algorithm will require good initial conditions. If the target is a plane, the camera axis is normal to the plane, and the image is roughly centered on the target, then the initial conditions are easy to obtain. Assume that the absolute coordinate system is set up so that the x and y axes are parallel to the camera axes. The initial conditions are:

$$\omega = \phi = \kappa = 0$$
$$x = \text{translation in } x \text{ from the origin}$$
$$y = \text{translation in } y \text{ from the origin}$$
$$z = \text{distance of the camera from the calibration plane}$$
$$f = \text{focal length of the lens}$$
$$x_p = y_p = 0$$
$$\kappa_1 = \kappa_2 = \kappa_3 = 0.$$

It is easy to build a target of dots using a laser printer. The uncorrected positions of the dots in the image can be found by computing the first moments of the connected components.

The disadvantage to nonlinear regression is that good initial values for the parameters are needed, but the advantage is that there is a body of literature on nonlinear regression with advice on solving nonlinear problems and methods for estimating errors in the parameter estimates.

12.11 Binocular Stereo Calibration

In this section we will discuss how the techniques presented in this chapter can be combined in a practical system for calibrating stereo cameras and using the stereo measurements. This provides a forum for reviewing the relationships between the various calibration problems.

There are several tasks in developing a practical system for binocular stereo:

1. Calibrate the intrinsic parameters for each camera.

2. Solve the relative orientation problem.

3. Resample the images so that the epipolar lines correspond to image rows.

4. Compute conjugate pairs by feature matching or correlation.

5. Solve the stereo intersection problem for each conjugate pair.

6. Determine baseline distance.

7. Solve the absolute orientation problem to transform point measurements from the coordinate system of the stereo cameras to an absolute coordinate system for the scene.

There are several ways to calibrate a binocular stereo system, corresponding to various paths through the diagram in Figure 12.6. To start, each camera must be calibrated to determine the camera constant, location of the

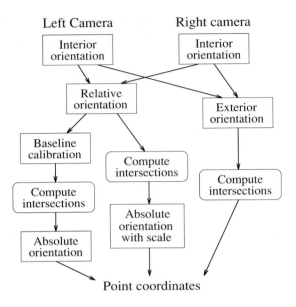

Figure 12.6: A diagram of the steps in various procedures for calibrating a binocular stereo system.

principal point, correction table for lens distortions, and other intrinsic parameters. Once the left and right stereo cameras have been calibrated, there are basically three approaches to using the cameras in a stereo system.

The first approach is to solve the relative orientation problem and determine the baseline by other means, such as using the stereo cameras to measure points that are a known distance apart. This fully calibrates the rigid body transformation between the two cameras. Point measurements can be gathered in the local coordinate system of the stereo cameras. Since the baseline has been calibrated, the point measurements will be in real units and the stereo system can be used to measure the relationships between points on objects in the scene. It is not necessary to solve the absolute orientation problem, unless the point measurements must be transformed into another coordinate system.

The second approach is to solve the relative orientation problem and obtain point measurements in the arbitrary system of measurement that results from assuming unit baseline distance. The point measurements will be correct, except for the unknown scale factor. Distance ratios and angles will be correct, even though the distances are in unknown units. If the baseline distance is obtained later, then the point coordinates can be multiplied by the baseline distance to get point measurements in known units. If it is necessary to transform the point measurements into another coordinate system, then solve the absolute orientation problem with scale (Section 12.7), since this will accomplish the calibration of the baseline distance and the conversion of point coordinates into known units without additional computation.

The third approach is to solve the exterior orientation problem for each stereo camera. This provides the transformation from the coordinate systems of the left and right camera into absolute coordinates. The point measurements obtained by intersecting rays using the methods of Section 12.6 will automatically be in absolute coordinates with known units, and no further transformations are necessary.

12.12   Active Triangulation

This section will cover methods for determining the coordinates of a point using an active sensor that projects a plane of light onto opaque surfaces in the scene. A method for calibrating an active triangulation system will be presented.

We will start with a simple geometry in camera-centered coordinates and proceed to the general case in absolute coordinates. Suppose that the plane of light rotates about an axis that is parallel to the y axis and displaced along the x axis by b_x. Let θ be the orientation of the plane relative to the z axis. With $\theta = 0$, the plane of light is parallel to the y–z plane and positive values for θ correspond to counterclockwise rotation about the y axis. In terms of vector geometry, the normal to the plane is

$$\mathbf{n} = (n_x, n_y, n_z) = (\cos(\theta), 0, \sin(\theta)), \tag{12.185}$$

the baseline is

$$\mathbf{b} = (b_x, b_y, b_z) = (b_x, 0, 0), \tag{12.186}$$

and point $\mathbf{p} = (x, y, z)$ lies in the plane if

$$(\mathbf{p} - \mathbf{b}) \cdot \mathbf{n} = 0. \tag{12.187}$$

The plane of light illuminates the scene and intersects an opaque surface to produce a curve in space that is imaged by the camera. A line detection operator is used to estimate the locations of points along the projected curve in the image plane. Suppose that an estimated line point has coordinates (x', y'). This corresponds to a ray in space represented by the equation

$$\begin{pmatrix} x \\ y \\ z \end{pmatrix} = t \begin{pmatrix} x' \\ y' \\ f \end{pmatrix}, \tag{12.188}$$

where f is the distance of the image plane from the center of projection. Replace $\mathbf{p}$ in Equation 12.187 with the equation for the ray and solve for t,

$$t = \frac{b_x \cos\theta}{x' \cos\theta - f \sin\theta}, \tag{12.189}$$

and plug into the equation for the ray to get the coordinates of the point in camera-centered coordinates:

$$\begin{pmatrix} x \\ y \\ z \end{pmatrix} = \begin{pmatrix} \frac{x' b_x \cos\theta}{x' \cos\theta - f \sin\theta} \\ \frac{y' b_x \cos\theta}{x' \cos\theta - f \sin\theta} \\ \frac{f b_x \cos\theta}{x' \cos\theta - f \sin\theta} \end{pmatrix}. \tag{12.190}$$

If the exterior orientation of the camera has been calibrated, then the ray can be represented in absolute coordinates; if the position and orientation of the plane are also represented in absolute coordinates, then the point measurements will be in absolute coordinates.

It is easy to generalize these equations to allow an arbitrary position and orientation for the plane in space. Suppose that the plane is rotated by θ counterclockwise about an axis ω and the normal $\mathbf{n}$ corresponds to the orientation with $\theta = 0$. Let $R(\theta)$ be the rotation matrix. A point $\mathbf{p}$ lies in the plane if

$$(\mathbf{p} - \mathbf{b}) \cdot R(\theta)\mathbf{n} = 0. \tag{12.191}$$

We can also change the position of the plane in space by varying $\mathbf{b}$,

$$(\mathbf{p} - \mathbf{b}(d)) \cdot R(\theta)\mathbf{n} = 0, \tag{12.192}$$

where d is the control parameter of a linear actuator.

12.13   Robust Methods

All of the calibration methods presented in this chapter have used least-squares regression, which is very sensitive to outliers due to mismatches in forming the conjugate pairs. There are two approaches to making calibration robust: use a resampling plan or change the regression procedure to use a robust norm.

To implement a resampling plan, it is necessary to consider all combinations of m conjugate pairs from the set of n conjugate pairs obtained from the calibration data. The size m of the subset must be large enough to get a good solution to the calibration problem. Choose the best set of parameters by comparing the fit of the transformation to all conjugate pairs according to the least median of squares criterion [207]. Use the best parameter estimates to remove outliers from the set of conjugate pairs, and repeat the calibration procedure using all of the remaining conjugate pairs.

The other approach involves replacing the square norm with a robust norm [197, pp. 558–565]. If the calibration procedure uses linear regression, then use a weighted least-squares solution method. Calculate the weights so that the weighted least-squares regression problem is equivalent to the unweighted robust regression problem. If the robust norm for residual r_i is $\rho(r_i)$, then solve

$$w_i r_i^2 = \rho(r_i) \tag{12.193}$$

for the weight w_i and use this weight for the conjugate pair. This leads to iterative reweighted least-squares: a sequence of regression problems is solved with the weights adjusted between iterations. For nonlinear regression, the solution method may allow the robust norm to be used directly.

12.14   Conclusions

Several methods for calibration have been presented in this chapter, including the basic problems in photogrammetry: absolute, relative, exterior, and interior orientation. The interior orientation problem should be solved for any camera to ensure that the camera obeys the assumptions of image formation assumed by most machine vision algorithms. The remaining calibration problems can be divided into two groups: methods used in image analysis and methods used in depth measurement. The exterior orientation problem must be solved for an image analysis application when it is necessary to relate image measurements to the geometry of the scene. The relative orientation problem is used to calibrate a pair of cameras for obtaining depth measurements with binocular stereo. The absolute orientation problem is used to calibrate the position and orientation of any system for depth measurement, including binocular stereo or active sensing, so that the depth measurements in camera coordinates can be translated into the coordinate system used in the application.

Further Reading

An excellent introduction to photogrammetry is provided by Horn [109], who has also published a closed-form solution to the absolute orientation problem using orthonormal matrices [113] and unit quaternions [110]. The American Society of Photogrammetry publishes a comprehensive collection of articles on photogrammetry [224], and there are several books on photogrammetry [17, 95, 170, 254]. Photogrammetry was originally developed for preparing topographic maps from aerial photographs—hence the name, which means making measurements in photographs. Terrestrial photogrammetry is covered in the *Handbook of Non-Topographic Photogrammetry* and includes methods of use in machine vision [137]. The camera calibration technique developed by Tsai [234] is widely used. Further work on camera calibration has been done by Lenz and Tsai [154]. Horn [111] published an excellent description of the solution to the relative orientation problem. Brown [51] published the algorithm for correcting radial lens distortions and decentering.

   Photogrammetry usually assumes that objects and cameras move by rigid body transformations. In many cases, an object will deform as it moves, in addition to undergoing translation, rotation, and change in scale. Bookstein has published an extensive body of literature on modeling nonrigid deformations in diverse fields using thin-plate splines [39, 40, 41, 42].

Exercises

12.1   How many types of orientations are possible in calibration? Consider some examples and illustrate the difference among them.

12.2 Why is photogrammetry concerned with the camera calibration problem? Illustrate using an example.

12.3 What is the effect of rectangular pixels on image measurements? How can you compensate for the rectangularity of pixels in measuring areas and centroids of regions? How will it affect measurements of relative locations and orientations of regions? What care would you take to get correct measurements in the real world?

12.4 Define affine transformation. Give three examples of objects that will undergo an affine transformation in an image at different time instants and three examples of objects that will not.

12.5 Define Euler angles. Where are they used? What are their strengths and weaknesses?

12.6 What are quaternions? Why are they considered a good representation for rotation in calibration? Demonstrate this considering the calibration of absolute orientation.

12.7 Define the coplanarity constraint. Where and how is it used in camera calibration?

12.8 Suppose that you are designing a hand–eye system. In this system a camera is mounted at a fixed position in the work space. The hand, a robot arm, is used to pick and place objects in the work space. What kind of calibration scheme will you require?

12.9 In the above problem, if the camera is mounted on the robot arm itself, what kind of calibration scheme will you require to find the location of objects? How many fixed known points will you require to solve the problem?

12.10 Now let's provide a stereo system on our robot arm so that we can compute the depth of points easily and perform more dexterous manipulations. What kind of calibration scheme is required in this case? What are the criteria for selecting scene points for calibration? What arrangements of points are bad for calibration?

12.11 What camera parameters should be determined for calibrating a camera? How can you determine these parameters without directly measuring them? What parameters are available from the data sheets provided by the camera manufacturer?

Computer Projects

12.1 You want to develop a camera calibration algorithm for preparing a three-dimensional model of a football field using lane markers. Assume that a sufficient number of lane markers, their intersections, and other similar features are visible in the field of view. Develop an approach to solve the camera calibration problem and use it to determine the location of each player, their height, and the location of the ball.

12.2 Develop an algorithm for the calibration of stereo cameras. Use this algorithm to calibrate a set of cameras mounted on a mobile robot. Develop an approach that will use known calibration points in an environment to determine the camera calibration and then use this to get the exact distance to all points.

Chapter 13

Curves and Surfaces

There are two fundamental problems with surfaces in machine vision: reconstruction and segmentation. Surfaces must be reconstructed from sparse depth measurements that may contain outliers. Once the surfaces are reconstructed onto a uniform grid, the surfaces must be segmented into different surface types for object recognition and refinement of the surface estimates.

This chapter begins with a discussion of the geometry of surfaces and includes sections on surface reconstruction and segmentation. The chapter will cover the following topics on surfaces:

Representations for surfaces such as polynomial surface patches and tensor-product cubic splines

Interpolation methods such as bilinear interpolation

Approximation of surfaces using variational methods and regression splines

Segmentation of point measurements into surface patches

Registration of surfaces with point measurements

Surface approximation is also called surface fitting, since it is like a regression problem where the model is the surface representation and the data are points sampled on the surface. The term *surface reconstruction* means estimating the continuous function for the surface from point samples, which can be implemented by interpolation or approximation.

There are many machine vision algorithms for working with curves and surfaces. This is a large area and cannot be covered completely in an introductory text. This chapter will cover the basic methods for converting point measurements from binocular stereo, active triangulation, and range cameras into simple surface representations. The basic methods include converting point measurements into a mesh of triangular facets, segmenting range measurements into simple surface patches, fitting a smooth surface to the point measurements, and matching a surface model to the point measurements. After studying the material in this chapter, the reader should have a good introduction to the terminology and notation of surface modeling and be prepared to continue the topic in other sources.

13.1   Fields

This chapter covers the problems of reconstructing surfaces from point samples and matching surface models to point measurements. Before a discussion of curves and surfaces, the terminology of fields of coordinates and measurements must be presented.

Measurements are a mapping from the coordinate space to the data space. The coordinate space specifies the locations at which measurements were made, and the data space specifies the measurement values. If the data space has only one dimension, then the data values are scalar measurements. If the data space has more than one dimension, then the data values are vector measurements. For example, weather data may include temperature and pressure (two-dimensional vector measurements) in the three-dimensional coordinate space of longitude, latitude, and elevation. An image is scalar measurements (image intensity) located on a two-dimensional grid of image plane positions. In Chapter 14 on motion, we will discuss the image flow velocity field, which is a two-dimensional space of measurements (velocity vectors) defined in the two-dimensonal space of image plane coordinates.

There are three types of fields: uniform, rectilinear, and irregular (scattered). In uniform fields, measurements are located on a rectangular grid with equal spacing between the rows and columns. Images are examples of uniform fields. As explained in Chapter 12 on calibration, the location of any grid point is determined by the position of the grid origin, the orientation of the grid, and the spacing between the rows and columns.

Rectilinear fields have orthogonal coordinate axes, like uniform fields, but the data samples are not equally spaced along the coordinate axes. The data samples are organized on a rectangular grid with various distances between the rows and columns. For example, in two dimensions a rectilinear field partitions a rectangular region of the plane into a set of rectangles of various sizes, but rectangles in the same row have the same height, and rectangles in the same column have the same width. Lists of coordinates, one list for each dimension, are needed to determine the position of the data samples in the coordinate space. For example, a two-dimensional rectilinear grid with coordinate axes labeled x and y will have a list of x coordinates $\{x_j\}, j = 1, 2, \ldots, m$ for the m grid columns and a list of y coordinates $\{y_i\}, i = 1, 2, \ldots, n$ for the n grid rows. The location of grid point $[i, j]$ is (x_j, y_i).

Irregular fields are used for scattered (randomly located) measurements or any pattern of measurements that do not correspond to a rectilinear structure. The coordinates (x_k, y_k) of each measurement must be provided explicitly in a list for $k = 1, \ldots, n$.

These concepts are important for understanding how to represent depth measurements from binocular stereo and active sensing. Depth measurements from binocular stereo can be represented as an irregular, scalar field of depth measurements z_k located at scattered locations (x_k, y_k) in the image plane or as an irregular field of point measurements located at scattered positions (x_k, y_k, z_k) in the coordinate system of the stereo camera with a null data part. Likewise, depth measurements from a range camera can be represented as distance measurements $z_{i,j}$ on a uniform grid of image plane locations (x_j, y_i) or as an irregular field of point measurements with a null data part. In other words, point samples of a graph surface $z = f(x, y)$ can be treated as displacement measurements from positions in the domain or as points in three-dimensional space.

13.2 Geometry of Curves

Before a discussion of surfaces, curves in three dimensions will be covered for two reasons: surfaces are described by using certain special curves, and representations for curves generalize to representations for surfaces. Curves can be represented in three forms: implicit, explicit, and parametric.

The parametric form for curves in space is

$$\mathbf{p} = (x, y, z) = (x(t), y(t), z(t)), \tag{13.1}$$

for $t_0 \leq t \leq t_1$, where a point along the curve is specified by three functions that describe the curve in terms of the parameter t. The curve starts at the point $(x(t_0), y(t_0), z(t_0))$ for the initial parameter value t_0 and ends at $(x(t_1), y(t_1), z(t_1))$ for the final parameter value t_1. The points corresponding to the initial and final parameter values are the start and end points of the curve, respectively. For example, Chapter 12 makes frequent use of the parametric form for a ray in space:

$$\begin{pmatrix} x \\ y \\ z \end{pmatrix} = t \begin{pmatrix} u_x \\ u_y \\ u_z \end{pmatrix} + \begin{pmatrix} x_0 \\ y_0 \\ z_0 \end{pmatrix}, \tag{13.2}$$

for $0 \leq t < \infty$, where the (unit) vector (u_x, u_y, u_z) represents the direction of the ray and (x_0, y_0, z_0) is the starting point of the ray.

The parametric equation for the line segment from point $\mathbf{p}_1 = (x_1, y_1, z_1)$ to point $\mathbf{p}_2 = (x_2, y_2, z_2)$ is

$$\begin{pmatrix} x \\ y \\ z \end{pmatrix} = t \begin{pmatrix} x_2 - x_1 \\ y_2 - y_1 \\ z_2 - z_1 \end{pmatrix} + \begin{pmatrix} x_1 \\ y_1 \\ z_1 \end{pmatrix}. \tag{13.3}$$

Curves can also be represented implicitly as the set of points (x, y, z) that satisfy some equation

$$f(x, y, z) = 0 \tag{13.4}$$

or set of equations.

13.3   Geometry of Surfaces

Like curves, surfaces can be represented in implicit, explicit, or parametric form. The parametric form for a surface in space is

$$(x, y, z) = (x(u, v), y(u, v), z(u, v)), \tag{13.5}$$

for $u_0 \leq u \leq u_1$ and $v_0 \leq v \leq v_1$. The domain can be defined more generally as $(u, v) \in D$.

The implicit form for a surface is the set of points (x, y, z) that satisfy some equation

$$f(x, y, z) = 0. \tag{13.6}$$

For example, a sphere with radius r centered at (x_0, y_0, z_0) is

$$f(x, y, z) = (x - x_0)^2 + (y - y_0)^2 + (z - z_0)^2 - r^2 = 0. \tag{13.7}$$

The explicit (functional) form

$$z = f(x, y) \tag{13.8}$$

is used in machine vision for range images. It is not as general and widely used as the parametric and implicit forms, because the explicit form is only useful for graph surfaces, which are surfaces that can be represented as displacements from some coordinate plane.

If a surface is a graph surface, then it can be represented as displacements normal to a plane in space. For example, a range image is a rectangular grid of samples of a surface

$$z = f(x, y), \tag{13.9}$$

where (x, y) are image plane coordinates and z is the distance parallel to the z axis in camera coordinates.

13.3.1 Planes

Three points define a plane in space and also define a triangular patch in space with corners corresponding to the points. Let $\mathbf{p}_0$, $\mathbf{p}_1$, and $\mathbf{p}_2$ be three points in space. Define the vectors $\mathbf{e}_1 = \mathbf{p}_1 - \mathbf{p}_0$ and $\mathbf{e}_2 = \mathbf{p}_2 - \mathbf{p}_0$. The normal to the plane is $\mathbf{n} = \mathbf{e}_1 \times \mathbf{e}_2$. A point $\mathbf{p}$ lies in the plane if

$$(\mathbf{p} - \mathbf{p}_0) \cdot \mathbf{n} = 0. \tag{13.10}$$

Equation 13.10 is one of the implicit forms for a plane and can be written in the form

$$ax + by + cz + d = 0, \tag{13.11}$$

where the coefficients a, b, and c are the elements of the normal to the plane and d is obtained by plugging the coordinates of any point on the plane into Equation 13.11 and solving for d.

The parametric form of a plane is created by mapping (u, v) coordinates into the plane using two vectors in the plane as the basis vectors for the coordinate system in the plane, such as vectors $\mathbf{e}_1$ and $\mathbf{e}_2$ above. Suppose that point $\mathbf{p}_0$ in the plane corresponds to the origin in u and v coordinates. Then the parametric equation for a point in the plane is

$$\begin{pmatrix} x \\ y \\ z \end{pmatrix} = u\mathbf{e}_1 + v\mathbf{e}_2 + \mathbf{p}_0. \tag{13.12}$$

If the coordinate system in the plane must be orthogonal, then force $\mathbf{e}_1$ and $\mathbf{e}_2$ to be orthogonal by computing $\mathbf{e}_2$ from the cross product between $\mathbf{e}_1$ and the normal vector $\mathbf{n}$,

$$\mathbf{e}_2 = \mathbf{n} \times \mathbf{e}_1. \tag{13.13}$$

The explicit form for the plane is obtained from Equation 13.11 by solving for z. Note that if coefficient c is close to zero, then the plane is close to being parallel to the z axis and the explicit form should not be used.

13.3.2   Differential Geometry

Differential geometry is the local analysis of how small changes in position (u, v) in the domain affect the position on the surface, $\mathbf{p}(u, v)$, the first derivatives, $\mathbf{p}_u(u, v)$ and $\mathbf{p}_v(u, v)$, and the surface normal, $\mathbf{n}(u, v)$. The geometry is illustrated in Figure 13.1.

The parameterization of a surface maps points (u, v) in the domain to points $\mathbf{p}$ in space:

$$\mathbf{p}(u, v) = (x(u, v), y(u, v), z(u, v)). \tag{13.14}$$

The first derivatives, $\mathbf{p}_u(u, v)$ and $\mathbf{p}_v(u, v)$, are vectors that span the tangent plane to the surface at point $(x, y, z) = \mathbf{p}(u, v)$. The surface normal n at point $\mathbf{p}$ is defined as the unit vector normal to the tangent plane at point $\mathbf{p}$ and is computed using the cross product of the partial derivatives of the surface parameterization,

$$\mathbf{n}(\mathbf{p}) = \frac{\mathbf{p}_u \times \mathbf{p}_v}{\|\mathbf{p}_u \times \mathbf{p}_v\|}. \tag{13.15}$$

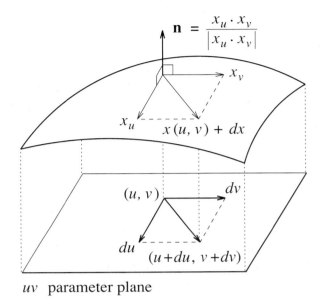

$$\mathbf{n} = \frac{x_u \cdot x_v}{|x_u \cdot x_v|}$$

uv parameter plane

Figure 13.1: A parametrically defined surface showing the tangent plane and surface normal.

The tangent vectors and the surface normal $\mathbf{n}(u,v)$ define an orthogonal coordinate system at point $\mathbf{p}(u,v)$ on the surface, which is the framework for describing the local shape of the surface.

The curvature of a surface is defined using the concept of the curvature of a planar curve. Suppose that we wish to know the curvature of the surface at some point $\mathbf{p}$ on the surface. Consider a plane that slices the surface at point $\mathbf{p}$ and is normal to the surface at point $\mathbf{p}$. In other words, the surface normal at point $\mathbf{p}$ lies in the slicing plane. The intersection of the plane with the surface is a planar curve that includes point $\mathbf{p}$. The normal curvature at point $\mathbf{p}$ is the curvature of this planar curve of intersection. Refer to Figure 13.2. The plane is spanned by the surface normal and a vector in the plane tangent to the surface at $\mathbf{p}$. The normal curvature is not unique, since it depends on the orientation of the plane about the surface normal. The minimum and maximum curvatures, κ_1 and κ_2, are called the

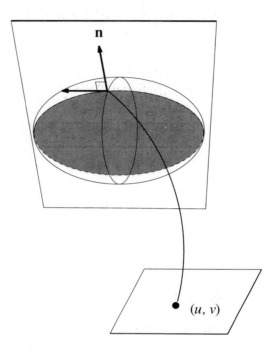

Figure 13.2: Diagram illustrating the geometry of the normal curvature at a point on a surface. The intersecting plane contains the surface normal.

principal curvatures, and the directions in the tangent plane corresponding to the minimum and maximum curvatures are the principal directions.

The Gaussian curvature is the product of the principal curvatures,

$$K = \kappa_1 \kappa_2, \qquad (13.16)$$

and the mean curvature is the average of the principal curvatures,

$$H = \frac{\kappa_1 + \kappa_2}{2}. \qquad (13.17)$$

The lines of curvature are curves in the surface obtained by following the principal directions. Umbilic points are locations on the surface, such as the end of an egg or any point on a sphere, where all normal curvatures (and hence the two principal curvatures) are equal.

13.4 Curve Representations

Curves can be represented in parametric form $x(t)$, $y(t)$, $z(t)$ with the position of each point on the curve specified by a function of the parameter t. Many functions can be used to specify curves in parametric form. For example, we can specify a line segment starting at (x_1, y_1, z_1) and ending at (x_2, y_2, z_2) as

$$x(t) = tx_2 + (1 - t)x_1 \tag{13.18}$$
$$y(t) = ty_2 + (1 - t)y_1 \tag{13.19}$$
$$z(t) = tz_2 + (1 - t)z_1, \tag{13.20}$$

for $0 \le t \le 1$. Cubic polynomials are used in a general representation of space curves described in the next section.

13.4.1 Cubic Spline Curves

A cubic spline is a sequence of cubic polynomial curves joined end to end to represent a complex curve. Each segment of the cubic spline is a parametric curve:

$$x(t) = a_x t^3 + b_x t^2 + c_x t + d_x \tag{13.21}$$
$$y(t) = a_y t^3 + b_y t^2 + c_y t + d_y \tag{13.22}$$
$$z(t) = a_z t^3 + b_z t^2 + c_z t + d_z, \tag{13.23}$$

for $0 \le t \le 1$. Cubic polynomials allow curves to pass through points with a specified tangent and are the lowest-order polynomials that allow nonplanar curves.

These equations can be represented more compactly by organizing the coefficients into vectors. If the coefficient vectors are written as

$$\mathbf{a} = (a_x, a_y, a_z) \tag{13.24}$$
$$\mathbf{b} = (b_x, b_y, b_z) \tag{13.25}$$
$$\mathbf{c} = (c_x, c_y, c_z) \tag{13.26}$$
$$\mathbf{d} = (d_x, d_y, d_z) \tag{13.27}$$

and $\mathbf{p}(t) = (x(t), y(t), z(t))$, then a cubic polynomial curve can be written as

$$\mathbf{p}(t) = \mathbf{a}t^3 + \mathbf{b}t^2 + \mathbf{c}t + \mathbf{d}. \tag{13.28}$$

More complex curves are represented as sequences of cubic polynomials joined at their end points:

$$\begin{aligned}
\mathbf{p}_1(t) &= \mathbf{a}_1 t^3 + \mathbf{b}_1 t^2 + \mathbf{c}_1 t + \mathbf{d}_1 \\
\mathbf{p}_2(t) &= \mathbf{a}_2 t^3 + \mathbf{b}_2 t^2 + \mathbf{c}_2 t + \mathbf{d}_2 \\
&\vdots \\
\mathbf{p}_n(t) &= \mathbf{a}_n t^3 + \mathbf{b}_n t^2 + \mathbf{c}_n t + \mathbf{d}_n
\end{aligned}$$

$$(13.29)$$

for $0 \leq t \leq 1$. Note that $\mathbf{a}_i$, $\mathbf{b}_i$, $\mathbf{c}_i$, and $\mathbf{d}_i$ are the vectors of coefficients for curve segment i. If we define cubic polynomial segment i on the unit interval $i - 1 \leq t \leq i$, then the entire sequence is defined on the interval $0 \leq t \leq n$ and the sequence of cubic polynomial segments can be treated as a single parametric curve starting at point $\mathbf{p}_1(0)$ and ending at point $\mathbf{p}_n(n)$. This sequence of cubic polynomials is called a cubic spline and is a common way of representing arbitrary curves in machine vision and computer graphics.

13.5   Surface Representations

This section will cover some simple surface representations that are used in machine vision.

13.5.1   Polygonal Meshes

Planar polygons, also called planar facets or faces, can used to model complex objects. Planes were covered in Section 13.3.1. In this section, we will show how many planar facets can be combined to model the surface of an object using a polygonal mesh.

In Chapter 6, we showed how a polyline can be represented by the list of coordinates for the vertices that connect the line segments. Likewise, a polygonal mesh is represented by the list of vertex coordinates for the vertices that define the planar polygons in the mesh. Since many polygons may share each vertex, we will use an indirect representation that allows each vertex to be listed once. Number the vertices from 1 to n and store the coordinates for each vertex once:

$$
\begin{aligned}
\mathbf{v}_1 &= (x_1, y_1, z_1) \\
\mathbf{v}_2 &= (x_2, y_2, z_2) \\
&\;\;\vdots \\
\mathbf{v}_n &= (x_n, y_n, z_n).
\end{aligned}
\tag{13.30}
$$

Represent each face by a list of the vertices in the polygon for the face. To ensure the consistency needed for using the vertex information, follow the convention of listing the vertices in the order in which they are encountered, moving clockwise around the face as seen from the outside of the object or from the same side of the surface (for example, above the surface). For outward-facing normals, this means that the vertex order follows the left-hand rule: the order of the vertex around the polygon is the same as the direction of the fingers of the left hand with the thumb pointing in the direction of the surface normal. This representation makes it very easy to find all of the vertices for a given face, and any change in the coordinates of a vertex automatically (indirectly) changes all faces that use the vertex. The list of vertices representation does not explicitly represent the edges between adjacent faces and does not provide an efficient way to find all faces that include a given vertex. These problems are resolved by using the winged edge data structure.

The winged edge data structure is a network with three types of records: vertices, edges, and faces. The data structure includes pointers that can be followed to find all neighboring elements without searching the entire mesh or storing a list of neighbors in the record for each element. There is one vertex record for every vertex in the polygonal mesh, one edge record for every edge in the polygonal mesh, and one face record for every face in the polygonal mesh. The size of the face, edge, or vertex record does not change with the size of the mesh or the number of neighboring elements. Both vertices at the ends of an edge and both faces on either side of an edge can be found directly; all of the faces (edges) that use a vertex can be found in time proportional to the number of faces (edges) that include the vertex; all of the vertices (edges) around a face can be found in time proportional to the number of vertices (edges) around the face.

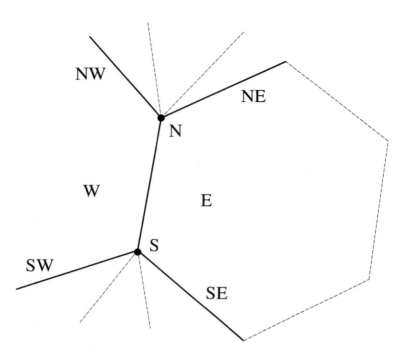

Figure 13.3: Each edge record contains pointers to the two vertices at the ends of the edge, the two faces on either side of the edge, and the four wing edges that allow the polygonal mesh to be traversed.

The winged edge data structure can handle polygons with many sides (the representation is not limited to polygonal meshes with triangular facets), and it is not necessary for all polygons in the mesh to have the same number of sides. The coordinates of vertices are only stored in the vertex records; the position of a face or edge is computed from the vertex coordinates.

Each face record points to the record for one of its edges, and each vertex record points to the record for one of the edges that end at that vertex. The edge records contain the pointers that connect the faces and vertices into a polygonal mesh and allow the polygonal mesh to be traversed efficiently. Each edge record contains a pointer to each of the vertices at its ends, a pointer to each of the faces on either side of the edge, and pointers to the four wing edges that are neighbors in the polygonal mesh. Figure 13.3 illustrates the information contained in an edge record. The faces, vertices, and wing edges are denoted by compass directions. This notation is just a convenience; in a polygonal mesh, there is no global sense of direction. Each wing allows the faces on either side of the edge to be traversed in either direction. For example, follow the northeast wing to continue traversing clockwise around the east face.

Whether a given face is east or west of an edge depends on the order in which faces were entered in the winged edge data structure. As a face is traversed, it is necessary to check if the face is east or west of each edge that is encountered. If the face is east of the edge, follow the northeast wing clockwise around the face or the southeast wing counterclockwise around the face. Otherwise, if the face is west of the edge, follow the southwest wing clockwise around the face or the northwest wing counterclockwise around the face.

Directions like clockwise and counterclockwise are viewer-centered; in other words, they assume that the polygonal mesh is seen from a particular position. When creating and using the polygonal mesh, it is necessary to adhere to some convention that provides an orientation for the viewer-centered directions relative to the polygonal mesh. Directions around the face are given from the side of the face with the surface normal pointing toward the viewer. Clockwise means the direction around the face indicated by how the fingers of the left hand point when the thumb points in the direction of the surface normal; counterclockwise means the direction around the face indicated by how the fingers of the right hand point when the thumb points in the direction of the surface normal. All face normals point to the outside if the polygonal mesh represents the boundary of a volume. If the polygonal mesh represents a surface, then the normals all point to the same side of the surface. If the surface is a graph surface, then the normals point in the direction of positive offsets from the domain. For example, if the polygonal mesh represents the graph surface $z = f(x, y)$, then the projection of the face normals onto the z axis is positive.

The algorithm for adding a face to a polygonal mesh is stated in Algorithm 13.1 and the algorithm for traversing the edges (or vertices) around a face is stated in Algorithm 13.2. Algorithm 13.1 assumes that the vertices are listed in clockwise order around the face, assuming the conventions described above. Algorithm 13.2 can be modified to find all edges (or faces) around a given vertex.

Algorithm 13.1 Adding a Face to a Winged Edge Data Structure
The input is a list of successive vertices for the face, including vertex numbers which are used to uniquely identify each vertex and the vertex coordinates, listed in clockwise order around the face.

1. *For each vertex in the list of vertices, add a record for the vertex to the winged edge data structure if the vertex is not already in the data structure.*

2. *For each pair of successive vertices (including the last and first vertices), add a record for the edge if an edge with those vertices is not currently in the data structure.*

3. *For each of the records for edges around the face, add the wings for clockwise and counterclockwise traversal around the face. The wing fields in each record that are affected depend on whether the new face being inserted in the data structure is the east or west face of each edge record.*

4. *Create a record for the face and add a pointer to one of the edges.*

13.5.2   Surface Patches

Portions of curved graph surfaces can be represented as surface patches modeled using bivariate (two-variable) polynomials. For example, a plane can be represented as

$$z = a_0 + a_1 x + a_2 y, \tag{13.31}$$

and curved surface patches can be modeled using higher-order polynomials.

Bilinear patches, so named because any cross section parallel to a coordinate axis is a line,

$$z = a_0 + a_1 x + a_2 y + a_3 xy, \tag{13.32}$$

biquadratic patches,

$$z = a_0 + a_1 x + a_2 y + a_3 xy + a_4 x^2 + a_5 y^2, \tag{13.33}$$

bicubic patches,

$$z = a_0 + a_1x + a_2y + a_3xy + a_4x^2 + a_5y^2 \\ + a_6x^3 + a_7x^2y + a_8xy^2 + a_9x^3, \quad\quad (13.34)$$

and biquartic patches,

$$z = a_0 + a_1x + a_2y + a_3xy + a_4x^2 + a_5y^2 \\ + a_6x^3 + a_7x^2y + a_8xy^2 + a_9y^3 \\ + a_{10}x^4 + a_{11}x^3y + a_{12}x^2y^2 + a_{13}xy^3 + a_{14}y^4, \quad (13.35)$$

are bivariate polynomials that are frequently used in machine vision to represent surface patches.

Polynomial surface patches are good for modeling portions of a surface, such as the neighborhood around a point, but surface patches are not convenient for modeling an entire surface and cannot model surfaces that are not graph surfaces. More complex surfaces can be modeled using cubic spline surfaces, presented in the next section.

Algorithm 13.2   Follow the Edges Clockwise Around a Face

The inputs are a pointer to the record for the face to traverse and a procedure to invoke for each edge that is visited.

1. *Get the first edge from the face record and make it the current edge.*

2. *Process the current edge: perform whatever operations must be done as each edge is visited. For example, to compile a list of vertices clockwise around the face, record the vertex at the end of the edge in the direction of traversal.*

3. *If the west face of the current edge is being circumnavigated, then the next edge is the southwest wing.*

4. *If the east face of the current edge is being circumnavigated, then the next edge is the northeast wing.*

5. *If the current edge is the first edge, then the traversal is finished.*

6. *Otherwise, go to step 2 to process the new edge.*

13.5.3   Tensor-Product Surfaces

In Section 13.4.1, we showed how a complex curve can be represented parametrically as a sequence of cubic polynomials. This representation can be extended to get a parametric representation for complex surfaces.

Write the equation for a parametric cubic polynomial curve in matrix notation:

$$\mathbf{p}(u) = \begin{bmatrix} \mathbf{a}_1 & \mathbf{a}_2 & \mathbf{a}_3 & \mathbf{a}_4 \end{bmatrix} \begin{bmatrix} u^3 \\ u^2 \\ u \\ 1 \end{bmatrix}. \tag{13.36}$$

Note that each coefficient is actually a three-element column vector.

A tensor-product surface representation is formed from the combination of two curve representations, one in each parametric coordinate,

$$\mathbf{p}(u,v) = \begin{bmatrix} u^3 & u^2 & u & 1 \end{bmatrix} \begin{bmatrix} \mathbf{a}_1 \\ \mathbf{a}_2 \\ \mathbf{a}_3 \\ \mathbf{a}_4 \end{bmatrix} \begin{bmatrix} \mathbf{b}_1 & \mathbf{b}_2 & \mathbf{b}_3 & \mathbf{b}_4 \end{bmatrix} \begin{bmatrix} v^3 \\ v^2 \\ v \\ 1 \end{bmatrix}, \tag{13.37}$$

where each $\mathbf{a}_i$ is a three-element row vector, each $\mathbf{b}_j$ is a three-element column vector, and each product $\mathbf{a}_i\mathbf{b}_j$ is the pairwise product of the coefficients for each coordinate. The parametric surface can be written as

$$\mathbf{p}(u,v) = U^T M V, \tag{13.38}$$

with

$$U = \begin{bmatrix} u^3 \\ u^2 \\ u \\ 1 \end{bmatrix} \tag{13.39}$$

and

$$V = \begin{bmatrix} v^3 \\ v^2 \\ v \\ 1 \end{bmatrix}. \tag{13.40}$$

The elements of the 4×4 matrix M are the vectors of coefficients for each coordinate in the parametric surface. In this notation, we can see that the tensor-product surface is literally the product of two curves: one curve in the u coordinate and another curve in the v coordinate. Any cross section through the tensor-product cubic polynomial surface, parallel to a coordinate axis in the parametric domain, is a parametric cubic polynomial curve. In other words, if one coordinate is fixed, then the result is a parametric cubic polynomial curve with the other coordinate as the curve parameter.

13.6 Surface Interpolation

This section shows how some of the surface representations presented above can be used to interpolate samples of a graph surface, such as depth measurements obtained with binocular stereo or active triangulation. Surface interpolation may be necessary when depth measurements do not conform to the uniform grid format required for image processing. It may be necessary to interpolate depth measurements onto a uniform grid before using image processing algorithms such as edge detection and segmentation.

13.6.1 Triangular Mesh Interpolation

Suppose that we have obtained samples z_k of a graph surface, $z = f(x, y)$, at scattered points (x_k, y_k) for $k = 1, \ldots, n$ using binocular stereo or active triangulation and we need to interpolate the depth measurements at grid locations $[i, j]$ in the image plane. The coordinates (x_j, y_i) at each location in the $n \times m$ grid (image array) are given by

$$x_j = j - \frac{m-1}{2} \tag{13.41}$$

$$y_i = -i + \frac{n-1}{2}. \tag{13.42}$$

We need to interpolate the z value at each point (x_j, y_i).

Use the scattered point coordinates and depth values (x_k, y_k, z_k) to create a triangular mesh. Since the depth measurements are from a graph surface, each triangle has an explicit planar equation

$$z = a_0 + a_1 x + a_2 y \tag{13.43}$$

with coefficients calculated from the formulas presented in Section 13.3.1. For each grid location, find the triangle that encloses point (x_j, y_i) and use the equation corresponding to this triangle to calculate the depth value at the grid location:

$$z_{ij} = a_0 + a_1 x_j + a_2 y_i. \tag{13.44}$$

13.6.2   Bilinear Interpolation

Sometimes it is necessary to interpolate values that already lie on a rectilinear grid. For example, rectification, discussed in Section 12.5, requires that the image intensity at a location between grid coordinates be interpolated from the pixel values at the four nearest grid locations. Bilinear interpolation is an easy method for interpolating values on a rectilinear grid.

The function

$$f(x, y) = a_1 + a_2 x + a_3 y + a_4 xy \tag{13.45}$$

is called bilinear because if one variable is set to a constant, then the function is linear in the other variable. In other words, each cross section of a bilinear surface patch taken parallel to a coordinate axis is a line segment. For any rectangle in the plane with sides parallel to the coordinate axes, there is a unique bilinear polynomial that interpolates the corner values.

Suppose that we need to interpolate the value at point (x, y) between four values on a rectilinear grid. The point is enclosed by a rectangle with sides parallel to the coordinate axes; the corners of the rectangle are the closest grid coordinates to the point at which the value is to be interpolated. The corner coordinates are (x_1, y_1), (x_1, y_2), (x_2, y_1), and (x_2, y_2) with values z_{11}, z_{12}, z_{21}, and z_{22}, respectively, as shown in Figure 13.4.

The coefficients of the bilinear interpolant are determined by the values at the four corners of the grid rectangle and are computed by plugging each grid coordinate and value into Equation 13.45:

$$z_{11} = a_1 + a_2 x_1 + a_3 y_1 + a_4 x_1 y_1 \tag{13.46}$$

$$z_{12} = a_1 + a_2 x_1 + a_3 y_2 + a_4 x_1 y_2 \tag{13.47}$$

$$z_{21} = a_1 + a_2 x_2 + a_3 y_1 + a_4 x_2 y_1 \tag{13.48}$$

$$z_{22} = a_1 + a_2 x_2 + a_3 y_2 + a_4 x_2 y_2 \tag{13.49}$$

and solving the four simultaneous equations for the coefficients of the interpolant:

$$a_1 = \frac{x_2 y_2 z_{11} - x_2 y_1 z_{12} - x_1 y_2 z_{21} + x_1 y_1 z_{22}}{(x_2 - x_1)(y_2 - y_1)} \tag{13.50}$$

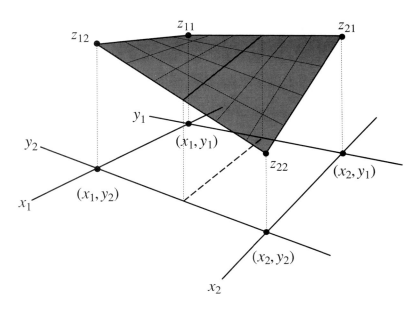

Figure 13.4: The bilinear interpolant is used to estimate the value between samples on a rectilinear grid. The values at the corners are known and uniquely determine the coefficients of the bilinear interpolant.

$$a_2 = \frac{-y_2 z_{11} + y_1 z_{12} + y_2 z_{21} - y_1 z_{22}}{(x_2 - x_1)(y_2 - y_1)} \qquad (13.51)$$

$$a_3 = \frac{-x_2 z_{11} + x_2 z_{12} + x_1 z_{21} - x_1 z_{22}}{(x_2 - x_1)(y_2 - y_1)} \qquad (13.52)$$

$$a_4 = \frac{z_{11} - z_{12} - z_{21} + z_{22}}{(x_2 - x_1)(y_2 - y_1)}. \qquad (13.53)$$

The bilinear interpolant has a very simple form for the special case where the rectilinear grid is a square with unit spacing between the rows and columns. Let the point (x, y) at which interpolation is to be performed be given as offsets $(\delta x, \delta y)$ from the upper left corner of a grid square. The bilinear interpolant is

$$f(\delta x, \delta y) = z_{11} + \delta x (z_{21} - z_{11}) + \delta y (z_{12} - z_{11}) + \delta x\, \delta y (z_{11} - z_{12} - z_{21} + z_{22}). \quad (13.54)$$

This formula can be used to interpolate the pixel value at an image plane point that is between the pixel locations.

13.6.3   Robust Interpolation

In Section 6.8.3, we presented the least-median-squares algorithm for fitting lines to edge points with outliers. Least-median-squares regression can be used to fit surface patches to depth measurements with outliers.

Least-median-squares is a robust regression algorithm with a breakdown point of 50% like the median on which it is based. The local least-median-squares estimator uses least-median-squares regression to fit a parametric model over local neighborhoods in the data. The algorithm finds the parameters that minimize the median of the squared residuals:

$$\min_a \left\{ \operatorname*{med}_{(x_i, y_i) \in N} \left[(z_i - f(x_i, y_i; a))^2 \right] \right\} \qquad (13.55)$$

where a is an estimated parameter vector and $f(x_i, y_i; a)$ is an estimate of the actual value of the graph surface at point measurement (x_i, y_i, z_i).

Least-median-squares can be used to grid surface samples that contain outliers. For example, binocular stereo depth measurements contain outliers due to mismatches, and range cameras may generate outliers at surface discontinuities. If least-median-squares is used to fit surface patches to depth measurements in local neighborhoods, then the surface fits will be immune to outliers. When the surface patches are used to grid the data, the outliers do not influence the grid values. Compare this algorithm with interpolation using a triangular mesh, presented in Section 13.6.1, which is very sensitive to outliers. Fitting surface patches with least-median-squares is very useful for processing sparse depth measurements that may be contaminated by outliers. We call this process cleaning and gridding the depth measurements.

A straightforward expression for least-median-squares is difficult to write, but the algorithm for implementing it is easy to explain. Assume that the surface patches are fit to the depth measurements in a local neighborhood about each grid point. The algorithm can be easily extended to fit higher-order surface patches. For each grid point, select the n depth measurements that are closest to the grid point. From this set, try all possible combinations of m data points, where m is the number of data points used to fit the surface patch. For each of the k subsets of the data points in the local neighborhood,

$$k = \binom{n}{m}, \qquad (13.56)$$

fit a surface patch to the points in the subset and denote the corresponding parameter vector by a_k. Compare all data points in the local neighborhood with the surface patch by computing the median of the squared residuals:

$$\chi_k^2 = \operatorname*{med}_i \left[(z_i - f(x_i, y_i, a_k))^2 \right]. \tag{13.57}$$

After surface patches have been fit to all possible subsets, pick the parameter vector a_k corresponding to the surface patch with the smallest median of squared residuals.

This procedure is computationally expensive since the model fit is repeated $\binom{n}{m}$ times for each local neighborhood; however, each surface fit is independent, and the procedure is highly parallelizable. Adjacent neighborhoods share data points and could share intermediate results. In practice, it may be necessary to try only a few of the possible combinations so that the probability of one of the subsets being free of outliers is close to 1.

13.7   Surface Approximation

Depth measurements are not free of errors, and it may be desirable to find a surface that approximates the data rather than requiring the surface to interpolate the data points.

If depth measurements are samples of a graph surface,

$$z = f(x, y), \tag{13.58}$$

then it may be desirable to reconstruct this surface from the samples. If we have a model for the surface,

$$z = f(x, y; a_1, a_2, \ldots, a_m), \tag{13.59}$$

with m parameters, then the surface reconstruction problem reduces to the regression problem of determining the parameters of the surface model that best fit the data:

$$\chi^2 = \sum_{i=1}^{n} (z_i - f(x_i, y_i; a_1, a_2, \ldots, a_m))^2. \tag{13.60}$$

If we do not have a parametric model for the surface and still need to reconstruct the surface from which the depth samples were obtained, then we must fit a generic (nonparametric) surface model to the data. The task is to find the graph surface $z = f(x, y)$ that best fits the data:

$$\chi^2 = \sum_{i=1}^{n} (z_i - f(x_i, y_i))^2. \qquad (13.61)$$

This is an ill-posed problem since there are many functions that can fit the data equally well; indeed, there are an infinite number of functions that interpolate the data. The term *ill-posed* means that the problem formulation does not lead to a clear choice, whereas a well-posed problem leads to a solution that is clearly the best from the set of candidates.

We need to augment the approximation norm in Equation 13.61 to constrain the selection of the approximating surface function to a single choice. There are many criteria for choosing a function to approximate the data. A popular criterion is to choose the function that both approximates the data and is a smooth surface. There are many measures of smoothness, but we will choose one measure and leave the rest for further reading. For the purposes of this discussion, the best approximation to the data points $\{(x_i, y_i, z_i)\}$ for $i = 1, \ldots, n$ is the function $z = f(x, y)$ that minimizes

$$\chi^2 = \sum_{i=1}^{n} (z_i - f(x_i, y_i))^2 + \alpha^2 \int \int \frac{\partial^2 f}{\partial x^2} + 2 \frac{\partial f}{\partial x} \frac{\partial f}{\partial y} + \frac{\partial^2 f}{\partial y^2} \, dx \, dy \qquad (13.62)$$

with $\alpha > 0$. This norm is the same as Equation 13.61 except for the addition of a smoothness term, weighted by α. The smoothness term is called the regularizing term or stabilizing functional, or stabilizer for short. The weight α is called the regularizing parameter and specifies the trade-off between achieving a close approximation to the data (small α) and forcing a smooth solution (large α). The first term in Equation 13.62 is called the problem constraint. The process of augmenting a problem constraint by adding a stabilizing functional to change an ill-posed problem into a well-posed problem is called regularization.

The notion of a well-posed problem is stronger than just requiring a unique solution. An approximation problem such as Equation 13.61 may have a unique solution, but the shape of the solution space is such that many other, very different solutions may be almost as good. A well-posed problem has a solution space where the solution corresponding to the minimum of the norm is not only unique, but is definitely better than the other solutions.

The solution to the regularized surface approximation problem in Equation 13.62 is obtained by using variational calculus to transform Equation 13.62 into a partial differential equation. Numerical methods are used to change the partial differential equation into a numerical algorithm for computing samples of the approximating surface from the depth measurements.

The partial differential equation for this problem in the variational calculus is

$$\alpha^2 \nabla^4 f(x, y) + f(x, y) - z(x, y) = 0, \tag{13.63}$$

which may be solved using finite difference methods as explained in Section A.4.

Another approach is to replace the partial derivatives in Equation 13.62 with finite difference approximations and differentiate with respect to the solution f to get a system of linear equations that can be solved using numerical methods.

13.7.1 Regression Splines

Another approach to surface approximation is to change Equation 13.61 into a regression problem by substituting a surface model for the approximating function. Of course, if we knew the surface model, then we would have formulated the regression problem rather than starting from Equation 13.61. If we do not know the surface model, we can still start with Equation 13.61 and substitute a generic surface representation, such as tensor-product splines, for the approximating function and solve the regression problem for the parameters of the generic surface representation. This technique is called regression splines.

Many surface representations, including tensor-product splines, can be represented as a linear combination of basis functions:

$$f(x, y; a_0, a_1, a_2, \ldots, a_m) = \sum_{i=0}^{m} a_i B_i(x, y) \tag{13.64}$$

where a_i are the scalar coefficients and B_i are the basis functions. With tensor-product splines, the basis functions (and their coefficients) are organized into a grid:

$$f(x, y) = \sum_{i=0}^{n} \sum_{j=0}^{m} a_{ij} B_{ij}(x, y). \tag{13.65}$$

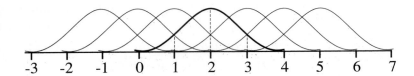

Figure 13.5: The B-spline curve in one dimension is a linear combination of basis functions located at integer positions in the interval $[0, m]$.

With either of these two equations, if we substitute the generic surface model into Equation 13.61, the result is a system of linear equations that can be solved for the regression parameters. Since the system of equations is sparse, it is better to use sparse matrix techniques, rather than singular value decomposition, for solving this regression problem.

The methods for calculating regression splines will be presented in detail, starting with the one-dimensional case. The B-spline in one dimension is a linear combination of basis functions,

$$\sum_{i=0}^{m} a_i B_i(x). \tag{13.66}$$

Assume that the B-spline basis functions are spaced uniformly at integer locations. The B-spline basis function $B_i(x)$ is nonzero over the interval $[i, i + 4)$. The basis functions are located at positions 0 through m, so there are $m+1$ coefficients that define the shape of the B-spline basis curve. (Refer to Figure 13.5.)

The B-spline curve is defined on the interval from $x = 3$ to $x = m + 1$. The basis function at $x = 0$ extends three intervals to the left of the curve and the basis function at $x = m + 1$ extends three intervals to the right of the curve. The extra intervals beyond the ends of the curve establish the boundary conditions for the ends of the curve and must be included so that the first and last curve segments on the intervals $[0, 1)$ and $[m, m + 1)$ are defined correctly.

Each B-spline basis function is nonzero over four integer intervals (segments). Since the basis functions overlap, each interval is covered by four basis functions. Each segment $b_i(x)$ of the basis function $B_i(x)$ is a cubic polynomial that is defined only on its integer interval, as illustrated in Figure 13.6. The cubic polynomials for the individual segments of the B-spline

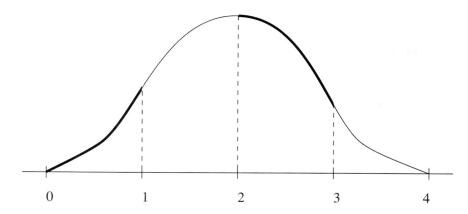

Figure 13.6: Each B-spline basis function consists of four cubic polynomials that are nonzero on adjacent integer intervals.

are

$$b_0(x) = \frac{x^3}{6} \tag{13.67}$$

$$b_1(x) = \frac{1 + 3x + 3x^2 - 3x^3}{6} \tag{13.68}$$

$$b_2(x) = \frac{4 - 6x^2 + 3x^3}{6} \tag{13.69}$$

$$b_3(x) = \frac{1 - 3x + 3x^2 - x^3}{6}. \tag{13.70}$$

It is not correct to simply add these cubic polynomials together to get a B-spline basis function since each segment is restricted to its interval; that is, each segment must be treated as nonzero outside of the interval over which it is intended to be used. To evaluate a particular B-spline basis function at a particular location x, it is necessary to determine in which segment x lies (in which piece of the B-spline basis function x is located) and evaluate the corresponding cubic polynomial.

Suppose that x is located in the interval $[i, i+1)$. The B-spline curve is the linear combination of $m + 1$ B-spline basis functions:

$$\sum_{i=0}^{m} a_i B_i(x). \tag{13.71}$$

Since each B-spline basis function covers four intervals, the interval $[i, i + 1)$ is covered by B-spline basis function $B_i(x)$ and the three basis functions to the left: $B_{i-1}(x)$, $B_{i-2}(x)$, and $B_{i-3}(x)$. So to evaluate the B-spline curve at x in the interval $[i, i + 1)$, it is only necessary to evaluate the portion of the linear combination that corresponds to the four B-spline basis functions that cover the interval:

$$a_{i-3}B_{i-3}(x) + a_{i-2}B_{i-2}(x) + a_{i-1}B_{i-1}(x) + a_i B_i(x). \tag{13.72}$$

Since each B-spline basis function consists of four cubic polynomial segments, with each segment defined over its own interval, it is only necessary to evaluate the cubic polynomial segments from each basis function that correspond to the interval containing the location at which the cubic spline curve is being evaluated. The value of the B-spline curve at location x in interval $[i, i + 1)$ is

$$\frac{(1 - 3x + 3x^2 - x^3)a_{i-3}}{6} + \frac{(4 - 6x^2 + 3x^3)a_{i-2}}{6}$$
$$+ \frac{(1 + 3x + 3x^2 - 3x^3)a_{i-1}}{6} + \frac{x^3 a_i}{6}. \tag{13.73}$$

To evaluate a B-spline curve at any location, it is necessary to determine the interval in which the location x lies and apply the formula in Equation 13.73, with the index i for the starting point of the interval, so that the correct coefficients for the four B-spline basis functions that cover the interval are used.

The formula for evaluating a B-spline curve in Equation 13.73 can be used for linear regression since the model is linear in the coefficients. It does not matter that the equation contains terms with powers of x. This model can be used to develop a regression algorithm for determining the coefficients of the B-spline from a set of data points (x_i, z_i). Each data point yields one equation that constrains four of the B-spline coefficients:

$$\frac{(1 - 3x_i + 3x_i^2 - x_i^3)a_{i-3}}{6} + \frac{(4 - 6x_i^2 + 3x_i^3)a_{i-2}}{6}$$
$$+ \frac{(1 + 3x_i + 3x_i^2 - 3x_i^3)a_{i-1}}{6} + \frac{x_i^3 a_i}{6} = z_i. \tag{13.74}$$

These equations are used to form a system of linear equations,

$$M\mathbf{a} = \mathbf{b}, \tag{13.75}$$

where the solution $\mathbf{a}$ is the column vector of B-spline coefficients a_i, the column vector on the right-hand side is the vector of data values z_i, and each row of the matrix M is zero except at the four elements corresponding to the coefficients of the B-spline basis functions that cover the interval that contains x_i. The system of linear equations can be solved using singular value decomposition, but since the system of equations is sparse, it is better to use sparse matrix techniques [197].

For surface approximation in three dimensions, the model is a tensor-product B-spline surface:

$$f(x,y) = \sum_{i=0}^{n} \sum_{j=0}^{m} a_{i,j} B_{i,j}(x,y) = \sum_{i=0}^{n} \sum_{j=0}^{m} a_{i,j} B_j(x) B_i(y). \tag{13.76}$$

The surface approximation problem is solved by minimizing the norm

$$\chi^2 = \sum_{k=1}^{N} (z(k) - f(x,y))^2. \tag{13.77}$$

Replace the general surface model with the expression for the tensor-product B-spline surface:

$$\chi^2 = \sum_{k=1}^{N} \left[z(k) - \left(\sum_{i=0}^{n} \sum_{j=0}^{m} a_{i,j} B_{i,j}(x,y) \right) \right]^2. \tag{13.78}$$

The data values range over a single index since the data can occur at scattered points in the plane. Each data value z_k is located at point (x_k, y_k) in the x–y plane.

The formulation of the regression problem for determining the coefficients of the tensor-product B-spline surface is similar to the one-dimensional case presented above. Each basis function $B_{ij}(x,y)$ covers 16 grid rectangles in $[i, i+4) \times [j, j+4)$. Since the basis functions overlap, each grid rectangle is covered by 16 basis functions. Each basis function consists of 16 bicubic polynomial patches. Each patch is defined on a grid rectangle and is nonzero outside of its grid rectangle.

Assume that the grid is uniform, so all grid rectangles are squares of the same size. The formulas for the bicubic patches in the B-spline basis functions are the same for each grid square. Each tensor-product basis function can be separated into the product of one-dimensional basis functions

$$B_{ij}(x,y) = B_j(x) B_i(y), \tag{13.79}$$

and each of the one-dimensional basis functions consists of four cubic polynomials. Each of the 16 polynomial patches defined on a grid rectangle is the product of two cubic polynomials, one in x and the other in y. The four cubic polynomials in x are

$$b_0(x) = \frac{x^3}{6} \tag{13.80}$$

$$b_1(x) = \frac{1 + 3x + 3x^2 - 3x^3}{6} \tag{13.81}$$

$$b_2(x) = \frac{4 - 6x^2 + 3x^3}{6} \tag{13.82}$$

$$b_3(x) = \frac{1 - 3x + 3x^2 - x^3}{6}. \tag{13.83}$$

The four cubic polynomials in y are identical except for the change of variable:

$$b_0(y) = \frac{y^3}{6} \tag{13.84}$$

$$b_1(y) = \frac{1 + 3y + 3y^2 - 3y^3}{6} \tag{13.85}$$

$$b_2(y) = \frac{4 - 6y^2 + 3y^3}{6} \tag{13.86}$$

$$b_3(y) = \frac{1 - 3y + 3y^2 - y^3}{6}. \tag{13.87}$$

The 16 polynomial patches in a B-spline basis function are formed from the pairwise products of the polynomials given above. The formula for evaluating the B-spline surface at point (x, y) in the grid is

$$\sum_{i=0}^{3} \sum_{j=0}^{3} a_{ij} b_j(x) b_i(y). \tag{13.88}$$

Substituting the formulas for the one-dimensional cubic polynomials yields the formula for evaluating the B-spline surface. The terms for each of the 16 coefficients are listed in the following table.

$a_{i,j}$	$x^3 y^3 / 18$
$a_{i,j-1}$	$x^3/18 + (x^3 y + x^3 y^2 - x^3 y^3)/6$

$a_{i,j-2}$ $2x^3/9 - x^3y^2/3 + x^3y^3/6$

$a_{i,j-3}$ $(x^3 - 3x^3y + 3x^3y^2 - x^3y^3)/18$

$a_{i-1,j}$ $(1 + 3x + 3x^2 - 3x^3)y^3/18$

$a_{i-1,j-1}$ $(1 + 3x + 3x^2 - 3x^3)/18 + (1 + 3x + 3x^2 - 3x^3)y/6$
$+ (1 + 3x + 3x^2 - 3x^3)y^2/6 - (1 + 3x + 3x^2 - 3x^3)y^3/6$

$a_{i-1,j-2}$ $2(1 + 3x + 3x^2 - 3x^3)/9 - (1 + 3x + 3x^2 - 3x^3)y^2/3$
$+ (1 + 3x + 3x^2 - 3x^3)y^3/6$

$a_{i-1,j-3}$ $(1 + 3x + 3x^2 - 3x^3)/18 - (1 + 3x + 3x^2 - 3x^3)y/6$
$+ (1 + 3x + 3x^2 - 3x^3)y^2/6 - (1 + 3x + 3x^2 - 3x^3)y^3/18$

$a_{i-2,j}$ $(4 - 6x^2 + 3x^3)y^3/18$

$a_{i-2,j-1}$ $(4 - 6x^2 + 3x^3)/18 + (4 - 6x^2 + 3x^3)y/6$
$+ (4 - 6x^2 + 3x^3)y^2/6 - (4 - 6x^2 + 3x^3)y^3/6$

$a_{i-2,j-2}$ $2(4 - 6x^2 + 3x^3)/9 - (4 - 6x^2 + 3x^3)y^2/3$
$+ (4 - 6x^2 + 3x^3)y^3/6$

$a_{i-2,j-3}$ $(4 - 6x^2 + 3x^3)/18 - (4 - 6x^2 + 3x^3)y/6$
$+ (4 - 6x^2 + 3x^3)y^2/6 - (4 - 6x^2 + 3x^3)y^3/18$

$a_{i-3,j}$ $(1 - 3x + 3x^2 - x^3)y^3/18$

$a_{i-3,j-1}$ $(1 - 3x + 3x^2 - x^3)/18 + (1 - 3x + 3x^2 - x^3)y/6$
$+ (1 - 3x + 3x^2 - x^3)y^2/6 - (1 - 3x + 3x^2 - x^3)y^3/6$

$a_{i-3,j-2}$ $2(1 - 3x + 3x^2 - x^3)/9 - (1 - 3x + 3x^2 - x^3)y^2/3$
$+ (1 - 3x + 3x^2 - x^3)y^3/6$

$a_{i-3,j-3}$ $(1 - 3x + 3x^2 - x^3)/18 - (1 - 3x + 3x^2 - x^3)y/6$
$+ (1 - 3x + 3x^2 - x^3)y^2/6 - (1 - 3x + 3x^2 - x^3)y^3/18$

These expressions are evaluated at the desired location (x, y), multiplied by the corresponding coefficient $a_{i,j}$, and summed together to get the value of the B-spline surface at (x, y).

The formula for evaluating a B-spline surface is also the model used in the regression problem for determining the coefficients of the B-spline surface. There are $(n + 1) \times (m + 1)$ coefficients in the B-spline surface. Each data point (x_k, y_k, z_k) yields one equation that constrains 16 of the coefficients. As in the one-dimensional case, the regression problem leads to a system of linear equations,

$$Ma = b, \tag{13.89}$$

where the solution a is the vector of B-spline coefficients (the two-dimensional grid of coefficients is unfolded into a column vector), the column vector b on the right-hand side is the vector of data values x_k for $k = 1, \ldots, N$, and each row of the matrix is zero except for the 16 elements corresponding to the coefficients of the basis functions that cover the grid square containing (x_k, y_k).

This method can be used to smooth images by fitting a B-spline surface to the uniform grid of pixels and sampling the spline surface. It is not necessary for the B-spline grid to correspond to the image grid; in fact, more smoothing is achieved if the B-spline grid has greater spacing. The B-spline basis function is like the Gaussian smoothing filter, covered in Chapter 4, and the widths of the B-spline basis functions are determined by the spacing of the B-spline grid. When using the formulas presented above for the B-spline regression problem, the image locations (x_j, y_i) for the pixel values z_{ij} in the image must be mapped into the grid coordinate system of the B-spline surface. After the B-spline coefficients have been determined, the B-spline surface can be sampled as needed, taking samples at the grid coordinates of the original image or at other coordinates, to calculate the pixels of the smoothed image.

The development of the equation for the B-spline surface shows that a B-spline is just a set of bicubic surface patches, which were presented in Section 13.5.2. The difference between a set of bicubic patches and the B-spline surface is that the bicubic patches in the B-spline surface are continuous up to second order, where the patches join, whereas an arbitrary collection of bicubic patches need not be continuous at all. The tensor-product B-spline surface is smooth, in the sense of having C^2 continuity, and can be used to model objects such as human bodies and organs, automobile body panels, and the skin of aircraft.

In many practical applications, the tensor-product B-spline surface must be in parametric form: $(x, y, z) = (x(u, v), y(u, v), z(u, v))$. As with the notation presented in Section 13.5.3, the coefficients for the spline surface in the regression problem will be three-element vectors, one element for each of the x, y, and z coordinates. The u–v domain is rectangular and is subdivided into a uniform grid (equal spacing between the basis functions). The most difficult part is constructing a formula for mapping (x, y, z) point measurements into the u–v domain so that the point measurements are correctly associated with the bicubic patches that make up the B-spline surface.

13.7.2 Variational Methods

The surface approximation problem can also be cast as a problem in variational calculus. The problem is to determine the surface $z = f(x, y)$ that approximates the set of depth measurements z_k at locations (x_k, y_k) in the image plane.

There are an infinite number of surfaces that will closely fit any set of depth measurements, but the best approximation can be defined as the surface that comes close to the depth measurements and is a smooth surface. The best approximation is obtained by minimizing the functional

$$\chi^2 = \sum_{k=1}^{n} (z_k - f(x_k, y_k))^2 + \int \int \left[\frac{\partial^2 f}{\partial x^2} + 2 \frac{\partial f}{\partial x} \frac{\partial f}{\partial y} + \frac{\partial^2 f}{\partial y^2} \right] dx\, dy. \quad (13.90)$$

This minimization problem is solved using variational calculus to obtain the partial differential equation

$$\nabla^4 f(x, y) = \sum_{k=1}^{n} (z_k - f(x_k, y_k)). \quad (13.91)$$

Appropriate boundary conditions must be chosen for the edges of the graph surface.

13.7.3 Weighted Spline Approximation

The problem with the surface approximation methods presented so far is that the solutions are smooth surfaces, even across discontinuities in the data that most likely correspond to surface boundaries in the scene. Smoothness is explicit in the smoothness functional used for the variational methods in Section 13.7.2 and is implicit in the choice of a smooth generic function for the regression splines presented in Section 13.7.1. If the approximating surface could be formulated as a set of piecewise smooth functions,

$$f(x, y) = \bigcup_{l=1}^{M} f_l(x, y), \quad (13.92)$$

and the depth measurements could be segmented into regions with one region
for each piecewise smooth function, then the solution would more accurately
reflect the structure of the surfaces in the scene. However, this leads to algorithms for manipulating surfaces that are beyond the scope of this book.
Some aspects of surface segmentation will be discussed in Section 13.8. We
can avoid some difficult topics in surface representation and still achieve a
good surface approximation that conforms to discontinuities by changing the
smoothness functional in Equation 13.62 to reduce smoothing across discontinuities in the depth data. This approach leads to weighted regularization.

In one dimension, the weighted regularization norm is

$$\sum_{i=1}^{n}(y_i - f(x_i))^2 + \int w(x)[f_{xx}(x)]^2 \, dx. \tag{13.93}$$

If the weight function $w(x)$ is choosen to be small at discontinuities in the
data and large elsewhere, then the weight function cancels the smoothness
criterion at discontinuities.

In two dimensions, the norm for weighted spline surface approximation is

$$\chi^2 = \sum_{i=1}^{n}(z_i - f(x_i, y_i))^2 + \int\int w(x, y)\left[\frac{\partial^2 f}{\partial x^2} + 2\frac{\partial f}{\partial x}\frac{\partial f}{\partial y} + \frac{\partial^2 f}{\partial y^2}\right] \, dx \, dy \tag{13.94}$$

with the weight function given by

$$w(x, y) = \frac{\alpha^2}{1 + \|\rho(x, y)\|^2}, \tag{13.95}$$

where $\rho(x, y)$ is the gradient of the surface from which the samples were
taken. Note that the regularizing parameter α from Equation 13.62 has been
absorbed into the weight function.

Variational methods can be used to transform Equation 13.94 into a partial differential equation, as shown in Section 13.7.2 and Appendix A.3, or
the method of regression splines can be used to replace Equation 13.94 with
a linear regression problem. The weight function in Equation 13.94 is computed from an approximation to the gradient by assuming that the weights
are constant over the span of the grid on which the tensor-product basis
functions are defined.

13.8 Surface Segmentation

A range image is a uniform grid of samples of a piecewise smooth graph surface

$$z = f(x, y). \tag{13.96}$$

This section will describe how a set of range samples, defined on a uniform grid, can be segmented into regions that have similar curvature and how to approximate each region with low-order bivariate polynomials. The variable-order surface segmentation algorithm, which is like the region-growing techniques described in Section 3.5, will be presented. Surface curvature properties are used to estimate core regions, which are grown to cover additional range samples. The regions are modeled by bivariate surface patches (see Section 13.5.2). The segmentation algorithm grows the regions by extending the surface patches to cover range samples that are neighbors of the region if the residual between the new range samples and the surface patch is low.

The surface segmentation problem is formally stated as follows. A piecewise smooth graph surface $z = f(x, y)$ can be partitioned into smooth surface primitives

$$z = f(x, y) = \sum_{l=1}^{n} f_l(x, y) \xi(x, y, l) \tag{13.97}$$

with one surface primitive for each region R_l. The characteristic function

$$\xi(x, y, l) = \begin{cases} 1 & \text{if } (x, y) \in R_l \\ 0 & \text{otherwise} \end{cases} \tag{13.98}$$

represents the segmentation of the range data into surface patches (regions). The surface for each region is approximated by a polynomial patch

$$f(x, y) = \sum_{i+j \leq m} a_{ij} x^i y^j \tag{13.99}$$

for points $(x, y) \in R_l$. The surface model includes planar, bilinear biquadratic, bicubic, and biquartic polynomial patches, which provide a nested hierarchy of surface representations.

The characteristic function $\xi(x, y, l)$ can be implemented as a list of the locations $[i, j]$ of the pixels that are in each region. Other representations such as bit masks or quad trees, described in Section 3.3, can be used.

$$K[i,j]$$

	+	0	−
−	Peak	Ridge	Saddle ridge
$H[i,j]$ 0		Flat	Minimal surface
+	Pit	Valley	Saddle valley

Table 13.1: The eight surface types defined by the signs of the mean and Gaussian curvatures.

13.8.1   Initial Segmentation

The core regions for the initial segmentation are estimated by computing the mean and Gaussian curvatures and using the signs of the mean and Gaussian curvatures to form initial region hypotheses. The portion of the regions near the region boundaries can include samples that do not belong in the regions since it can be difficult to accurately estimate the curvature in the vicinity of the transitions between regions. The regions are shrunk to core regions (see Section 2.5.12) to ensure that the initial region hypotheses do not include any falsely labeled samples. There are eight surface types corresponding to the signs of the mean and Gaussian curvatures, listed in Table 13.1. These surface types are used to form core regions.

Assume that the range data are a square grid of range samples that may be processed like an image. The first and second partial derivatives of the range image, f_x, f_y, f_{xx}, f_{xy}, f_{yy}, are estimated by convolving the range image with separable filters to estimate the partial derivatives. Compute the mean curvature

$$H[i,j] = \frac{(1 + f_y^2[i,j]) f_{xx}[i,j] + (1 + f_x^2[i,j]) f_{yy} - 2 f_x[i,j] f_y[i,j] f_{xy}[i,j]}{2 \left(\sqrt{1 + f_x^2[i,j] + f_y^2[i,j]} \right)^3}$$

(13.100)

and the Gaussian curvature

$$K[i,j] = \frac{f_{xx}[i,j] f_{yy}[i,j] - f_{xy}^2[i,j]}{\left(1 + f_x^2[i,j] + f_y^2[i,j]\right)^2}$$

using the estimates of the first and second partial derivatives of the range image. Compute an integer label that encodes the signs of the mean and Gaussian curvatures using the formula

$$T[i,j] = 1 + 3(1 + \mathrm{sgn}(H[i,j])) + (1 + \mathrm{sgn}(K[i,j])).$$ (13.101)

Use the sequential connected components algorithm (Section 2.5.2) to group range samples with identical labels into connected regions and reduce the size of the regions to a conservative core of range samples using a shrink or erosion operator. A size filter can be used to discard regions that are too small to correspond to scene surfaces. The regions that remain form the seeds that initialize the region-growing process.

13.8.2 Extending Surface Patches

Bivariate polynomials are fit to the range samples in the core regions. The surface fit starts with a planar patch, $m = 1$, and the order of the surface patch is increased until a good fit (root-mean-square error below a threshold) is achieved. If a good fit is not achieved, then the region is discarded.

The rest of the segmentation algorithm tries to extend the regions to cover neighboring range samples. Unassigned range samples are added to regions using a region-growing process. The similarity predicate for deciding if a range sample should be added to a region is based on comparing the value of the range sample to the bivariate surface patch evaluated at the same location as the range sample. If the residual between the surface patch and range sample is below a threshold, then the range sample is included in the set of candidates for inclusion in the region; otherwise, the range sample is not added to the region. The bivariate surface patch is refit to all range samples in the region, including the samples that are currently in the region and the set of candidates, increasing the order of the surface patch if necessary, up to the maximum order. If the fitting error is below a threshold, then the candidates are permanently added to the region; otherwise, the entire set of candidates is discarded. Region growing continues until no regions are enlarged. The segmentation algorithm is summarized in Algorithm 13.3.

Algorithm 13.3   Surface Segmentation

Segment a range image into regions represented as a surface patches.

1. *Compute the first and second partial derivatives of the range image using separable filters.*

2. *Compute the mean and Gaussian curvature at each image location.*

3. *Label each range pixel with the surface type.*

4. *Shrink the regions to eliminate false labels near the region boundaries.*

5. *Identify core regions using the sequential connected components algorithm.*

6. *Remove core regions that are too small.*

7. *Fit a bivariate patch to each region.*

8. *For each region, determine the set of neighboring pixels with values that are close to the value of the surface patch evaluated at the location of the neighbor. This set of pixels are candidates for inclusion in the region.*

9. *Refit the surface patch to the union of the set of pixels in the region and the set of candidates. If the fit is acceptable, add the set of candidates to the region; otherwise, discard the set of candidates.*

10. *Repeat steps 8 through 9 until no region is changed.*

Note that it is possible for a set of candidates to be discarded because some of the candidates should be included in another region. Later, some of the candidates may be retested and included in the region. Once a pixel is assigned to a region, the assignment cannot be reversed.

13.9 Surface Registration

The transformation between corresponding points in a set of conjugate pairs can be determined by solving the absolute orientation problem (Section 12.3), but there are applications where it is necessary to align two shapes without point correspondences. For example, it may be necessary to align a set of range samples to an object model without advance knowledge of how the point samples correspond to points on the object. This section describes the iterative closest point algorithm for determining the rigid body transformation between two views of an object without using point correspondences.

The iterative closest point algorithm can be adapted to a wide variety of object models including sets of points, curves in two or three dimensions, and various surface representations. The curves may be represented as polylines, implicit curves, or parametric curves, and the surfaces may be represented as faceted surfaces (polygonal meshes), implicit or parametric surfaces, or tensor-product B-splines. It is not necessary for both views of an object to have the same representation. For example, one view could be a set of space curves obtained with active triangulation, and the other view could be a tensor-product cubic spline surface. The key idea is that points sampled from the representation for one view can be mapped to the closest points on the representation for the other view to approximate a set of conjugate pairs. Solving the absolute orientation problem with these correspondences will improve the alignment between views. The process of finding closest point correspondences and solving the absolute orientation problem with these approximate conjugate pairs can be repeated until the two views are aligned.

The iterative closest point algorithm will be presented for the case of registering a set of points with a surface model. Given a set of points P sampled over the surface of an object and a model M for the object, determine the set of points on the model that are the closest points to each sample point. The distance between a point p in the sample set and the model M is

$$d(p, M) = \min_{q \in M} \|q - p\|. \tag{13.102}$$

Compute $Q = \{q_1, q_2, \ldots, q_n\}$, the set of points $q \in M$ that are the closest points to each point in the sample set $P = \{p_1, p_2, \ldots, p_n\}$. Pair each sample point p with its closest point q in the model to form a set of conjugate pairs $\{(p_1, q_1), (p_2, q_2), \ldots, (p_n, q_n)\}$. Solve the absolute orientation problem using this set of conjugate pairs. Even though the point sets are not true conjugate pairs, an approximate solution to the transformation between the views will be obtained. Apply the transformation to the point set P and recompute the set of closest points Q. Now the point set is closer to the model and the pairs of points in P and Q will be closer to correct conjugate pairs. The absolute orientation problem can be solved again and the new transformation applied to point set P. This procedure is repeated until the sum of the squared distances between closest points (the root-mean-square closest point distance) is below a threshold.

In the early iterations, the closest point pairs may be very poor approximations to true conjugate pairs, but applying the rigid body transformation obtained by solving the absolute orientation problem with these approximate conjugate pairs brings the point set closer to the model. As the iterations continue, the closest point pairs become more like valid conjugate pairs. For example, if the point set is initially very far from the model, then all points may be mapped to the same point on the model. Clearly, such a many-to-one mapping cannot be a valid set of conjugate pairs, but the first iteration pulls the point set to the model so that the points are centered on the matching model point and subsequent iterations align the center of the point set with the center of the model. The final iterations rotate the point set and adjust its position so that the point set is aligned with the model. The steps of iterative closest point registration are listed in Algorithm 13.4.

Algorithm 13.4 Iterative Closest Point Registration
Register a set of points to a surface modeled as a polygonal mesh.

1. *Compute the set of closest points.*

2. *Compute the registration between the point sets.*

3. *Apply the transform to register the point sets.*

4. *Return to step 1 unless the registration error is within tolerance.*

Further Reading

An excellent introduction to the differential geometry of curves and surfaces is the book by do Carmo [67]. Bartels, Beatty, and Barsky [22] have written an excellent introduction to spline curves and surfaces. The series of books, called *Graphics Gems*, contains much useful information on geometry, curves, and surfaces [89]. The most recent volume includes a disk with code for all of the algorithms presented in the series.

Surface reconstruction is a common problem in computer vision [92, 232, 233, 35]. Surface reconstruction is required for fitting a surface to sparse depth values from binocular stereo and range sensors [28, 92, 232]. Sinha and Schunck [223] have proposed a two-stage algorithm for surface reconstruction. The first stage removes outliers from the depth measurements and interpolates the sparse depth values onto a uniform grid. The second stage fits weighted bicubic splines to the output from the second stage.

The section on surface segmentation is based on the work of Besl and Jain [28]. The iterative closest point algorithm was developed by Besl and McKay [29].

Exercises

13.1 Define implicit, explicit, and parametric representations of a surface. When we consider an image as a two-dimensional mathematical function, can we consider the image as a surface? If so, which of the three representations above corresponds to an image?

13.2 Given four points in space, how will you determine whether they are coplanar?

13.3 Define principal curvature, Gaussian curvature, and mean curvature for a surface. What information about the nature of a surface do you get by knowing these curvature values at a point? Do you get local information or global information about the surface from these curvature values?

13.4 What is a spline curve? Why are cubic splines frequently used to represent curves?

13.5   Polygonal meshes are commonly used to represent surfaces. What are polygonal meshes? Why are they used to represent arbitrary curved surfaces? How closely can you represent an arbitrary curved surface using a polygonal mesh?

13.6   What is a winged edge data structure? Where and why is one used?

13.7   What is a graph surface? Show that a graph surface can be modeled using a bivariate polynomial surface representation. Discuss the limitations of polynomial patch representations for surfaces.

13.8   Define tensor product surface. Compare it with the polynomial surface representation for applications in machine vision. You should consider three concrete applications in inspection, measurement, and object recognition and study the requirements of these applications. Then compare these representations for the specific task requirements.

13.9   What is the difference between surface approximation and surface interpolation? What is segmentation: approximation or interpolation? Explain your answer.

13.10 What is regularization? When should it be used in computer vision? Explain your answer using an example where regularization may be used to solve a problem effectively.

13.11 Give the mathematical formulation of a B-spline surface. How do you evaluate the quality of fit for these surfaces?

13.12 How can you use B-splines for smoothing imges? Which type of applications will be suitable for spline smoothing?

13.13 Using differential geometric characteristics, it is possible to classify the local nature of a surface as one of the eight surface types. Which characteristics are used and what are the classes?

Computer Projects

13.1 Write a program to segment images, including range images, starting with a seed that contains local regions of similar type based on the differential geometric characteristics. Use a region growing method to fit biquadratic and bicubic surfaces in an image. Select suitable region growing and region termination criterion. Apply this program to several images. Study the error characteristics of surfaces approximated by your program.

13.2 Implement an algorithm to take scattered three-dimensional points and convert them to a uniform grid representation using an interpolation algorithm. Test the algorithm using the output of a stereo algorithm.

Chapter 14

Dynamic Vision

Most biological vision systems have evolved to cope with the changing world. Machine vision systems have developed in the same way. For a computer vision system, the ability to cope with moving and changing objects, changing illumination, and changing viewpoints is essential to perform several tasks. Although early computer vision systems were concerned primarily with static scenes, computer vision systems for analyzing dynamic scenes are being designed for different applications.

The input to a dynamic scene analysis system is a sequence of image frames taken from a changing world. The camera used to acquire the image sequence may also be in motion. Each frame represents an image of the scene at a particular instant in time. The changes in a scene may be due to the motion of the camera, the motion of objects, illumination changes, or changes in the structure, size, or shape of an object. It is usually assumed that the changes in a scene are due to camera and/or object motion, and that the objects are either rigid or quasi-rigid; other changes are not allowed. The system must detect changes, determine the motion characteristics of the observer and the objects, characterize the motion using high-level abstraction, recover the structure of the objects, and recognize moving objects. In applications such as video editing and video databases, it may be required to detect *macro* changes in a sequence. These changes will partition the segment into many related segments exhibiting similar camera motion or a similar scene in a sequence.

A scene usually contains several objects. An image of the scene at a given time represents a projection of the scene, which depends on the position of the camera. There are four possibilities for the dynamic nature of the camera and world setup:

1. Stationary camera, stationary objects (SCSO)

2. Stationary camera, moving objects (SCMO)

3. Moving camera, stationary objects (MCSO)

4. Moving camera, moving objects (MCMO)

For analyzing image sequences, different techniques are required in each of the above cases. The first case is simply static-scene analysis.

In many applications, it may be necessary to process a single image to obtain the required information. This type of analysis has been the topic of discussion in the earlier chapters in this book.

Some applications require information extracted from a dynamic environment; in some cases a vision system must understand a dynamic process from a single viewpoint. In applications such as mobile robots or autonomous vehicles, a vision system must analyze an image sequence acquired while in motion. As we will see, recovering information from a mobile camera requires different techniques than those useful when the camera remains stationary.

A sequence of image frames offers much more information to aid in understanding a scene but significantly increases the amount of data to be processed by the system. The application of static-scene analysis techniques to each frame of a sequence requires an enormous amount of computation, in addition to all of the difficulties of static-scene analysis. Fortunately, research in dynamic-scene analysis has shown that the recovery of information in many cases is easier in dynamic scenes than in static scenes.

In dynamic-scene analysis, SCMO scenes have received the most attention. In analyzing such scenes, the goal is usually to detect motion, to extract masks of moving objects for recognizing them, and to compute their motion characteristics. MCSO and MCMO scenes are very important in navigation applications. MCMO is the most general and possibly the most difficult situation in dynamic scene analysis, but it is also the least developed area of computer vision.

Dynamic scene analysis has three phases:

- Peripheral

- Attentive

- Cognitive

The peripheral phase is concerned with extraction of approximate information which is very helpful in later phases of analysis. This information indicates the activity in a scene and is used to decide which parts of the scene need careful analysis. The attentive phase concentrates analysis on the active parts of the scene and extracts information which may be used for recognition of objects, analysis of object motion, preparation of a history of events taking place in the scene, or other related activities. The cognitive phase applies knowledge about objects, motion verbs, and other application-dependent concepts to analyze the scene in terms of the objects present and the events taking place.

The input to a dynamic scene analysis system is a frame sequence, represented by $F(x, y, t)$ where x and y are the spatial coordinates in the frame representing the scene at time t. The value of the function represents the intensity of the pixel. It is assumed that the image is obtained using a camera located at the origin of a three-dimensional coordinate system. The projection used in this observer-centered system may be either perspective or orthogonal.

Since the frames are usually taken at regular intervals, we will assume that t represents the tth frame of the sequence, rather than the frame taken at absolute time t.

14.1   Change Detection

Detection of changes in two successive frames of a sequence is a very important step for many applications. Any perceptible motion in a scene results in some change in the sequence of frames of the scene. Motion characteristics can be analyzed if such changes are detected. A good quantitative estimate of the motion components of an object may be obtained if the motion is restricted to a plane that is parallel to the image plane; for three-dimensional motion, only qualitative estimates are possible. Any illumination change in

a scene will also result in changes in intensity values, as will scene changes
in a TV broadcast or a movie.

Most techniques for dynamic-scene analysis are based on the detection of
changes in a frame sequence. Starting with frame-to-frame changes, a global
analysis of the sequence may be performed. Changes can be detected at
different levels: pixel, edge, or region. Changes detected at the pixel level
can be aggregated to obtain useful information with which the computational
requirements of later phases can be constrained.

In this section, we will discuss different techniques for change detection.
We will start with one of the simplest, yet one of the most useful change
detection techniques, *difference pictures*, and then discuss change detection
for edges and regions.

14.1.1 Difference Pictures

The most obvious method of detecting change between two frames is to
directly compare the corresponding pixels of the two frames to determine
whether they are the same. In the simplest form, a binary difference picture
$DP_{jk}(x, y)$ between frames $F(x, y, j)$ and $F(x, y, k)$ is obtained by:

$$DP_{jk}(x, y) = \begin{cases} 1 & \text{if } |F(x, y, j) - F(x, y, k)| > \tau \\ 0 & \text{otherwise} \end{cases} \tag{14.1}$$

where τ is a threshold.

In a difference picture, pixels which have value 1 may be considered to be
the result of object motion or illumination changes. This assumes that the
frames are properly registered. In Figures 14.1 and 14.2 we show two cases
of change detection, one due to illumination changes in a part of image and
the other due to motion of an object.

A concept discussed in Chapter 3, thresholding, will play a very important
role here also. Slow-moving objects and slowly varying intensity changes may
not be detected for a given threshold value.

Size Filter

A difference picture obtained using the above simple test on real scenes usually results in too many noisy pixels. A simple size filter is effective in elimi-

(a)   (b)   (c)

Figure 14.1: Two frames from a sequence, (a) and (b), and their difference picture, (c). Notice that the changed areas (shown in black) are due to the motion of objects. We used $\tau = 25$.

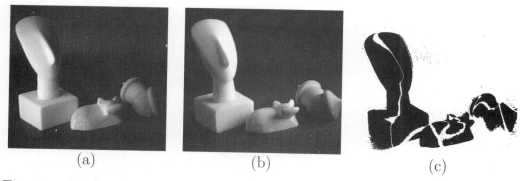

(a)   (b)   (c)

Figure 14.2: Two frames from a sequence, (a) and (b), and their difference picture, (c). Notice that the changed areas are due to the changes in the illumination in the part of the scene. We used $\tau = 25$.

nating many noisy areas in the difference picture. Pixels that do not belong to a connected cluster of a minimum size are usually due to noise and can be filtered out. Only pixels in a difference picture that belong to a 4-connected (or 8-connected) component larger than some threshold size are retained for further analysis. For motion detection, this filter is very effective, but unfortunately it also filters some desirable signals, such as those from slow or small moving objects. In Figure 14.3 we show the difference picture for the frames shown in Figure 14.1 with $\tau = 10$ and the result of the size filtering.

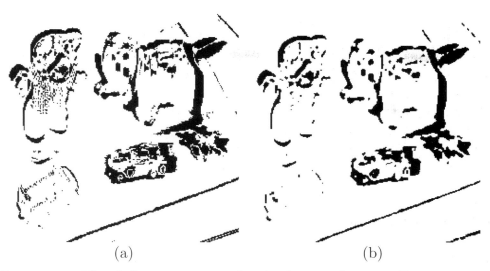

Figure 14.3: The difference picture for the frames shown in Figure 14.1 with $\tau = 10$ and the result of the size filtering are shown in (a) and (b), respectively. Notice that size filtering has removed many noisy regions in the image. All regions below 10 pixels were filtered out.

Robust Change Detection

To make change detection more robust, intensity characteristics of regions or groups of pixels at the same location in two frames may be compared using either a statistical approach or an approach based on the local approximation of intensity distributions. Such comparisons will result in more reliable change detection at the cost of added computation.

A straightforward domain-independent method for comparing regions in images is to consider corresponding areas of the frames. These corresponding areas may be the superpixels formed by pixels in nonoverlapping rectangular areas comprising m rows and n columns. The values of m and n are selected to compensate for the aspect ratio of the camera. Thus, a frame partitioned into disjoint superpixels, as shown in Figure 14.4(a), may be considered. Another possibility is to use a local mask, as in all convolutions, and compare the intensity distributions around the pixel, as shown in Figure 14.4(b).

One such method is based on comparing the frames using the likelihood

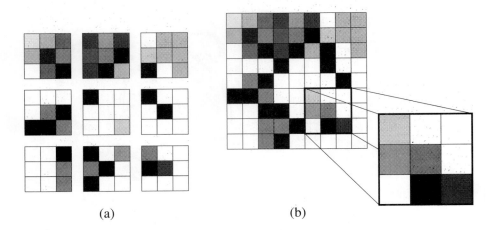

(a)   (b)

Figure 14.4: Partitioning frames for applying the likelihood ratio test. In (a) we show nonoverlapping areas, called superpixels, and (b) shows regular masks representing the local area of a pixel.

ratio. Thus, we may compute

$$\lambda = \frac{\left[\frac{\sigma_1^2 + \sigma_2^2}{2} + \left(\frac{\mu_1 - \mu_2}{2} \right)^2 \right]^2}{\sigma_1^2 * \sigma_2^2} \qquad (14.2)$$

(where μ and σ^2 denote the mean gray value and the variance for the sample areas from the frames) and then use

$$DP_{jk}(x, y) = \begin{cases} 1 & \text{if } \lambda > \tau \\ 0 & \text{otherwise} \end{cases} \qquad (14.3)$$

where τ is a threshold. The likelihood ratio test combined with the size filter works quite well for many real world scenes. In Figure 14.5, we show the results of change detection using the likelihood ratio test.

The likelihood test discussed above was based on the assumption of uniform second-order statistics over a region. The performance of the likelihood ratio test can be improved significantly by using facets and quadratic surfaces to approximate the intensity values of pixels belonging to superpixels. These-higher order approximations allow for better characterization of intensity values and result in more robust change detection.

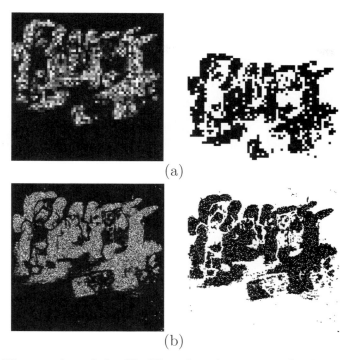

Figure 14.5: The results of the likelihood ratio test on the image pair in Fig. 14.1 for (a) superpixels and (b) regular masks.

Note that the likelihood ratio test results in detecting dissimilarities at the superpixel level. Since the tests use the likelihood ratio, they can only determine whether or not the areas under consideration have similar gray-level characteristics; information about the relative intensities of the areas is not retained. As shown later, the sign of the changes can also provide useful information for the analysis of motion.

Accumulative Difference Pictures

Small or slow-moving objects will usually result in a small number of changes using differencing approaches. A size filter may eliminate such pixels as noise. This problem of detecting small and slow-moving objects is exacerbated when using robust differencing approaches since superpixels effectively raise the size threshold for the detection of such objects.

By analyzing the changes over a sequence of frames, instead of just be-

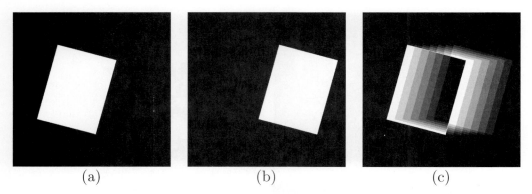

Figure 14.6: Results for change detection using accumulative difference pictures. In (a) and (b) we show the first and last frames, respectively, of a synthetic object in motion. The accumulative difference picture is given in (c).

tween two frames, this problem may be solved. An accumulative difference picture may be used to detect the motion of small and slow-moving objects more reliably. An accumulative difference picture is formed by comparing every frame of an image sequence to a reference frame and increasing the entry in the accumulative difference picture by 1 whenever the difference for the pixel, or some measure for the superpixel, exceeds the threshold. Thus, an accumulative difference picture ADP_k is computed over k frames by

$$ADP_0(x, y) = 0 \tag{14.4}$$
$$ADP_k(x, y) = ADP_{k-1}(x, y) + DP_{1k}(x, y). \tag{14.5}$$

The first frame of a sequence is usually the reference frame, and the accumulative difference picture ADP_0 is initialized to 0. Small and slow-moving objects can be detected using an ADP. In Figure 14.6, results of detecting changes using accumulative difference pictures are shown.

Difference Pictures in Motion Detection

The most attractive aspect of the difference picture for motion detection is its simplicity. In its simplest form, the difference picture is noise-prone. Changes in illumination and registration of the camera, in addition to electronic noise of the camera, can result in many false alarms. A likelihood ratio in conjunction with a size filter can eliminate most of the camera noise. Changes in illumination will create problems for all intensity-based approaches and can be handled only at a symbolic level. Misregistration of frames results in the assignment of false motion components. If the misregistration is not severe, accumulative difference pictures can eliminate it.

It should be emphasized that measuring dissimilarities at the pixel level can only detect intensity changes. In dynamic scene analysis, this is the lowest level of analysis. After such changes have been detected, other processes are required to interpret these changes. Experience has shown that the most efficient use of the difference picture is to have peripheral processes direct the attention of interpretation processes to areas of the scene where some *activity* is taking place. Approximate information about events in a scene may be extracted using some features of difference pictures.

14.1.2   Static Segmentation and Matching

Segmentation is the task of identifying the semantically meaningful components of an image and grouping the pixels belonging to such components. Segmentation need not be performed in terms of objects; some predicates based on intensity characteristics may also be used. Predicates based on intensity characteristics are usually called *features.* If an object or feature appears in two or more images, segmentation may be necessary in order to identify the object in the images. The process of identifying the same object or feature in two or more frames is called the correspondence process.

Static-scene analysis techniques can be used to segment, or at least partially segment, each frame of a dynamic sequence. Matching can then be used to determine correspondences and detect changes in the location of corresponding segments. Cross-correlation and Fourier domain features have been used to detect cloud motion. Several systems have been developed which segment each frame of a sequence to find regions, corners, edges, or other features in the frames. The features are then matched in consecutive frames to detect any displacement. Some restriction on the possible matches for a feature can be achieved by predicting the new location of the feature based on the displacements in previous frames.

The major difficulty in the approaches described above is in segmentation. Segmentation of an image of a static scene has been a difficult problem. In most cases, a segmentation algorithm results in a large number of features in each frame, which makes the correspondence problem computationally quite expensive. Moreover, it is widely believed that motion detection can be used to produce better segmentation, as opposed to the techniques of segmentation and matching used to determine motion above.

14.2   Segmentation Using Motion

The goal of many dynamic-scene analysis systems is to recognize moving objects and to find their motion characteristics. If the system uses a stationary camera, segmentation generally involves separating moving components in the scene from stationary components. Then the individual moving objects are identified based either on their velocity or on other characteristics. For systems using a moving camera, the segmentation task may be the same as above or may include further segmentation of the scene's stationary components by exploiting the camera motion. Most research efforts for the segmentation of dynamic scenes have assumed a stationary camera.

Researchers in perception have known for a long time that motion cues are helpful for segmentation. Computer vision techniques for segmenting SCMO scenes perform well compared to techniques for segmenting stationary scenes. Segmentation into stationary and nonstationary components, in a system using a moving camera, has only recently received attention. One problem in segmenting moving-observer scenes is that every surface in the scene has image plane motion. This is precisely what can be used to aid the separation of moving and stationary objects in stationary-camera scenes. For segmenting moving-camera scenes, the motion assigned to components in the images due to the motion of the camera should be removed. The fact that the image motion of a surface depends both on the surface's distance from the camera and on the structure of the surface complicates the situation.

Segmentation may be performed using either region-based or edge-based approaches. In this section, some approaches for the segmenting of dynamic scenes are discussed.

14.2.1   Time-Varying Edge Detection

As a result of the importance of edge detection in static scenes, it is reasonable to expect that time-varying edge detection may be very important in dynamic-scene analysis. In segment-and-match approaches, efforts are

Figure 14.7: In (a) and (b), two frames of a sequence are shown. In (c), edges are detected using the time-varying edge detector.

wasted on attempting to match static features to moving features. These static features are obstacles to extracting motion information. If only moving features are detected, the computation needed to perform matching may be substantially reduced.

A moving edge in a frame is an edge, *and* moves. Moving edges can be detected by combining the temporal and spatial gradients using a logical AND operator. This AND can be implemented through multiplication. Thus, the time-varying edginess of a point in a frame $E(x, y, t)$ is given by

$$E_t(x, y, t) = \frac{dF(x, y, t)}{dS} \cdot \frac{dF(x, y, t)}{dt} \tag{14.6}$$
$$= E(x, y, t) \cdot D(x, y) \tag{14.7}$$

where $dF(x, y, t)/dS$ and $dF(x, y, t)/dt$ are, respectively, the spatial and temporal gradients of the intensity at point (x, y, t). Various conventional edge detectors can be used to compute the spatial gradient, and a simple difference can be used to compute the temporal gradient. In most cases, this edge detector works effectively. By applying a threshold to the product, rather than first differencing and then applying an edge detector or first detecting edges and then computing their temporal gradient, this method overcomes the problem of missing slow-moving or weak edges. See Figures 14.7 and 14.8.

As shown in Figure 14.8, this edge detector will respond to slow-moving edges that have good edginess and to poor edges that are moving with appreciable speed. Another important fact about this detector is that it does

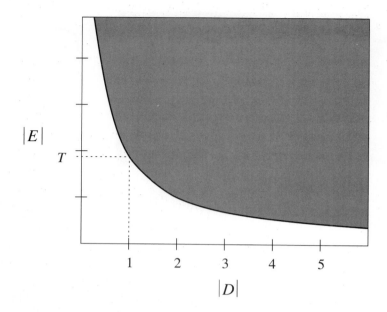

Figure 14.8: A plot showing the performance of the edge detector. Note that slow-moving edges will be detected if they have good contrast, and that poor-contrast edges will be detected if they move well.

not assume any displacement size. The performance of the detector is satisfactory even when the motion of an edge is very large.

14.2.2   Stationary Camera

Using Difference Pictures

Difference and accumulative difference pictures find the areas in a scene which are changing. An area is usually changing due to the movement of an object. Although change detection results based on difference pictures are sensitive to noise, the areas produced by a difference picture are a good place from which to start segmentation. In fact, it is possible to segment a scene with very little computation using accumulative difference pictures. In this section, such an approach is discussed.

Let us define absolute, positive, and negative difference pictures and accumulative difference pictures as follows:

$$DP_{12}(x, y) = \begin{cases} 1 & \text{if } |F(x, y, 1) - F(x, y, 2)| > T \\ 0 & \text{otherwise} \end{cases} \quad (14.8)$$

$$PDP_{12}(x, y) = \begin{cases} 1 & \text{if } F(x, y, 1) - F(x, y, 2) > T \\ 0 & \text{otherwise} \end{cases} \quad (14.9)$$

$$NDP_{12}(x, y) = \begin{cases} 1 & \text{if } F(x, y, 1) - F(x, y, 2) < T \\ 0 & \text{otherwise} \end{cases} \quad (14.10)$$

$$AADP_n(x, y) = AADP_{n-1}(x, y) + DP_{1n}(x, y) \quad (14.11)$$

$$PADP_n(x, y) = PADP_{n-1}(x, y) + PDP_{1n}(x, y) \quad (14.12)$$

$$NADP_n(x, y) = NADP_{n-1}(x, y) + NDP_{1n}(x, y). \quad (14.13)$$

Depending on the relative intensity values of the moving object and the background being covered and uncovered, PADP and NADP provide complementary information. In either the PADP or NADP, the region due to the motion of an object continues to grow after the object has been completely displaced from its projection in the reference frame, while in the other, it stops growing. The area in the PADP or NADP corresponds to the area covered by the object image in the reference frame. The entries in this area continue to increase in value, but the region stops growing in size. Accumulative difference pictures for a synthetic scene are shown in Figure 14.9. A test to determine whether or not a region is still growing is needed in order to obtain a mask of an object. The mask can then be obtained from the accumulative difference picture where the object's area stops growing after the object is displaced from its projection. In its simplest form, this approach has one obvious limitation. Masks of moving objects can be extracted only after the object has been completely displaced from its projection in the reference frame. However, it appears that properties of difference and accumulative difference pictures can be used to segment images in complex situations, such as running occlusion. To prevent running occlusions from disrupting segmentation, the segmentation process should not wait for an object's current position to be completely displaced from its projection in the reference frame. Regions in the accumulative difference pictures can be monitored as opposed to monitoring the reference frame projections of objects. Simple tests on the rate of region growth and on the presence of *stale* entries allow a system to determine which regions are eventually going to mature and result

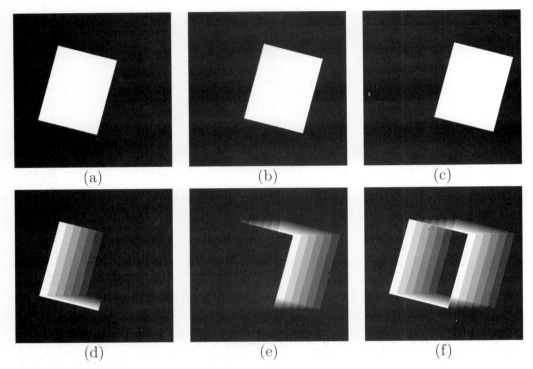

Figure 14.9: (a)–(c) show frames 1, 5, and 7 of a scene containing a moving object. The intensity-coded positive, negative, and absolute ADPs are shown in parts (d), (e), and (f), respectively.

in a mask for an object in the reference frame. Early determination of reference frame positions of objects, and hence, extraction of masks for objects, allows a system to take the action necessary to prevent running occlusion.

14.3   Motion Correspondence

Given two frames of a sequence, one can analyze them to determine features in each frame. To determine the motion of objects, one can establish the correspondence among these features. The correspondence problem in motion is similar to the correspondence problem in stereo. In stereo, the major constraint used is the epipolar constraint. However, in motion, other constraints must be used. In the following we describe a constraint propagation approach to solve the correspondence problem. Since this approach is similar to the problem for stereo, and for historical reasons, we present the formulation similar to that for stereo.

Relaxation Labeling

In many applications, we are given a set of labels that may be assigned to objects that may appear in the "world." Possible relationships among different objects in this world and the conditions under which a certain set of labels may or may not be applied to a set of objects is also known. The relationships among the objects in the image may be found using techniques discussed in earlier chapters. Now, based on the knowledge about the labels in the domain, proper labels must be assigned to the objects in the image. This problem is called the *labeling problem*.

The labeling problem may be represented as shown in Figure 14.10. Each node represents an object or entity which should be assigned a label. The arcs connecting nodes represent relations between objects. This figure represents the observed entities and relations among them in a given situation. We have to assign labels to each entity based on the label set and the constraints among the labels for the given domain.

Assume that we have a processor at each node. We define sets R, C, L, and P for each node. The set R contains all possible relations among the nodes. The set C represents the compatibility among these relations. The compatability among relations helps in constraining the relationships and labels for each entity in an image. The set L contains all labels that can be assigned to nodes, and the set P represents the set of possible levels that can be assigned to a node at any instant in computation. Each processor knows the label of its node and all nodes which are connected to it. It also knows all relations involving its node and the sets R and C. Assume that in the first iteration the possible label set P_i^1 of node i is L for all i. In other words, all nodes are assigned all possible labels initially. The labeling process should then iteratively remove invalid labels from P_i^k to give P_i^{k+1}. Since at any stage labels are discarded considering only the current labels of the node, its relations with other nodes, and the constraints, each processor has sufficient information to refine its label set P_i^k. Thus, it is possible for all processors to work synchronously. Note that at any time a processor uses only information directly available to it, that is, information pertaining

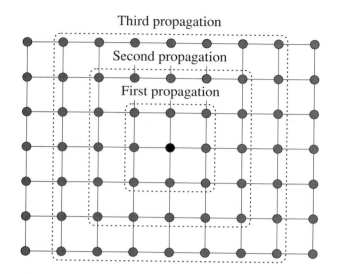

Figure 14.10: Parallel propagation in graphs.

only to its object. Each iteration, however, propagates the effect through its neighbor or related nodes, to other nodes that are not directly related. The circle of influence of a node increases with each iteration. This propagation of influence results in global consistency through direct local effects.

In most applications, some knowledge about the objects is available at the start of the labeling process. Segmentation, or some other process which takes place before labeling, often gives information that can be used to refine the initial set P_i for a node. This knowledge can be used to refine the initial label sets for objects. The labeling process is then used to further refine these sets to yield a unique label for each object. The labeling problem now may be considered in a slightly different form than the above. Based on some unary relations, a label set P_i^1 can be assigned to an object. The correct label is uncertain. However, a confidence value can be assigned to each label $l_k \in P_i^1$. The confidence value, like a subjective probability, indicates a belief that the entity may be assigned this label based on the available evidence in the image. Thus, for each element $l_k \in P_i$, a nonnegative probability p_{ik} represents the confidence that the label l_k is the correct label for node i. This confidence value may be considered the membership value if approaches from fuzzy set theory are used in constraint propogation.

The task of the labeling process is to use the constraints to refine the confidence value for each label. The confidence value p_{ik} is influenced by the confidence values in the labels of the connected nodes. Thus, in the tth iteration, the confidence value p_{ik}^t for the label l_k at node i is the function of the confidence value p_{ik}^{t-1} and the confidence values of the labels of all directly related nodes. In each iteration a node looks at the labels of all its related nodes and then uses the known constraints to update the confidence in its labels. The process may terminate either when each node has been assigned a unique label or when the confidence values achieve a steady state. Note that in place of just the presence or absence of a label, there is now a continuum of confidence values in a label for an object.

The above process, commonly called the *relaxation labeling process*, attempts to decide which of the possible interpretations is correct on the basis of local evidence. Interestingly, though, the final interpretation is globally correct. In each iteration, the confidence in a label is directly influenced only by directly related nodes. However, this influence is propagated to other nodes in later iterations. The sphere of influence increases with the number of iterations. In relaxation labeling, the constraints are specified in terms of compatibility functions. Suppose that objects O_i and O_j are related by R_{ij}, and under this relation labels L_{i_k} and L_{j_l} are highly "likely" to occur. The knowledge about the likelihood of these labels can be expressed in terms of a function that will increase the confidence in these labels for the objects under consideration. In such a situation, the presence of L_{i_k} at O_i encourages the assignment of L_{j_l} to O_j. It is also possible that the incompatibility of certain labels can be used to discourage labels by decreasing their confidence values.

In the following, we discuss an algorithm that uses relaxation labeling to determine disparity values in images. The algorithm to determine optical flow, discussed later in this chapter, also is an example of relaxation labeling.

Disparity Computations as Relaxation Labeling

The matching problem is to pair a point $p_i = (x_i, y_i)$ in the first image with a point $p_j = (x_j, y_j)$ in the second image. The disparity between these points is the displacement vector between the two points:

$$d_{ij} = (x_i - x_j, y_i - y_j). \tag{14.14}$$

The result of matching is a set of conjugate pairs.

In any kind of matching problem, there are two questions that must be answered:

- How are points selected for matching? In other words, what are the features that are matched?

- How are the correct matches chosen? What constraints, if any, are placed on the displacement vectors?

Three properties guide matching:

Discreteness, which is a measure of the distinctiveness of individual points

Similarity, which is a measure of how closely two points resemble one another

Consistency, which is a measure of how well a match conforms to nearby matches

The property of discreteness means that features should be isolated points. For example, line segments would not make good features since a point can be matched to many points along a line segment. Discreteness also minimizes expensive searching by reducing the problem of analyzing image disparities to the problem of matching a finite number of points.

The set of potential matches form a bipartite graph, and the matching problem is to choose a (partial) covering of this graph. As shown in Figure 14.11, initially each node can be considered as a match for each node in the other partition. Using some criterion, the goal of the correspondence problem is to remove all other connections except one for each node. The property of similarity indicates how close two potential matching points are to one another; it is a measure of affinity. Similarity could be based on any property of the features that are selected to implement discreteness.

The property of consistency is implied by the spatial continuity of surfaces in the scene and assumes that the motion is well behaved. Consistency allows the obvious matches to improve the analysis of more difficult matches. Some points are sufficiently distinct and similar that it is easy to match them; this match can assist in matching nearby points.

The discrete feature points can be selected using any corner detector or a feature detector. One such feature detector is the Moravec interest operator.

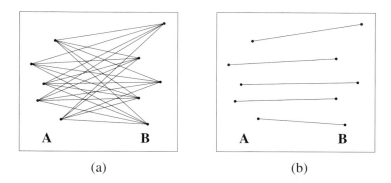

Figure 14.11: (a) A complete bipartite graph. Here each node in group A has a connection with each node in group B. Using a characteristic of nodes (points) and some other knowledge, a correspondence algorithm must remove all but one connection for each node, as shown in (b).

This operator detects points at which intensity values are varying quickly in at least one direction. This operator can be implemented in the following steps:

1. Compute sums of the squares of pixel differences in four directions (horizontal, vertical, and both diagonals) over a 5×5 window.

2. Compute the minimum value of these variances.

3. Suppress all values that are not local maxima.

4. Apply a threshold to remove weak feature points.

Any feature detector can be used in place of the above operator. One can use a corner detector or computed curvature values at every point and select high curvature points as features.

Next one must pair each feature point in the first image with all points in the second image within some maximum distance. This will eliminate many connections from the complete bipartite graph. The connections removed are those that are between points far away in two images and hence unlikely to be candidate matches. Each node a_i has position (x_i, y_i) in the first image and a set of possible labels (disparity vectors). The disparity labels are displacement vectors or the undefined disparity, which allows some feature points to remain unmatched.

The initial probabilities of a match are computed using a measure of similarity between the feature points in the two images. A good measure is the sum of the squares of the pixel differences, s_i, in corresponding windows. The following approach may be used for assigning these probabilities.

Let l be a candidate label at a point. This label represents a disparity vector at the point. First we compute $w_i(l)$, which represents the similarity between the point (x_i, y_i) and its potential match at disparity l.

$$w_i(l) = \frac{1}{1 + cs_i(l)}, \tag{14.15}$$

where $s_i(l)$ is the sum of the squared differences corresponding to label l, and c is some positive constant. The probability that this point has undefined disparity is obtained by first defining

$$p_i^0(\text{undefined}) = 1 - \max(w_i(l)). \tag{14.16}$$

This probability is determined based on the strength of the most similar point for (x_i, y_i). If there are no strongly similar points, then it is likely that the point has no match in this image. The probabilities of the various matches (labels) are

$$p_i(l|i) = \frac{w_i(l)}{\sum w_i(l')}, \tag{14.17}$$

where $p_i(l|i)$ is the conditional probability that a_i has label l given a_i is matchable, and the sum is over all labels l' excluding the "undefined" label. The probability estimates are refined using the consistency property and iterative relaxation algorithm. In this approach, the labels at each node are strengthened or weakened based on the labels of the neighboring nodes in that iteration. The most important property used here is that all disparities should be similar in a given neighborhood. Thus, similar disparities of nodes in a neighborhood should strengthen each other and dissimilar ones should be weakened. This is accomplished using the following approach.

Let us consider probability for disparity vectors of all neighbors of a_i. For each neighbor, sum the probability of labels (disparities) that are close to, or similar to, the disparity of a_i:

$$q_i^k(l) = \sum_{\substack{\text{Neighbors} \\ \text{of } a_i}} \sum_{\substack{\text{Nearby} \\ \text{disparities}}} p_j^k(l'). \tag{14.18}$$

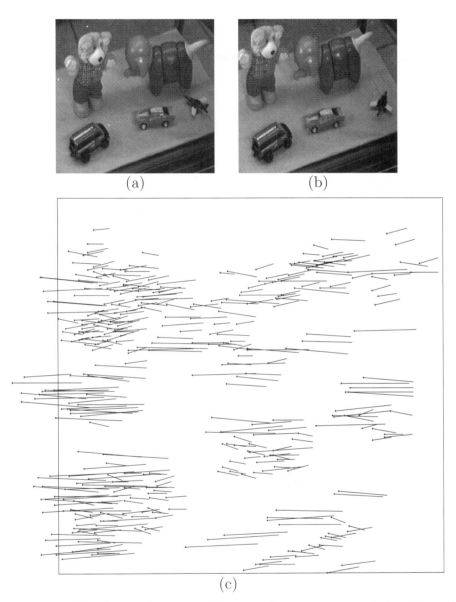

Figure 14.12: This figure shows two frames of a sequence and the disparities of the matched feature points (shown magnified by a factor of 5) after applying the relaxation labeling algorithm.

The probabilities are now refined with the iterative calculation:

$$p_i^{k+1}(l) = p_i^k(A + Bq_i^k(l)) \qquad (14.19)$$

for constants A and B. Constants A and B are selected to control the rate of convergence of the algorithm. The updated probabilities must be normalized. Usually, a good solution is obtained after only a few iterations. To speed up the algorithm, matches with low probability are removed.

In Figure 14.12, we show two frames of a sequence and disparities found using the above algorithm. Interested readers should see [21] for an in-depth analysis of disparity calculations.

14.4   Image Flow

Image flow is the distribution of velocity, relative to the observer, over the points of an image. Image flow carries information which is valuable for analyzing dynamic scenes. Several methods for dynamic-scene analysis have been proposed which assume that image flow information is available. Unfortunately, however, although image flow has received a significant amount of attention from researchers, the techniques developed for computing image flow do not produce results of the quality which will allow the valuable information to be recovered. Current methods for computing image flow, information which is critical in optical flow, and the recovery of such information are discussed in this section.

Definition 14.1 *Image flow is the velocity field in the image plane due to the motion of the observer, the motion of objects in the scene, or apparent motion which is a change in the image intensity between frames that mimics object or observer motion.*

14.4.1   Computing Image Flow

Image flow is determined by the velocity vector of each pixel in an image. Several schemes have been devised for calculating image flow based on two or more frames of a sequence. These schemes can be classified into two general categories: feature-based and gradient-based. If a stationary camera is used, most of the points in an image frame will have zero velocity. This is assuming that a very small subset of the scene is in motion, which is usually true. Thus, most applications for image flow involve a moving camera.

14.4.2 Feature-Based Methods

Feature-based methods for computing image flow first select some features in the image frames and then match these features and calculate the disparities between frames. As discussed in an earlier section, the correspondence may be solved on a stereo image pair using relaxation. The same approach may be used to solve the correspondence problem in dynamic scenes. However, the problem of selecting features and establishing correspondence is not easy. Moreover, this method only produces velocity vectors at sparse points. This approach was discussed above as disparity analysis.

14.4.3 Gradient-Based Methods

Gradient-based methods exploit the relationship between the spatial and temporal gradients of intensity. This relationship can be used to segment images based on the velocity of points.

Suppose the image intensity at a point in the image plane is given as $E(x, y, t)$. Assuming small motion, the intensity at this point will remain constant, so that

$$\frac{dE}{dt} = 0. \tag{14.20}$$

Using the chain rule for differentiation, we see that

$$\frac{\partial E}{\partial x}\frac{dx}{dt} + \frac{\partial E}{\partial y}\frac{dy}{dt} + \frac{\partial E}{\partial t} = 0. \tag{14.21}$$

Using

$$u = \frac{dx}{dt} \tag{14.22}$$

and

$$v = \frac{dy}{dt}, \tag{14.23}$$

the relationship between the spatial and temporal gradients and the velocity components is:

$$E_x u + E_y v + E_t = 0. \tag{14.24}$$

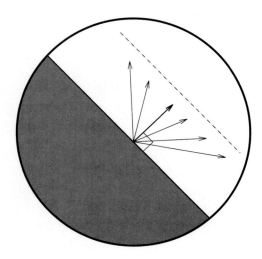

Figure 14.13: If one sees a point using a tube such that only one point is visible, then motion of the point cannot be determined. One can only get the sense of the motion, not the components of the motion vector. This problem is commonly called the *aperture problem.*

In the above equation, E_x, E_y, and E_t can be computed directly from the image. Thus, at every point in an image, there are two unknowns, u and v, and only one equation. Using information only at a point, image flow cannot be determined. This can be explained using Figure 14.13. This is known as the *aperture problem.* The velocity components at a point cannot be determined using the information at only one point in the image without making further assumptions.

It can be assumed that the velocity field varies smoothly over an image. Under this assumption, an iterative approach for computing image flow using two or more frames can be developed. The following iterative equations are used for the computation of image flow. These equations can be derived using the variational approach discussed below.

$$u = u_{\text{average}} - E_x \frac{P}{D} \tag{14.25}$$

$$v = v_{\text{average}} - E_y \frac{P}{D} \tag{14.26}$$

where

$$P = E_x u_{\text{average}} + E_y v_{\text{average}} + E_t \tag{14.27}$$

and

$$D = \lambda^2 + E_x^2 + E_y^2. \tag{14.28}$$

In the above equations E_x, E_y, E_t, and λ represent the spatial gradients in the x and y directions, the temporal gradient, and a constant multiplier, respectively. When only two frames are used, the computation is iterated over the same frames many times. For more than two frames, each iteration uses a new frame.

An important fact to remember about gradient-based methods is that they assume a linear variation of intensities and compute the point-wise velocities under this assumption. It is typically expected that this assumption is satisfied at edge points in images and, hence, the velocity can be computed at these points. The smoothness constraint is not satisfied at the boundaries of objects because the surfaces of objects may be at different depths. When overlapping objects are moving in different directions, the constraint will also be violated. These abrupt changes in the velocity field at the boundaries cause problems. To remove these problems, some other information must be used to refine the optical flow determined by the above method.

14.4.4 Variational Methods for Image Flow

Recall that the aperture problem says that the image-flow velocity at a point in the image plane cannot be computed by only using the changes in the image at that point without using information from other sources. Image flow can be computed using variational methods that combine the image flow constraint equation with an assumption about the smoothness of the image-flow velocity field. The image-flow constraint equation is

$$E_x u + E_y v + E_t = 0 \tag{14.29}$$

where u and v are the x and y components of the image-flow, respectively, and E_x, E_y, and E_t are the spatial and temporal derivatives of the image intensity. For a smoothness measure, use the sum of the squared magnitudes of each image flow component as the integrand in the regularization term:

$$\iint \left[\left(\frac{\partial u}{\partial x}\right)^2 + \left(\frac{\partial u}{\partial y}\right)^2 + \left(\frac{\partial v}{\partial x}\right)^2 + \left(\frac{\partial v}{\partial y}\right)^2 \right] dx \, dy. \tag{14.30}$$

Combine the smoothness measure with a measure of deviations from the problem constraint weighted by a parameter that controls the balance between deviations from the image-flow constraint and deviations from smoothness:

$$\int\int\left\{(E_xu+E_yv+E_t)^2 + \nu^2\left[\left(\frac{\partial u}{\partial x}\right)^2 + \left(\frac{\partial u}{\partial y}\right)^2 + \left(\frac{\partial v}{\partial x}\right)^2 + \left(\frac{\partial v}{\partial y}\right)^2\right]\right\}\,dx\,dy.$$

$$(14.31)$$

Use the calculus of variations to transform this norm into a pair of partial differential equations

$$\nu^2\nabla^2u \;=\; E_x^2u + E_xE_yv + E_xE_t \qquad\qquad (14.32)$$

$$\nu^2\nabla^2v \;=\; E_xE_yu + E_y^2v + E_yE_t. \qquad\qquad (14.33)$$

Use finite difference methods to replace the Laplacian in each equation with a weighted sum of the flow vectors in a local neighborhood, and use iterative methods to solve the difference equations.

14.4.5   Robust Computation of Image Flow

The motion information can be unreliable at motion boundaries, since the process of occluding or disoccluding the background does not obey the image-flow constraints of Equation 14.24. The incorrect motion constraints are outliers. Robust methods avoid the problems caused by incorrect motion constraints at boundaries.

Image flow can be computed using robust regression with least-median-squares regression. The least-median-squares algorithm is applied over successive neighborhoods. Within each neighborhood, the algorithm tries all possible pairs of constraint lines. The intersection of each pair of constraint lines is computed and the median of the square of the residuals is computed to assign a cost to each estimate. Each intersection and its cost are stored. After all possible pairs have been tried, the intersection corresponding to the minimum cost is used as the estimate for the image-flow velocity for the center of the neighborhood.

There are several steps in computing the intersection of the constraint lines and the residuals. The constraint lines are represented in polar form using the distance d of the constraint line from the origin in velocity space and the angle α of the image gradient:

$$d = \rho \cos(\alpha - \beta), \qquad (14.34)$$

where $\rho(x, y)$ and $\beta(x, y)$ are the speed and direction of motion, respectively. Let the coordinates of the first constraint line be d_1 and α_1, and the coordinates of the second constraint line be d_2 and α_2. The position of the intersection in rectangular coordinates is

$$x = \frac{d_1 \sin \alpha_2 - d_2 \sin \alpha_1}{\sin(\alpha_1 - \alpha_2)} \qquad (14.35)$$

$$y = \frac{d_2 \cos \alpha_1 - d_1 \cos \alpha_2}{\sin(\alpha_1 - \alpha_2)}. \qquad (14.36)$$

The fit of the model to the constraint lines is the median of the squared residuals:

$$\operatorname*{med}_i (r_i^2). \qquad (14.37)$$

The r residual between the motion estimate and each constraint line is the perpendicular distance of the constraint line from the estimate x and y. The residual is given by

$$r = x \cos \alpha + y \sin \alpha. \qquad (14.38)$$

The position of the intersection of the pair of constraint lines, given by Equation 14.36, is a candidate solution. The median of the squared residuals of the constraint lines with respect to the candidate is computed and saved, along with the candidate, as a potential solution. The median of the squared residuals is the median of the square of the perpendicular distance of each constraint line in the neighborhood from the candidate.

The typical neighborhood size is 5×5. An $n \times n$ neighborhood contains n^2 constraint lines. The number of possible pairs of constraint lines in an $n \times n$ neighborhood would be

$$\frac{n^2(n^2 - 1)}{2}. \qquad (14.39)$$

A 5×5 neighborhood would yield 300 pairs. It is not necessary to try all possible pairs if computation time is restricted. Rousseeuw and Leroy [207, p. 198] provide a table showing the number of trials that must be run to fit a model with p parameters and 95% confidence to data sets with various percentages of outliers. Assume that at most 50% of the constraints

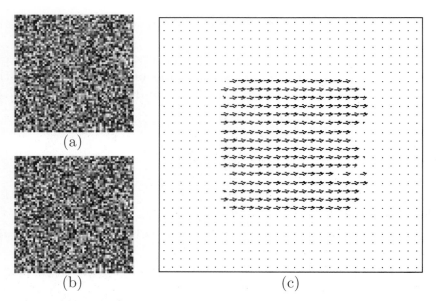

Figure 14.14: Image flow computed using the least-median-squares algorithm. Two frames of a synthetic image sequence were computed by filling a 64×64 background image and a 32×32 foreground image with pixels from a uniform random number generator. The foreground image was overlayed on the center of the background image to create the first frame and overlayed one pixel to the right to create the second frame.

in the neighborhood will be outliers. The local estimate of the image-flow velocity field requires only two constraint lines. From the table published by Rousseeuw and Leroy, only 11 pairs of constraints would have to be tried to provide a consistent estimate with 95% confidence. Using more pairs would increase the odds of finding a consistent estimate. If fewer than all possible pairs of constraint lines are used, the pairs should be selected so that the constraints in each pair are far apart. This reduces the problems with ill-conditioning caused by intersecting constraint lines that have nearly the same orientation. A preprogrammed scheme could be used for selecting the constraint line pairs in each neighborhood. The results using this approach are shown in Figure 14.14.

14.4.6 Information in Image Flow

Many researchers have studied the types of information that can be extracted from an image-flow field, assuming that high-quality image flow has been computed. Let us assume an environment which contains rigid, stationary surfaces at known depths, and that the observer, the camera, moves through this world. The image flow can be derived from the known structure. Thus, the structure of the environment can be obtained, in principle, from the computed image-flow field.

Areas with smooth velocity gradients correspond to single surfaces in the image and contain information about the structure of the surface. Areas with large gradients contain information about occlusion and boundaries, since only different objects at different depths can move at different speeds relative to the camera. Using an observer-based coordinate system, a relationship between the surface orientation and the smooth velocity gradients can be derived. The orientation is specified with respect to the direction of motion of the observer.

The translational component of object motion is directed toward a point in the image called the focus of expansion (FOE) when the observer is approaching or the focus of contraction when the observer is receding; see Figure 14.15. This point is the intersection of the direction of object motion in the image plane. Surface structure can be recovered from the first and second spatial derivatives of the translational component. The angular velocity is fully determined by the rotational component.

The importance of the FOE for recovering structure from the translational components of image flow encouraged several researchers to develop methods for determining the FOE. If the FOE is correctly determined, it may be used for computing the translational components of image flow. Since all flow vectors meet at the FOE, their direction is already known and only their magnitude remains to be computed. Thus, the two-dimensional problem of the computation of image flow is reduced to a one-dimensional problem. This fact has been noted by many researchers. However, it has not been applied, possibly due to the uncertainty in the proposed approaches for locating the FOE in real scenes.

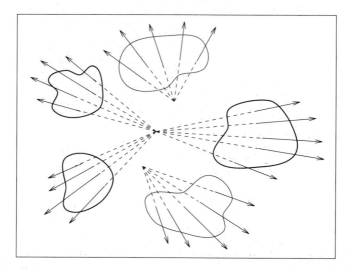

Figure 14.15: The velocity vectors for the stationary components of a scene, as seen by a translating observer, meet at the focus of expansion (FOE).

14.5   Segmentation Using a Moving Camera

If the camera is moving, then every point in the image has nonzero velocity relative to it.[1] The velocity of points relative to the camera depends both on their own velocity and on their distance from the camera. Difference picture–based approaches may be extended for segmenting moving camera scenes. Additional information will be required, however, to decide whether the motion at a point is due solely to its depth or is due to a combination of its depth and its motion. Gradient-based approaches will also require additional information.

If the camera's direction of motion is known, then the FOE with respect to the stationary components in a scene can easily be computed. The FOE will have coordinates

$$x'_f = \frac{dx}{dz} \tag{14.40}$$

and

$$y'_f = \frac{dy}{dz} \tag{14.41}$$

---

[1]With the exception of the pathological case where object points are moving with the same velocity as the camera.

in the image plane, where dx, dy, dz is the camera displacement between frames. As discussed in a later section, the velocity vectors of all the stationary points in a scene project onto the image plane so that they intersect at the FOE. A transformation with respect to the FOE may be used to simplify the task of segmentation. The ego-motion polar (EMP) transformation of an image transforms a frame $F(x', y', t)$ into $E(r', \theta, t)$ using

$$E(r', \theta, t) = F(x', y', t) \tag{14.42}$$

where

$$r' = \sqrt{(x' - x'_f)^2 + (y' - y'_f)^2} \tag{14.43}$$

and

$$\theta = \tan^{-1} \left(\frac{y' - y'_f}{x' - x'_f} \right). \tag{14.44}$$

In EMP space, stationary points are displaced only along the θ axis between the frames of an image sequence, while points on moving objects are displaced along the r' axis as well as the θ axis. Thus, the displacement in the EMP space may be used to segment a scene into its stationary and nonstationary components. In Figure 14.16, three frames of a sequence acquired by a moving camera are shown. The results of the segmentation are shown in Figure 14.17.

A method for representing object motion in images, acquired by a moving camera, is derived from the fact that all of the velocity vectors of stationary objects in a scene acquired by a translating observer intersect at the FOE. An image frame may be transformed, with respect to the FOE, to a second frame in which the abscissa is r' and the ordinate is θ. Under this transformation, it is possible to segment a dynamic scene into its moving and stationary components, as discussed earlier.

14.5.1 Ego-Motion Complex Log Mapping

More information about moving, as well as stationary, objects may be extracted using a complex logarithmic mapping (CLM) rather than the simple polar mapping. Let us define

$$\omega = \log \alpha, \tag{14.45}$$

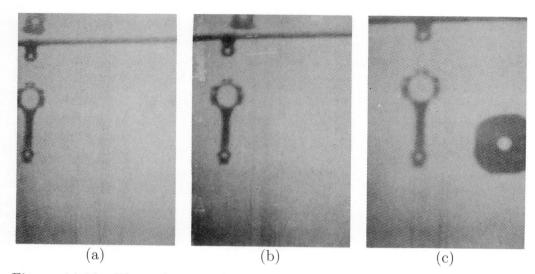

(a)   (b)   (c)

Figure 14.16: Three frames of a scene acquired by a moving camera are shown in (a), (b), and (c).

Figure 14.17: Moving objects are segmented from stationary objects, in the frames shown in Figure 14.16, using EMP segmentation. Moving objects appear brighter.

where ω and α are complex variables:

$$\alpha = x' + jy' = r'(\cos\theta + j\sin\theta) = r'e^{j\theta} \qquad (14.46)$$

and

$$\omega = u(z) + jv(z). \qquad (14.47)$$

Under this transformation, it can be shown that

$$u(r', \theta) = \log r' \qquad (14.48)$$

$$v(r', \theta) = \theta. \qquad (14.49)$$

The above results state that if the observer is moving, the horizontal displacement of a stationary point in CLM space depends only on the depth of the point and, furthermore, the vertical displacement is zero. This fact is very useful, not only in segmenting dynamic scenes into moving and stationary components, but also in determining the depth of stationary points. It has been shown that the retino-striate mapping can be approximated by a complex log function. This mapping is considered to be responsible for size, rotation, and projection invariances in biological systems. In the following discussion it is shown that if the mapping is obtained with respect to the FOE, rather than the center of the image, some other advantages can be achieved.

14.5.2 Depth Determination

First assume a camera is translating along its optical axis in a static world. For a stationary point in the environment, with real-world coordinates (x, y, z) relative to the observer at a time instant, the perspective projection, (x', y'), of this point onto the image plane is given by

$$x' = \frac{x}{z} \qquad (14.50)$$

$$y' = \frac{y}{z} \qquad (14.51)$$

assuming that the projection plane is parallel to the x–y plane at $z = 1$. The latter assumption simplifies the derivation without loss of generality. For translational motion along the direction of the gaze of the observer, the relationship between the distance r' of the projection of the point from the center of the image and the distance z of the scene point from the observer is

$$\frac{dr'}{dz} = \frac{d\sqrt{x'^2 + y'^2}}{dz} = -\frac{r'}{z}.$$ (14.52)

By the chain rule for differentiation,

$$\frac{du}{dz} = \frac{du}{dr'} \frac{dr'}{dz},$$ (14.53)

and from Equation 14.48,

$$\frac{du}{dr'} = \frac{1}{r'}.$$ (14.54)

Therefore, we have

$$\frac{du}{dz} = \frac{1}{r'} \cdot \left(-\frac{r'}{z}\right) = -\frac{1}{z}.$$ (14.55)

Similarly, to find dv/dz,

$$\frac{d\theta}{dz} = \frac{d(\tan^{-1}\frac{y'}{x'})}{dz} = 0$$ (14.56)

and

$$\frac{dv}{dz} = \frac{dv}{d\theta} \frac{d\theta}{dz} = 0.$$ (14.57)

In Equation 14.55 we see that the depth, z, of a point can be determined from the horizontal displacement, du, in CLM space for that point and from the velocity, dz, of the observer. This is a formalization of the observable phenomenon that near objects appear to get bigger faster than far objects, as you approach them. Furthermore, the axial movement of the observer will result only in a horizontal change in the mapping of the image points, since $dv/dz = 0$. There will be no vertical movement of the mapped points. Thus, the correspondence of points between the two mapped images will become easier, since there is only a horizontal change in the points' locations over time. Now, assuming that there is sufficient control of the camera to be able to determine the amount of its movement, both variables necessary to determine image depths are readily available. Thus, it is possible to recover depth, in principle, if the camera motion is along its optical axis.

In real-life applications of machine vision it is not always possible to constrain the motion of the camera to be along the optical axis. Thus, the approach must be extended to arbitrary camera motion. To see that the depth can be recovered for an arbitrary translational motion of the camera, let us assume that the transform is taken with respect to the point (a, b) in the image plane. Then

$$r' = \sqrt{(x' - a)^2 + (y' - b)^2} \tag{14.58}$$

$$u = \log r' = \log \sqrt{(x' - a)^2 + (y' - b)^2}. \tag{14.59}$$

Now

$$\frac{du}{dz} = \frac{d}{dz} \log r' = \frac{1}{r'} \frac{dr'}{dz}. \tag{14.60}$$

Let us substitute for x' and y' from Equations 14.50 and 14.51, and evaluate dr'/dz.

$$\frac{dr'}{dz} = \frac{d\sqrt{(\frac{x}{z} - a)^2 + (\frac{y}{z} - b)^2}}{dz} \tag{14.61}$$

$$= \frac{1}{2\sqrt{(\frac{x}{z} - a)^2 + (\frac{y}{z} - b)^2}}$$
$$\cdot \left[2\left(\frac{x}{z} - a\right) \cdot \frac{z\frac{dx}{dz} - z}{z^2} + 2\left(\frac{y}{z} - b\right) \cdot \frac{z\frac{dy}{dz} - y}{z^2} \right] \tag{14.62}$$

$$= \frac{1}{\sqrt{(\frac{x}{z} - a)^2 + (\frac{y}{z} - b)^2}}$$
$$\cdot \frac{1}{z} \cdot \left[\left(\frac{x}{z} - a\right)\left(\frac{dx}{dz} - \frac{x}{z}\right) + \left(\frac{y}{z} - b\right)\left(\frac{dy}{dz} - \frac{y}{z}\right) \right]. \tag{14.63}$$

Hence

$$\frac{du}{dz} = \frac{1}{(\frac{x}{z} - a)^2 + (\frac{y}{z} - b)^2} \cdot \frac{1}{z}$$
$$\cdot \left[\left(\frac{x}{z} - a\right)\left(\frac{dx}{dz} - \frac{x}{z}\right) + \left(\frac{y}{z} - b\right)\left(\frac{dy}{dz} - \frac{y}{z}\right) \right]. \tag{14.64}$$

If (a, b) is an arbitrary point in the image, the above equations are complicated and require detailed knowledge of the relationship between the camera and the objects in the scene. However, if we let (a, b) be the focus of expansion, the equations become greatly simplified. The focus of expansion (FOE) is an important point on the image plane. If a camera moves toward the objects in the scene, the objects appear to get bigger. If the vectors that represent this expansion of the objects in the image plane are extended, they will all meet in a single point, the FOE. If the camera is moving away from the scene, the objects will seem to get smaller. In this case the point at which the extended vectors meet is called the focus of contraction, the FOC. For either case, it is the pixel where the path of the translating camera pierces the image plane. If (a, b) is the FOE, then

$$a \; = \; \frac{dx}{dz} \quad \text{and} \quad b \; = \; \frac{dy}{dz}. \tag{14.65}$$

Substituting for dx/dz and dy/dz

$$\frac{du}{dz} \; = \; \frac{1}{(\frac{x}{z} - a)^2 + (\frac{y}{z} - b)^2} \cdot \frac{1}{z} \cdot \left[-\left(\frac{x}{z} - a\right)^2 - \left(\frac{y}{z} - b\right)^2 \right] \tag{14.66}$$

$$= \; -\frac{1}{z}. \tag{14.67}$$

Now let us examine dv/dz, when v is calculated with respect to the FOE (a, b):

$$v \; = \; \theta \; = \; \tan^{-1}\left(\frac{y' - b}{x' - a}\right) \tag{14.68}$$

$$\frac{dv}{dz} \; = \; \frac{1}{1 + \left(\frac{y'-b}{x'-a}\right)^2} \cdot \frac{d}{dz}\left(\frac{y' - b}{x' - a}\right). \tag{14.69}$$

Considering only the second factor of this equation, and substituting for x' and y',

$$\frac{d}{dz}\left(\frac{y' - b}{x' - a}\right) \; = \; \frac{d}{dz}\frac{(\frac{y}{z} - b)}{(\frac{x}{z} - a)} \tag{14.70}$$

$$= \; \frac{(\frac{x}{z} - a)(z\frac{dy}{dz} - y)\frac{1}{z^2} - (\frac{y}{z} - b)(z\frac{dx}{dz} - x)\frac{1}{z^2}}{(\frac{x}{z} - a)^2} \tag{14.71}$$

$$= \; \frac{(\frac{x}{z} - a)(\frac{dy}{dz} - \frac{y}{z}) - (\frac{y}{z} - b)(\frac{dx}{dz} - \frac{x}{z})}{z(\frac{x}{z} - a)^2}. \tag{14.72}$$

Remembering that $dx/dz = a$ and $dy/dz = b$,

$$\frac{d}{dz}\left(\frac{y'-b}{x'-a}\right) = \frac{(\frac{x}{z}-a)(b-\frac{y}{z}) - (\frac{x}{z}-a)(b-\frac{y}{z})}{z(\frac{x}{z}-a)^2} \qquad (14.73)$$

$$= 0. \qquad (14.74)$$

Therefore,

$$\frac{dv}{dz} = 0. \qquad (14.75)$$

Note that when the mapping is done with respect to the FOE, then the displacement in the u direction depends only on the z coordinate of the point. For other values of (a, b) the above property will not be true. Thus, if the FOE is known or can be computed, the approach can be applied to an arbitrarily translating camera, as long as there is some movement in depth (dz is in the denominator and so cannot be zero). This extension is called the ego-motion complex logarithmic mapping (ECLM) since it is based on the motion parameters of the camera itself.

14.6 Tracking

In many applications, an entity, a feature or an object, must be tracked over a sequence of frames. If there is only one entity in the sequence, the problem is easy to solve. In the presence of many entities moving independently in a scene, tracking requires the use of constraints based on the nature of objects and their motion. Due to inertia, the motion of a physical entity cannot change instantaneously. If a frame sequence is acquired at a rate such that no dramatic change takes place between two consecutive frames, then for most physical objects, no abrupt change in motion can be observed. The projection of a smooth three-dimensional trajectory is also smooth in the two-dimensional image plane. This allows us to make the smoothness assumption in images. This property is used to formulate *path coherence*. Path coherence implies that the motion of an object at any point in a frame sequence will not change abruptly.

We can combine the solution of the correspondence problem for stereopsis and motion. The following three assumptions help in formulating an approach to solve the correspondence problem:

- The location of the given point will be relatively unchanged from one frame to the next frame.

- The scalar velocity of a given point will be relatively unchanged from one frame to the next.

- The direction of motion of a given point will be relatively unchanged from one frame to the next frame.

We can also use the smoothness of image motion in monocular image sequences. To formalize the idea of the smoothness of motion, let us first formalize the idea of path coherence which states that the motion of an object at any time instant cannot change abruptly.

14.6.1   Deviation Function for Path Coherence

To use the above properties in an algorithm, we formulate a function to implement the above ideas concretely and use it to evaluate motion properties in a frame sequence. The path coherence function should follow these four guiding principles:

1. The function value is always positive.

2. It should consider the amount of angular deviation without any effect of the sign of the direction of motion.

3. The function should respond equally to the incremental speed.

4. The function should be normalized in the range 0.0 to 1.0.

Trajectories of two point-tokens are shown in Figure 14.18. Let the trajectory be represented as

$$T_i = \left\langle P_i^1, P_i^2, P_i^3, \ldots, P_i^n \right\rangle \tag{14.76}$$

where T_i is trajectory and P_i^k represents a point in the kth image. If the coordinates of the point are given by the vector X_i in the frame, the coordinates in the kth frame can be represented as X_{ik}. Representing the trajectory in vector form,

$$T_i = \left\langle X_{i1}, X_{i2}, X_{i3}, \ldots, X_{in} \right\rangle . \tag{14.77}$$

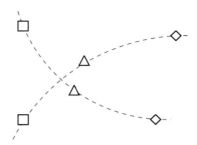

Figure 14.18: The trajectories of two points. The points in the first, second, and third frames are labeled $\square$, $\triangle$, and $\diamond$, respectively.

Now let us consider the deviation $d_i{}^k$ in the path of the point in the kth frame. The deviation in the path is a measure of path coherence, given by

$$d_i{}^k = \phi(\overline{X_{ik-1}X_{ik}}, \overline{X_{ik}X_{ik+1}}), \qquad (14.78)$$

where ϕ is a path coherence function. The deviation for the complete trajectory is defined as

$$D_i = \sum_{k=2}^{n-1} d_i{}^k. \qquad (14.79)$$

If there are m points in a sequence of n frames resulting in m trajectories, the deviation of all trajectories should be considered, which is given by

$$D(T_1, T_2, T_3, \ldots, T_m) = \sum_{i=k}^{m} \sum_{k=2}^{n-1} d_i{}^k. \qquad (14.80)$$

Thus, the correspondence problem is solved by maximizing the smoothness of motion; that is, the total deviation D is minimized to find the set of correct trajectories.

14.6.2 Path Coherence Function

After understanding the trajectory function, let us define a constraint function for path coherence. If the sampling rate of the camera is high enough, then change in the direction and velocity of any moving point in consecutive time frames is smooth.

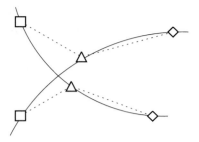

Figure 14.19: Deviations in a trajectory.

This is described by the deviation function:

$$\phi(P_i^{k-1}, P_i^{k}, P_i^{k+1}) = w_1(1 - \cos\theta) + w_2\left(1 - 2\frac{\sqrt{d_1 d_2}}{d_1 + d_2}\right). \qquad (14.81)$$

In the vector form it can represented as

$$\phi(P_i^{k-1}, P_i^{k}, P_i^{k+1}) = w_1\left(1 - \frac{\overline{X_{ik-1}X_{ik}} \cdot \overline{X_{ik}X_{ik+1}}}{\|\overline{X_{ik-1}X_{ik}}\| \|\overline{X_{ik}X_{ik+1}}\|}\right)$$

$$+ w_2\left(1 - 2\frac{\sqrt{\|\overline{X_{ik-1}X_{ik}}\| \|\overline{X_{ik}X_{ik+1}}\|}}{\|\overline{X_{ik-1}X_{ik}}\| + \|\overline{X_{ik}X_{ik+1}}\|}\right) \quad (14.82)$$

where w_1 and w_2 are weights that are selected to assign differing importance to direction and velocity changes (see Figure 14.19).

Note that the first term is the dot product of displacement vectors and the second considers the geometric and arithmetic mean of the magnitude. The first term in the above expression can be considered as *direction coherence*, and the second term can be considered as *speed coherence*. The weight can be selected in the range 0.00 to 1.00 such that their sum is 1.

One of the main difficulties with the use of multiple frames is the problem of occlusion. When working with a large sequence of frames, it is possible that some objects may disappear totally or partially. Similarly, some new objects may appear in the frame sequence from some intermediate frame onward. In addition, the changing geometry of objects due to motion and the changes in scene illumination over the frame sequence can cause significant changes in the feature point data that is to be matched over the frames.

All these changes in the feature point data set lead to incorrect correspondence, if such a correspondence is obtained by performing minimization to extract a globally smooth set of complete trajectories. By forcing the trajectories to satisfy some local constraints and allowing them to be incomplete, if necessary, trajectories in the presence of occlusion can be obtained.

14.6.3   Path Coherence in the Presence of Occlusion

Two limitations of the path coherence algorithm are:

- It assumes that the same number of feature points are available in every frame.

- It assumes that the same set of feature points are being extracted in every frame.

In practice, the number of feature points can drastically change from frame to frame. Even when the feature point count is the same over all the frames, it may not necessarily be the same set of feature points that was extracted in the earlier frames. Thus, while searching for a smooth set of trajectories, we must allow the possibility of obtaining incomplete trajectories indicating either occlusion, appearance of a new object, or simply the absence of the corresponding feature points in the subsequent or earlier frames due to poor feature detection. Moreover, some constraints in the form of maximum possible displacement and maximum local smoothness deviation must be placed on the acceptable trajectories to avoid chance groupings of distant feature points that may happen to form a smooth path.

Given a set P^j of m_j feature points for each of the n frames, the algorithm below finds the maximal set of complete or partially complete trajectories that minimizes the sum of local smoothness deviations for all the trajectories obtained subject to the conditions that the local smoothness deviation for none of the trajectories exceeds $\phi_{\max}$ and the displacement between any two successive frames for any trajectory is always less than $d_{\max}$. The above local constraints limit the acceptable location of a feature point in the next frame given its location in two previous frames.

14.6.4   Modified Greedy Exchange Algorithm

To account for missing points due to occlusion, *phantom points* are used. Phantom feature points are hypothetical points which are used as fillers to extend all trajectories over the given frame set. These points are introduced during the trajectory initialization phase as described later. The notion of phantom feature points serves two purposes: it allows us to satisfy the local constraints, and it provides a way to deal with incomplete trajectories. For the notion of phantom feature points to be useful, we must define the displacement and local smoothness deviation values for a trajectory that has some phantom feature points assigned to it. For computing the frame-to-frame displacement for a trajectory T_i, a displacement function $\mathrm{Disp}(P_i^k, P_i^{k+1})$ is defined as follows:

$$\mathrm{Disp}(P_i^k, P_i^{k+1}) = \begin{cases} \text{Euclidean } \mathrm{Disp}(P_i^k, P_i^{k+1}) & \text{if both points are true feature} \\ d_{\max} & \text{points otherwise.} \end{cases}$$

This definition of frame-to-frame displacement for a trajectory implies that a phantom feature point always moves by a fixed amount $d_{\max}$. For computing the local smoothness deviation for a trajectory T_i in the kth frame function, $\mathrm{DEV}(P_i^{k-1}, P_i^k, P_i^{k+1})$ is defined as follows:

$$\mathrm{DEV}(P_i^{k-1}, P_i^k, P_i^{k+1}) = \begin{cases} 0 & \text{if } P_i^{k-1} \text{ is a phantom point} \\ \phi(P_i^{k-1}, P_i^k, P_i^{k+1}) & \text{if all three are true feature points} \\ \phi_{\max} & \text{otherwise.} \end{cases}$$

The above definition of the local smoothness deviation function is equivalent to the path coherence function if all three feature points are true feature points. It introduces a penalty of $\phi_{\max}$ for not having a true feature for T_i in the subsequent frame or missing a true feature point in the current frame, the kth frame. There is no penalty if a trajectory begins from the current frame under consideration. However, as the greedy exchange algorithm is applied alternately in the forward and backward directions, it is clear that the assignment of phantom feature points in the subsequent frames or earlier frames is equally discouraged. With the above definitions for computing the displacement and local smoothness deviation values, the steps of the modified greedy exchange algorithm are as follows:

Initialization:

1. For each of the m_k true feature points in $P^k, k = 1, 2, \ldots, n-1$, determine the nearest neighbor in P^{k+1} that is within the distance $d_{\max}$. Resolve arbitrarily in case of multiple choices.

2. Form initial trajectories by linking the nearest neighbors in the successive frames. Extend all incomplete trajectories using phantom feature points to span n frames.

3. For every trajectory of step 2 above, form an additional trajectory consisting only of phantom feature points.

Exchange Loop:

forward-flag = on;
backward-flag = on;
For each frame index $k = 2$ to $n - 1$:
    while (forward-flag == on or backward-flag = on)
    do
        if (forward-flag == on) then
        begin
            for $i = 1$ to $m - 1$
                for $j = i + 1$ to m
                    if within constraints of $d_{\max}$
                    calculate:

$$G_{ij}{}^k = [\phi(P_i^{k-1}, P_i^k, P_i^{k+1}) + \phi(P_j^{k-1}, P_j^k, P_j^{k+1})]$$
$$-[\phi(P_i^{k-1}, P_i^k, P_j^{k+1}) + \phi(P_j^{k-1}, P_j^k, P_i^{k+1})]. \qquad (14.83)$$

        Pick the ij pair with maximum gain.
        if (gain is greater than 0)
        begin
            Exchange the point in $(k + 1)$th frame.
            Set backward-flag on.
        end (if)
        else
            Set forward-flag off.
      end (if)

else if (backward-flag == on) then
begin
    for $i = 1$ to $m - 1$
       for $j = i + 1$ to m
          if within constraints of d_{max}
             calculate $G_{ij}{}^{k}$ from Equation 14.83.
    Pick the ij pair with maximum gain.
    if (gain is greater than 0)
    begin
       Exchange the point in $(k + 1)$th frame.
       Set backward-flag on.
    end (if)
    else
       Set forward-flag off.
    end (if)
  end (while)
end (for)

Termination:

Repeat the exchange loop until there are no more frames.

The exchange operation of the above algorithm must be applied alternately in the forward and backward directions. The number of trajectories formed during step 2 of the initialization depends on the quality of the data. In an ideal case, where the same set of m feature points are consistently present in all the frames, only m trajectories will be formed and none of these will have any phantom points. In the worst case, when the feature points of any one frame do not have any correlation with the feature points of any other frame, the number of trajectories formed can be as large as the total count of true feature points from all the frames. In general, the number of trajectories formed will be at least m_{max}, where $m_{max} = \max(m_1, m_2, \ldots, m_n)$, and different trajectories will have different numbers of phantom feature points.

It should be noted that the introduction of phantom feature points does not require the locations of these points. Also, the introduction of trajectories with only phantom feature points during step 3 of the initialization phase does not affect the overall sum of local smoothness deviation values, the criterion being minimized to determine the correspondence. The introduction of these trajectories ensures that no final trajectory will have a local smoothness deviation value greater than ϕ_{max} in any frame. The checking of the *ij* pairs in the exchange loop ensures that the d_{max} constraint will not be violated.

14.7 Shape from Motion

One of the major goals of dynamic scene analysis is to get three-dimensional information about the structure of objects in the scene and their three-dimensional motion characteristics. By the structure of an object, we mean the relative locations of points on the object. Thus, if we know the locations of points on an object with respect to a *reference point*, with a scale factor, then we say that the structure of the object is known. To get this information, however, image plane information about objects must be converted to scene information. As we saw in stereo, by using multiple views of an object, or by using multiple locations of a camera, this information may be recovered. The interpretation of two-dimensional displacements in terms of three-dimensional motion is complicated.

An assumption about the *rigidity* of objects is helpful in recovering the structure of objects. The rigidity assumption states that any set of elements undergoing a two-dimensional transformation, which has a unique interpretation as a rigid body moving in space, should be so interpreted. This assumption provides a much-needed constraint in the scene that can be used to determine the relative locations of points. The *structure-from-motion* theorem states that given three distinct orthographic projections of four noncoplanar points in a rigid configuration, the structure and motion compatible with the three views are uniquely determined up to a reflection about the image plane.

The structure-from-motion theorem provides a precise statement about constraints and possible solutions. By changing constraints on the types of objects and types of projections, and by using different mathematical approaches, many different formulations of the above problem are possible. Approaches to recovering three-dimensional structures from image sequences may be divided into the following two general classes.

Token-Based Methods

Suppose that we have several views of an object available and that these views can be considered orthographic. An interest operator is applied to consecutive frames of this sequence and some interesting points or tokens, such as corners, are extracted. Also suppose that the correspondence problem between interesting points has been solved, using a method discussed earlier. If token correspondence has been established, the structure-from-motion theorem states that it is possible to recover the three-dimensional location of four noncoplanar points and their motion from their orthogonal projections. This gives an implicit three-dimensional structure for an object. Here we will consider an approach to recover this structure.

Suppose that we have four points in space and their orthographic projections are known in three frames. Let the points be $P_0, P_1, P_2,$ and P_3 and the frames be taken at time instants $t_1, t_2,$ and t_3. We represent three-dimensional coordinates of points as (x_{ij}, y_{ij}, z_{ij}) and two-dimensional coordinates as (x'_{ij}, y'_{ij}), where i and j represent the point number, $i = 0, 1, 2, 3$, and frame number, $j = 1, 2, 3$, respectively. Now we can write the transformation of three-dimensional points as

$$\begin{bmatrix} x_{i,2} \\ y_{i,2} \\ z_{i,2} \end{bmatrix} = R_{12} \begin{bmatrix} x_{i,1} \\ y_{i,1} \\ z_{i,1} \end{bmatrix} + T_{12} \qquad (14.84)$$

and

$$\begin{bmatrix} x_{i,3} \\ y_{i,3} \\ z_{i,3} \end{bmatrix} = R_{23} \begin{bmatrix} z_{i,2} \\ y_{i,2} \\ z_{i,2} \end{bmatrix} + T_{23}. \qquad (14.85)$$

In the above, R_{ij} and T_{ij} are rotation matrices and translation vectors from time t_i to time t_j. We want to determine rotations matrices, translation vectors, and three-dimensional coordinates of points to determine the motion and structure. To solve for these quantities, we have image plane coordinates of four points in three frames. Now since we are using orthographic projections, we know that

$$x'_{ij} = x_{ij} \quad \text{and} \quad y'_{ij} = y_{ij}. \qquad (14.86)$$

Upon substituting this information in the above equations, we see that this is an underconstrained situation; we have too few equations to determine the required variables.

Many different approaches have been proposed to solve this undercon- strained problem. Most of these methods introduce an assumption that al- lows the formulation of the problem in such a way that one of the known techniques for determining the unknowns may be applied. We do not dis- cuss any specific method here. Interested readers may want to study the references given in the Further Readings section.

Feature-based methods for the recovery of three-dimensional structure or for the estimation of motion parameters require two steps: determination of the precise location of tokens or points, and the correspondence between tokens or points. If interest operators are applied based on small neighbor- hoods, then the number of tokens extracted in a real image is very large, mak- ing correspondence a difficult problem. Interest operators based on a large neighborhood and higher-order gray-level characteristics result in a more rea- sonable number of tokens for determining correspondences, but the location of tokens may not be precise.

Trajectory-Based Methods

The above methods depend on a set of points in two or three frames. If a token is tracked over several frames by solving correspondences, a two- dimensional trajectory of the token is obtained. Methods for recovering three-dimensional structure and motion from trajectories are more reliable than methods based on sets of features in new frames. A trajectory may be interpolated to obtain better resolution of the two-dimensional path by using curve fitting techniques. Moreover, the correspondence problem may be simplified by considering more than two frames and extending relaxation across the frames.

Further Reading

Nagel [174] proposed the use of the likelihood ratio for motion detection. Much of the work presented here on difference and accumulative difference pictures was done by Jain with other researchers [129, 130, 122, 125].

Fennema and Thompson proposed use of the gradient method to com- pute optical flow for use in segmentation based on motion [79]. Horn and Schunck [114] developed the optical flow determination algorithm. Schunck

[212] developed robust approaches to compute optical flow. Nagel and Enkelmann [176] formulate and evaluate an oriented smoothness constraint for computing optical flow. Determination of optical flow has been a very active research area. For some recent approaches see [4, 106, 169, 175, 9, 222]. Clocksin [58] and Longuet-Higgins and Prazdny [157] describe the significance of optical flow and its potential in extracting information about surface orientation and structure. Subbarao [229] has presented a rigorous approach for recovering information from optical flow.

Barnard and Thompson [21] introduced a feature-based matching algorithm to estimate geometrical disparity between images. Recovery of three-dimensional motion parameters and the three-dimensional structure of objects has been an active research area. Ullman published a book containing an extensive discussion of the correspondence problem from the viewpoint of natural vision [241]. He popularized the *structure-from-motion* problem. Later Huang, with his students, did pioneering research in recovering motion characteristics of objects under various conditions [237, 77, 76, 249]. Jerian and Jain showed the sensitivity of various approaches to noise and initial estimates and provided a polynomial systems solution [134, 133]. Chellappa and Broida developed a framework for estimating motion in a sequence of frames using Kalman filters [47, 48].

Ego-motion complex logarithmic mapping was developed by Jain [124, 123, 127]. Schwartz [214, 213, 215, 216] and Cavanaugh [54, 55] have studied this mapping in the context of biological vision systems.

Jenkin [132] proposed a novel approach for combining the solution of the correspondence problem for stereopsis and motion. Sethi and Jain [219] developed a tracking algorithm based on smoothness of motion. The algorithm for path coherence in the presence of occlusion was developed by by Salari and Sethi [209]. A large number of image frames taken at short intervals helps minimize the correspondence problem since the amount of change in successive images is expected to be very small. This concept has led to the so-called epipolar plane image analysis in which the images are acquired using a moving camera. Explicit representation of both the spatial and temporal structure of such image sequences is captured in a spatiotemporal surface. Tracking mechanisms that operate locally on these evolving surfaces to obtain three-dimensional scene reconstruction are described by Baker and Bolles in [15]. In the last few years, there have been several other approaches based on the spatiotemporal image solid. Some of these are [106, 247, 156].

Determination of optical flow and solving the correspondence problem have been two difficult problems in dynamic vision. In the last few years, several techniques have been proposed for direct computation of motion properties, bypassing optical flow and correspondence [126, 7, 6, 184]. These approaches appear promising.

Zhang, Faugeras, and Ayache [262] approach the problem of motion determination as a stereo matching and motion estimation problem. Stereo and motion are also used by Grosso, Sandini, and Tistarelli in their system described in [94]. However, in their system, the objects remain stationary, and multiple views are obtained by moving the cameras around the objects while maintaining the direction of gaze fixed toward a point in space. A survey of many of the methods for dynamic-scene analysis is given by Aggarwal and Nandhakumar [2]. A framework for the abstraction of motion concepts from image sequences is described by Tsotos et al. [238]. It includes semantic nets for knowledge representation and associated algorithms operating in a competing and cooperating feedback mode.

Exercises

14.1 What are the three phases in a dynamic-scene analysis system? Consider the task of a driver on a freeway. What are the operations performed in the three phases of dynamic-scene analysis for this task?

14.2 Define a difference picture. How would you select an appropriate threshold in a particular application of a difference picture?

14.3 How can you form a difference picture using a statistical approach to determine dissimilarity of intensity values in an image? How is this approach better than the one based on the straightforward difference picture approach?

14.4 What is an accumulative difference picture? What limitations of difference pictures does it overcome? Does sign of difference help in motion analysis? How?

14.5 What is the difference between a time-varying edge and a moving edge? How will you detect moving edges in a dynamic scene? Can you use three-dimensional edge detectors to detect motion in an image sequence?

14.6 Define the correspondence problem in motion analysis. How can you solve this problem? List various approaches to solve this problem. What do you think is the problem that makes determination of corresponding points so difficult?

14.7 To obtain the feature point locations to subpixel resolution, compute the x and y moments from the intermediate image of the minimum variance obtained in step 2 of the Moravec interest operator.

14.8 What is image flow? Where can it be used?

14.9 Derive the motion constraint equation. How can you use this equation to segment a dynamic scene into moving and stationary objects?

14.10 Define the aperture problem. What problems does it cause in computing image flow? How can you overcome the aperture problem? Suggest at least two different approaches.

14.11 Define focus of expansion. Where and how it can be used? Is it related to vanishing points?

14.12 What is complex logarithmic mapping? What properties of this mapping make it attractive in dynamic-scene analysis?

14.13 When a camera is moving, all points in the environment are getting displaced in an image. How can you segment a dynamic scene to determine moving objects in case of a moving camera?

14.14 What is path coherence? How can you use it for tracking objects in a scene?

14.15 What is structure-from-motion? Under what conditions can you realistically determine the structure of an object using three frames?

14.16 The top left corner of a 6×8-pixel object is initially located at $(0, 0)$. It moves to the location $(4, 8)$ after four frames of uniform velocity. Find the absolute, positive, and negative accumulative difference pictures.

14.17 Consider a point object located at *world* coordinates $(10, 0, 10)$ at time $t = 0$. A camera system with a focal length of 2 is located such that its *lens center* is at the origin of the world coordinate system and its optical axis is looking directly at the object at time $t = 0$. The object is moving with a uniform velocity of $(5, 0, 0)$.

    a. What are the coordinates of the point object in the image plane at time $t = 0$?

    b. Find the image coordinates of the point at $t = 1$ if

        (i) The camera is stationary.

        (ii) The camera translates with a uniform velocity of $(0, 0, 5)$.

14.18 A family decides to enjoy a nice Spring weekend outdoors along with their pets, Cilly the cat and Billy the dog. Just when they had their new video camera set up on a tripod and leveled so that its axis is horizontal, Cilly decides to go after a songbird feasting on some bread crumbs. Immediately, Billy decides to chase Cilly. After four seconds of running at constant speed, both Billy and Cilly reach their destination only to find that the bird just took off in the direction $(0, 0, 200)$. Initial positions of all animals and the camera are as shown in Figure P14.18. The camera height is 5 feet and the focal length of its lens is 0.25 feet. Assume that the animals are small enough so that they can be represented as points on the ground. Also, assume that Billy's instantaneous direction of travel is always toward the current position of Cilly and that he also travels at constant speed. Find the image plane coordinates and the direction of optical flow for all animals at time instances $t = 0^+$ (just after start), $t = 2$ (an estimate is enough for Billy's position), and $t = 4^-$ (just before end) seconds. Also mark the direction of optical flow for the bird after it starts its flight. Mark all distances in the image plane. Note that all distances are measured from the center of the lens for convenience.

Computer Projects

14.1 Design a system to count how many people entered the coffee room and what percentage took coffee. Assume that you can use multiple cameras if required and you are free to determine the location and orientation of cameras.

14.2 Implement an accumulative difference picture–based approach for motion analysis. Use this approach to extract images of all people who came to your coffee room and took coffee.

14.3 Develop a program to establish correspondence. Use this to determine all corresponding points in a scene containing at least three objects moving in different directions.

14.4 Implement a tracking algorithm using path coherence. Test it using a basketball sequence to track the ball. Modify this algorithm to work with a mobile camera. Assume that you want to design a mobile robot that will track a particular moving object (pointed to it interactively). Apply your tracking algorithm to this problem.

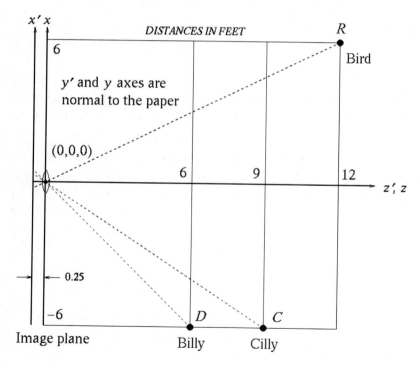

Figure P14.18

Chapter 15

Object Recognition

An object recognition system finds objects in the real world from an image of the world, using *object models* which are known a priori. This task is surprisingly difficult. Humans perform object recognition effortlessly and instantaneously. Algorithmic description of this task for implementation on machines has been very difficult. In this chapter we will discuss different steps in object recognition and introduce some techniques that have been used for object recognition in many applications. We will discuss the different types of recognition tasks that a vision system may need to perform. We will analyze the complexity of these tasks and present approaches useful in different phases of the recognition task.

The object recognition problem can be defined as a labeling problem based on models of known objects. Formally, given an image containing one or more objects of interest (and background) and a set of labels corresponding to a set of models *known* to the system, the system should assign correct labels to regions, or a set of regions, in the image. The object recognition problem is closely tied to the segmentation problem: without at least a partial recognition of objects, segmentation cannot be done, and without segmentation, object recognition is not possible.

In this chapter, we discuss basic aspects of object recognition. We present the architecture and main components of object recognition and discuss their role in object recognition systems of varying complexity.

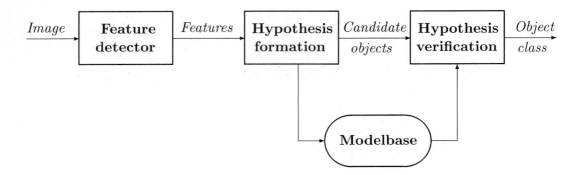

Figure 15.1: Different components of an object recognition system are shown.

15.1   System Component

An object recognition system must have the following components to perform the task:

- Model database (also called modelbase)

- Feature detector

- Hypothesizer

- Hypothesis verifier

A block diagram showing interactions and information flow among different components of the system is given in Figure 15.1.

The model database contains all the models known to the system. The information in the model database depends on the approach used for the recognition. It can vary from a qualitative or functional description to precise geometric surface information. In many cases, the models of objects are abstract feature vectors, as discussed later in this section. A feature is some attribute of the object that is considered important in describing and recognizing the object in relation to other objects. Size, color, and shape are some commonly used features.

The feature detector applies operators to images and identifies locations of features that help in forming object hypotheses. The features used by a system depend on the types of objects to be recognized and the organization of the model database. Using the detected features in the image, the hypothesizer assigns likelihoods to objects present in the scene. This step is used to reduce the search space for the recognizer using certain features. The modelbase is organized using some type of indexing scheme to facilitate elimination of unlikely object candidates from possible consideration. The verifier then uses object models to verify the hypotheses and refines the likelihood of objects. The system then selects the object with the highest likelihood, based on all the evidence, as the correct object.

All object recognition systems use models either explicitly or implicitly and employ feature detectors based on these object models. The hypothesis formation and verification components vary in their importance in different approaches to object recognition. Some systems use only hypothesis formation and then select the object with highest likelihood as the correct object. Pattern classification approaches are a good example of this approach. Many artificial intelligence systems, on the other hand, rely little on the hypothesis formation and do more work in the verification phases. In fact, one of the classical approaches, template matching, bypasses the hypothesis formation stage entirely.

An object recognition system must select appropriate tools and techniques for the steps discussed above. Many factors must be considered in the selection of appropriate methods for a particular application. The central issues that should be considered in designing an object recognition system are:

- *Object or model representation:* How should objects be represented in the model database? What are the important attributes or features of objects that must be captured in these models? For some objects, geometric descriptions may be available and may also be efficient, while for another class one may have to rely on generic or functional features. The representation of an object should capture all relevant information without any redundancies and should organize this information in a form that allows easy access by different components of the object recognition system.

- *Feature extraction:* Which features should be detected, and how can they be detected reliably? Most features can be computed in two-dimensional images but they are related to three-dimensional characteristics of objects. Due to the nature of the image formation process,

some features are easy to compute reliably while others are very difficult. Feature detection issues were discussed in many chapters in this
book.

- *Feature-model matching:* How can features in images be matched to
  models in the database? In most object recognition tasks, there are
  many features and numerous objects. An exhaustive matching approach will solve the recognition problem but may be too slow to be
  useful. Effectiveness of features and efficiency of a matching technique
  must be considered in developing a matching approach.

- *Hypotheses formation:* How can a set of likely objects based on the
  feature matching be selected, and how can probabilities be assigned
  to each possible object? The hypothesis formation step is basically a
  heuristic to reduce the size of the search space. This step uses knowledge of the application domain to assign some kind of probability or
  confidence measure to different objects in the domain. This measure
  reflects the likelihood of the presence of objects based on the detected
  features.

- *Object verification:* How can object models be used to select the most
  likely object from the set of probable objects in a given image? The
  presence of each likely object can be verified by using their models. One
  must examine each plausible hypothesis to verify the presence of the
  object or ignore it. If the models are geometric, it is easy to precisely
  verify objects using camera location and other scene parameters. In
  other cases, it may not be possible to verify a hypothesis.

Depending on the complexity of the problem, one or more modules in
Figure 15.1 may become trivial. For example, pattern recognition-based
object recognition systems do not use any feature-model matching or object
verification; they directly assign probabilities to objects and select the object
with the highest probability.

15.2   Complexity of Object Recognition

As we studied in earlier chapters in this book, images of scenes depend on
illumination, camera parameters, and camera location. Since an object must

be recognized from images of a scene containing multiple entities, the complexity of object recognition depends on several factors. A qualitative way to consider the complexity of the object recognition task would consider the following factors:

- *Scene constancy:* The scene complexity will depend on whether the images are acquired in similar conditions (illumination, background, camera parameters, and viewpoint) as the models. As seen in earlier chapters, scene conditions affect images of the same object dramatically. Under different scene conditions, the performance of different feature detectors will be significantly different. The nature of the background, other objects, and illumination must be considered to determine what kind of features can be efficiently and reliably detected.

- *Image-models spaces:* In some applications, images may be obtained such that three-dimensional objects can be considered two-dimensional. The models in such cases can be represented using two-dimensional characteristics. If models are three-dimensional and perspective effects cannot be ignored, then the situation becomes more complex. In this case, the features are detected in two-dimensional image space, while the models of objects may be in three-dimensional space. Thus, the same three-dimensional feature may appear as a different feature in an image. This may also happen in dynamic images due to the motion of objects.

- *Number of objects in the model database:* If the number of objects is very small, one may not need the hypothesis formation stage. A sequential exhaustive matching may be acceptable. Hypothesis formation becomes important for a large number of objects. The amount of effort spent in selecting appropriate features for object recognition also increases rapidly with an increase in the number of objects.

- *Number of objects in an image and possibility of occlusion:* If there is only one object in an image, it may be completely visible. With an increase in the number of objects in the image, the probability of occlusion increases. Occlusion is a serious problem in many basic image computations. Occlusion results in the absence of expected features and the generation of unexpected features. Occlusion should also be considered in the hypothesis verification stage. Generally, the difficulty in the recognition task increases with the number of objects in an image. Difficulties in image segmentation are due to the presence of multiple occluding objects in images.

The object recognition task is affected by several factors. We classify the object recognition problem into the following classes.

Two-dimensional

In many applications, images are acquired from a distance sufficient to consider the projection to be orthographic. If the objects are always in one stable position in the scene, then they can be considered two-dimensional. In these applications, one can use a two-dimensional modelbase. There are two possible cases:

- Objects will not be occluded, as in remote sensing and many industrial applications.

- Objects may be occluded by other objects of interest or be partially visible, as in the bin of parts problem.

In some cases, though the objects may be far away, they may appear in different positions resulting in multiple stable views. In such cases also, the problem may be considered inherently as two-dimensional object recognition.

Three-dimensional

If the images of objects can be obtained from arbitrary viewpoints, then an object may appear very different in its two views. For object recognition using three-dimensional models, the perspective effect and viewpoint of the image have to be considered. The fact that the models are three-dimensional and the images contain only two-dimensional information affects object recognition approaches. Again, the two factors to be considered are whether objects are separated from other objects or not.

For three-dimensional cases, one should consider the information used in the object recognition task. Two different cases are:

- *Intensity:* There is no surface information available explicitly in intensity images. Using intensity values, features corresponding to the three-dimensional structure of objects should be recognized.

- 2.5-*dimensional images:* In many applications, surface representations with viewer-centered coordinates are available, or can be computed, from images. This information can be used in object recognition. Range images are also 2.5-dimensional. These images give the distance to different points in an image from a particular view point.

Segmented

The images have been segmented to separate objects from the background. As discussed in Chapter 3 on segmentation, object recognition and segmentation problems are closely linked in most cases. In some applications, it is possible to segment out an object easily. In cases when the objects have not been segmented, the recognition problem is closely linked with the segmentation problem.

15.3   Object Representation

Images represent a scene from a camera's perspective. It appears natural to represent objects in a camera-centric, or viewer-centered, coordinate system. Another possibility is to represent objects in an object-centered coordinate system. Of course, one may represent objects in a world coordinate system also. Since it is easy to transform from one coordinate system to another using their relative positions, the central issue in selecting the proper coordinate system to represent objects is the ease of representation to allow the most efficient representation for feature detection and subsequent processes.

A representation allows certain operations to be efficient at the cost of other operations. Representations for object recognition are no exception. Designers must consider the parameters in their design problems to select the best representation for the task. The following are commonly used representations in object recognition.

15.3.1   Observer-Centered Representations

If objects usually appear in a relatively few stable positions with respect to the camera, then they can be represented efficiently in an observer-centered coordinate system. If a camera is located at a fixed position and objects move such that they present only some aspects to the camera, then one can represent objects based on only those views. If the camera is far away from objects, as in remote sensing, then three-dimensionality of objects can be ignored. In such cases, the objects can be represented only by a limited set of views—in fact, only one view in most cases. Finally, if the objects in a domain of applications are significantly different from each other, then observer-centered representations may be enough.

Observer-centered representations are defined in image space. These representations capture characteristics and details of the images of objects in their relative camera positions.

One of the earliest and most rigorous approaches for object recognition is based on characterizing objects using a feature vector. This feature vector captures essential characteristics that help in distinguishing objects in a domain of application. The features selected in this approach are usually global features of the images of objects. These features are selected either based on the experience of a designer or by analyzing the efficacy of a feature in grouping together objects of the same class while discriminating it from the members of other classes. Many feature selection techniques have been developed in pattern classification. These techniques study the probabilistic distribution of features of known objects from different classes and use these distributions to determine whether a feature has sufficient discrimination power for classification.

In Figure 15.2 we show a two-dimensional version of a feature space. An object is represented as a point in this space. It is possible that different features have different importance and that their units are different. These problems are usually solved by assigning different weights to the features and by normalizing the features.

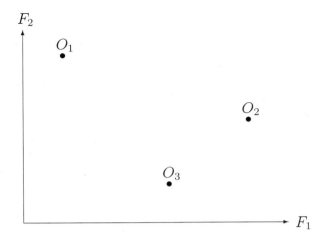

Figure 15.2: Two-dimensional feature space for object recognition. Each object in this space is a point. Features must be normalized to have uniform units so that one may define a distance measure for the feature space.

Most so-called approaches for two-dimensional object recognition in the literature are the approaches based on the image features of objects. These approaches try to partition an image into several local features and then represent an object as image features and relations among them. This representation of objects allows partial matching also. In the presence of occlusion in images, this representation is more powerful than feature space. In Figure 15.3 we show local features for an object and how they will be represented.

15.3.2 Object-Centered Representations

An object-centered representation uses description of objects in a coordinate system attached to objects. This description is usually based on three-dimensional features or description of objects.

Object-centered representations are independent of the camera parameters and location. Thus, to make them useful for object recognition, the representation should have enough information to produce object images or object features in images for a known camera and viewpoint. This requirement suggests that object-centered representations should capture aspects of the geometry of objects explicitly. Some commonly used object-centered representations are discussed here.

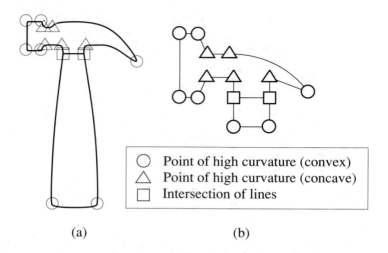

(a)   (b)

Figure 15.3: In (a) an object is shown with its prominent local features highlighted. A graph representation of the object is shown in (b). This representation is used for object recognition using a graph matching approach.

Constructive Solid Geometry

A CSG representation of an object uses simple volumetric primitives, such as blocks, cones, cylinders, and spheres, and a set of boolean operations: union, intersection, and difference. Since arbitrarily curved objects cannot be represented using just a few chosen primitives, CSG approaches are not very useful in object recognition. These representations are used in object representation in CAD/CAM applications. In Figure 15.4, a CSG representation for a simple object is shown.

Spatial Occupancy

An object in three-dimensional space may be represented by using nonoverlapping subregions of the three-dimensional space occupied by an object. In addition to simple occupancy, one may consider representing other properties of objects at points in space. There are many variants of this representation such as voxel representation, octree, and tetrahedral cell decomposition. In Figure 15.5, we show a voxel representation of an object.

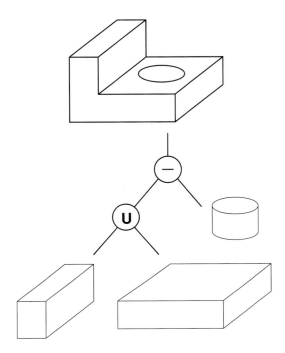

Figure 15.4: A CSG representation of an object uses some basic primitives and operations among them to represent an object. Here we show an object and its CSG representation.

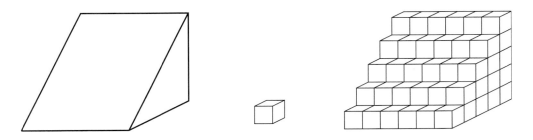

Figure 15.5: A voxel representation of an object.

A spatial occupancy representation contains a detailed description of an object, but it is a very low-level description. This type of representation must be processed to find specific features of objects to enable the hypothesis formation process.

Multiple-View Representation

Since objects must be recognized from images, one may represent a three-dimensional object using several views obtained either from regularly spaced viewpoints in space or from some strategically selected viewpoints. For a limited set of objects, one may consider arbitrarily many views of the object and then represent each view in an observer-centered representation.

A three-dimensional object can be represented using its aspect graph. An aspect graph represents all stable views of an object. Thus, an aspect graph is obtained by partitioning the view-space into areas in which the object has stable views. The aspect graph for an object represents a relationship among all the stable views. In Figure 15.6 we show a simple object and its aspect graph.

Surface-Boundary Representation

A solid object can be represented by defining the surfaces that bound the object. The bounding surfaces can be represented using one of several methods popular in computer graphics. These representations vary from triangular patches to nonuniform rational B-splines (NURBS). Some of these representations were discussed in Chapter 13.

Sweep Representations: Generalized Cylinders

Object shapes can be represented by a three-dimensional space curve that acts as the spine or axis of the cylinder, a two-dimensional cross-sectional figure, and a sweeping rule that defines how the cross section is to be swept along the space curve. The cross section can vary smoothly along the axis. This representation is shown in Figure 15.7.

For many industrial and other objects, the cross section of objects varies smoothly along an axis in space, and in such cases this representation is satisfactory. For arbitrarily shaped objects, this condition is usually not satisfied, making this representation unsuitable.

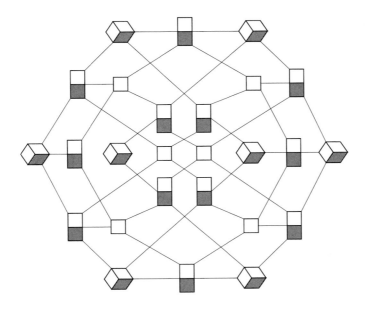

Figure 15.6: An object and its aspect graph. Each node in the aspect graph represents a stable view. The branches show how one can go from one stable view to other stable views through accidental views.

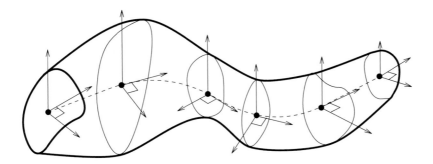

Figure 15.7: An object and its generalized cylinder representation. Note the axis of the cylinder is shown as a dashed line, the coordinate axes are drawn with respect to the cylinder's central axis, and the cross sections at each point are orthogonal to the cylinder's central axis.

15.4   Feature Detection

Many types of features are used for object recognition. Most features are based on either regions or boundaries in an image. It is assumed that a region or a closed boundary corresponds to an entity that is either an object or a part of an object. Some of the commonly used features are as follows.

Global Features

Global features usually are some characteristics of regions in images such as area (size), perimeter, Fourier descriptors, and moments. Global features can be obtained either for a region by considering all points within a region, or only for those points on the boundary of a region. In each case, the intent is to find descriptors that are obtained by considering all points, their locations, intensity characteristics, and spatial relations. These features were discussed at different places in the book.

Local Features

Local features are usually on the boundary of an object or represent a distinguishable small area of a region. Curvature and related properties are commonly used as local features. The curvature may be the curvature on a boundary or may be computed on a surface. The surface may be an intensity surface or a surface in 2.5-dimensional space. High curvature points are commonly called corners and play an important role in object recognition. Local features can contain a specific shape of a small boundary segment or a surface patch. Some commonly used local features are *curvature, boundary segments,* and *corners.*

Relational Features

Relational features are based on the relative positions of different entities, either regions, closed contours, or local features. These features usually include distance between features and relative orientation measurements. These features are very useful in defining composite objects using many regions or local features in images. In most cases, the relative position of entities is what defines objects. The exact same feature, in slightly different relationships, may represent entirely different objects.

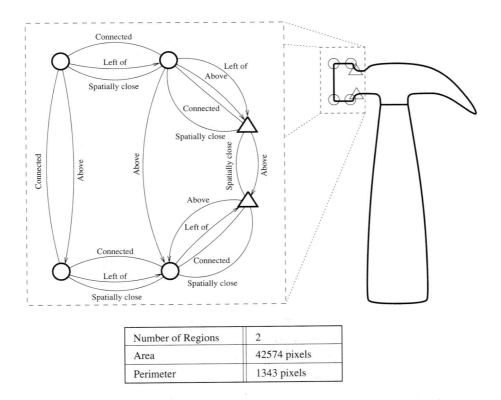

Number of Regions	2
Area	42574 pixels
Perimeter	1343 pixels

Figure 15.8: An object and its partial representation using multiple local and global features.

In Figure 15.8, an object and its description using features are shown. Both local and global features can be used to describe an object. The relations among objects can be used to form composite features.

15.5 Recognition Strategies

Object recognition is the sequence of steps that must be performed after appropriate features have been detected. As discussed earlier, based on the detected features in an image, one must formulate hypotheses about possible objects in the image. These hypotheses must be verified using models of objects. Not all object recognition techniques require strong hypothesis formation and verification steps. Most recognition strategies have evolved to

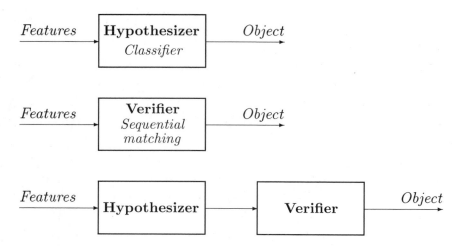

Figure 15.9: Depending on the complexity of the problem, a recognition strategy may need to use either or both the hypothesis formation and verification steps.

combine these two steps in varying amounts. As shown in Figure 15.9, one may use three different possible combinations of these two steps. Even in these, the application contest, characterized by the factors discussed earlier in this section, determines how one or both steps are implemented. In the following, we discuss a few basic recognition strategies used for recognizing objects in different situations.

15.5.1   Classification

The basic idea in classification is to recognize objects based on features. Pattern recognition approaches fall in this category, and their potential has been demonstrated in many applications. Neural net-based approaches also fall in this class. Some commonly used classification techniques are discussed briefly here. All techniques in this class assume that N features have been detected in images and that these features have been normalized so that they can be represented in the same metric space. We will briefly discuss techniques to normalize these features after classification. In the following discussion, it will be assumed that the features for an object can be represented as a point in the *N-dimensional* feature space defined for that particular object recognition task.

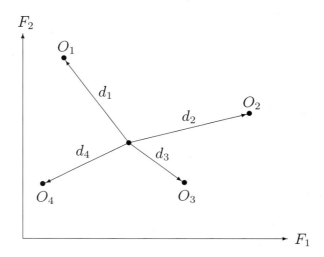

Figure 15.10: The prototypes of each class are represented as points in the feature space. An unknown object is assigned to the closest class by using a distance measure in this space.

Nearest Neighbor Classifiers

Suppose that a model object (ideal feature values) for each class is known and is represented for class i as f_{ij}, $j = 1, \ldots, N$, $i = 1, \ldots, M$ where M is the number of object classes. Now suppose that we detect and measure features of the unknown object U and represent them as u_j, $j = 1, \ldots, N$. For a 2-dimensional feature space, this situation is shown in Figure 15.10. To decide the class of the object, we measure its similarity with each class by computing its distance from the points representing each class in the feature space and assign it to the nearest class. The distance may be either Euclidean or any weighted combination of features. In general, we compute the distance d_j of the unknown object from class j as given by

$$d_i = \left[\sum_{j=1}^{N} (u_j - f_{ij})^2 \right]^{1/2}, \tag{15.1}$$

then the object is assigned to the class R such that

$$d_R = \min_{i=1}^{M} [d_i]. \tag{15.2}$$

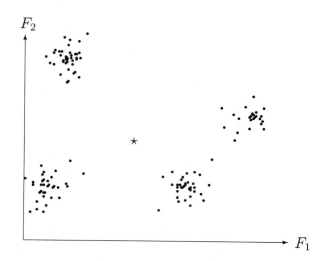

Figure 15.11: All known objects of each class are represented as points in the feature space. Each class is thus represented by a cluster of points in the feature space. Either the centroid of the cluster representing the class or the closest point of each class is considered the prototype for classification.

In the above, the distance to a class was computed by considering distance to the feature point representing a prototype object. In practice, it may be difficult to find a prototype object. Many objects may be known to belong to a class. In this case, one must consider feature values for all known objects of a class. This situation is shown in Figure 15.11. Two common approaches in such a situation are

1. Consider the centroid of the cluster as the prototype object's feature point, and compute the distance to this.

2. Consider the distance to the closest point of each class.

Bayesian Classifier

A Bayesian approach has been used for recognizing objects when the distribution of objects is not as straightforward as shown in the cases above. In general, there is a significant overlap in feature values of different objects.

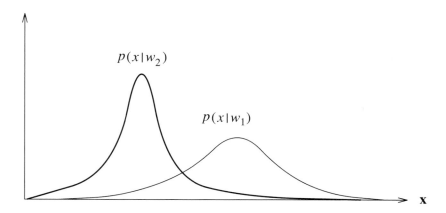

Figure 15.12: The conditional density function for $p(x|w_j)$. This shows the probability of the feature values for each class.

Thus, as shown for the one-dimensional feature space in Figure 15.12, several objects can have same feature value. For an observation in the feature space, multiple-object classes are equally good candidates. To make a decision in such a case, one may use a Bayesian approach to decision making.

In the Bayesian approach, probabilistic knowledge about the features for objects and the frequency of the objects is used. Suppose that we know that the probability of objects of class j is $P(w_j)$. This means that a priori we know that the probability that an object of class j will appear is $P(w_j)$, and hence in absence of any other knowledge we can minimize the probability of error by assigning the unknown object to the class for which $P(w_j)$ is maximum.

Decisions about the class of an object are usually made based on feature observations. Suppose that the probability $p(x|w_j)$ is given and is as shown in Figure 15.12. The conditional probability $p(x|w_j)$ tells us that based on the probabilistic information provided, we know that if the feature value is observed to be x, then the probability that the object belongs to class j is $p(x|w_j)$. Based on this knowledge, we can compute the a posteriori probability $p(w_j|x)$ for the object. The a posteriori probability is the probability that, for the given information and observations, the unknown object belongs to class j. Using Bayes' rule, this probability is given as:

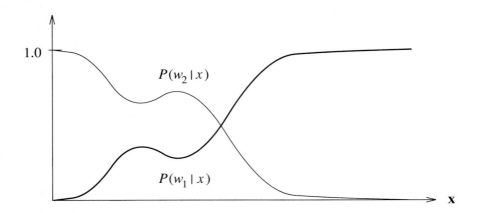

Figure 15.13: A posteriori probabilities for two different values of a priori probabilities for objects.

$$P(w_j|x) = \frac{p(x|w_j)P(w_j)}{p(x)} \tag{15.3}$$

where

$$p(x) = \sum_{j=1}^{N} p(x|w_j)P(w_j). \tag{15.4}$$

The unknown object should be assigned to the class with the highest a posteriori probability $P(w_j|x)$. As can be seen from the above equations, and as shown in Figure 15.13, a posteriori probability depends on prior knowledge about the objects. If a priori probability of the object changes, so will the result.

We discussed the Bayesian approach above for one feature. It can be easily extended to multiple features by considering conditional density functions for multiple features.

Off-Line Computations

The above classification approaches consider the feature space, and then, based on the knowledge of the feature characteristics of objects, a method is used to partition the feature space so that a class decision is assigned to each point in the feature space. To assign a class to each point in the feature space, all computations are done before the recognition of unknown objects begins.

This is called *off-line computation*. These off-line computations reduce the computations at the run time. The recognition process can be effectively converted to a *look-up table* and hence can be implemented very quickly.

Neural Nets

Neural nets have been proposed for object recognition tasks. Neural nets implement a classification approach. Their attraction lies in their ability to partition the feature space using nonlinear boundaries for classes. These boundaries are obtained by using training of the net. During the training phase, many instances of objects to be recognized are shown. If the training set is carefully selected to represent all objects encountered later during the recognition phase, then the net may learn the classification boundaries in its feature space. During the recognition phase, the net works like any other classifier.

The most attractive feature of neural nets is their ability to use nonlinear classification boundaries and learning abilities. The most serious limitations have been the inability to introduce known facts about the application domain and difficulty in debugging their performance.

15.5.2 Matching

Classification approaches use effective features and knowledge of the application. In many applications, a priori knowledge about the feature probabilities and the class probabilities is not available or not enough data is available to design a classifier. In such cases one may use direct matching of the model to the unknown object and select the best-matching model to classify the object. These approaches consider each model in sequence and fit the model to image data to determine the similarity of the model to the image component. This is usually done after the segmentation has been done. In the following we discuss basic matching approaches.

Feature Matching

Suppose that each object class is represented by its features. As above, let us assume that the jth feature's value for the ith class is denoted by f_{ij}. For an unknown object the features are denoted by u_j. The similarity of the object with the ith class is given by

$$S_i = \sum_{j=1}^{N} w_j s_j \qquad (15.5)$$

where w_j is the weight for the jth feature. The weight is selected based on the relative importance of the feature. The similarity value of the jth feature is s_j. This could be the absolute difference, normalized difference, or any other distance measure. The most common method is to use

$$s_j = |u_j - f_{ij}| \qquad (15.6)$$

and to account for normalization in the weight used with the feature.

The object is labeled as belonging to class k if S_k is the highest similarity value. Note that in this approach, we use features that may be local or global. We do not use any relations among the features.

Symbolic Matching

An object could be represented not only by its features but also by the relations among features. The relations among features may be spatial or some other type. An object in such cases may be represented as a graph. As shown in Figure 15.8, each node of the graph represents a feature, and arcs connecting nodes represent relations among the objects. The object recognition problem then is considered as a graph matching problem.

A graph matching problem can be defined as follows. Given two graphs G_1 and G_2 containing nodes N_{ij}, where i and j denote the graph number and the node number, respectively, the relations among nodes j and k is represented by R_{ijk}. Define a similarity measure for the graphs that considers the similarities of all nodes and functions.

In most applications of machine vision, objects to be recognized may be partially visible. A recognition system must recognize objects from their partial views. Recognition techniques that use global features and must have all features present are not suitable in these applications. In a way, the partial-view object recognition problem is similar to the graph embedding problem studied in graph theory. The problem in object recognition becomes different when we start considering the *similarity* of nodes and relations among them.

We discuss this type of matching in more detail later, in the section on verification.

15.5.3 Feature Indexing

If the number of objects is very large and the problem cannot be solved using feature space partitioning, then indexing techniques become attractive. The symbolic matching approach discussed above is a sequential approach and requires that the unknown object be compared with all objects. This sequential nature of the approach makes it unsuitable with a number of objects. In such a case, one should be able to use a hypothesizer that reduces the search space significantly. The next step is to compare the models of each object in the reduced set with the image to recognize the object.

Feature indexing approaches use features of objects to structure the modelbase. When a feature from the indexing set is detected in an image, this feature is used to reduce the search space. More than one feature from the indexing set may be detected and used to reduce the search space and in turn reduce the total time spent on object recognition.

The features in the indexing set must be determined using the knowledge of the modelbase. If such knowledge is not available, a learning scheme should be used. This scheme will analyze the frequency of each feature from the feature set and, based on the frequency of features, form the indexing set, which will be used for structuring the database.

In the indexed database, in addition to the names of the objects and their models, information about the orientation and pose of the object in which the indexing feature appears should always be kept. This information helps in the verification stage.

Once the candidate object set has been formed, the verification phase should be used for selecting the best object candidate.

15.6 Verification

Suppose that we are given an image of an object and we need to find how many times and where this object appears in an image. Such a problem is essentially a verification, rather than an object recognition, problem. Obviously a verification algorithm can be used to exhaustively verify the presence of each model from a large modelbase, but such an exhaustive approach will not be a very effective method. A verification approach is desirable if one, or at most a few, objects are possible candidates. There are many approaches for verification. Here we discuss some commonly used approaches.

15.6.1   Template Matching

Suppose that we have a template $g[i,j]$ and we wish to detect its instances in an image $f[i,j]$. An obvious thing to do is to place the template at a location in an image and to detect its presence at that point by comparing intensity values in the template with the corresponding values in the image. Since it is rare that intensity values will match exactly, we require a measure of dissimilarity between the intensity values of the template and the corresponding values of the image. Several measures may be defined:

$$\max_{[i,j]\,\in\,R} |f-g| \tag{15.7}$$

$$\sum_{[i,j]\,\in\,R} |f-g| \tag{15.8}$$

$$\sum_{[i,j]\,\in\,R} (f-g)^2 \tag{15.9}$$

where R is the region of the template.

The sum of the squared errors is the most popular measure. In the case of template matching, this measure can be computed indirectly and computational cost can be reduced. We can simplify:

$$\sum_{[i,j]\,\in\,R} (f-g)^2 = \sum_{[i,j]\,\in\,R} f^2 + \sum_{[i,j]\,\in\,R} g^2 - 2\sum_{[i,j]\,\in\,R} fg. \tag{15.10}$$

Now if we assume that f and g are fixed, then $\sum fg$ gives a measure of mismatch. A reasonable strategy for obtaining all locations and instances of the template is to shift the template and use the match measure at every point in the image. Thus, for an $m \times n$ template, we compute

$$M[i,j] = \sum_{k=1}^{m}\sum_{l=1}^{n} g[k,l]\,f[i+k,j+l] \tag{15.11}$$

where k and l are the displacements with respect to the template in the image.[1]

Our aim will be to find the locations that are local maxima and are above a certain threshold value. However, a minor problem in the above computation

---

[1]This operation is called the *cross-correlation* between f and g.

was introduced when we assumed that f and g are constant. When applying this computation to images, the template g is constant, but the value of f will be varying. The value of M will then depend on f and hence will not give a correct indication of the match at different locations. This problem can be solved by using normalized cross-correlation. The match measure M then can be computed using

$$C_{fg}[i,j] = \sum_{k=1}^{m} \sum_{l=1}^{n} g[k,l]\, f[i+k, j+l] \tag{15.12}$$

$$M[i,j] = \frac{C_{fg}[i,j]}{\{\sum_{k=1}^{m} \sum_{l=1}^{n} f^2[i+k, j+l]\}^{1/2}}. \tag{15.13}$$

It can be shown that M takes maximum value for $[i,j]$ at which $g = cf$. In Figure 15.14, we show an image, a template, and the result of the above computation. Notice that at the location of the template, we get local maxima.

The above computations can be simplified significantly in binary images. Template matching approaches have been quite popular in optical computing: frequency domain characteristics of convolution are used to simplify the computation.

A major limitation of template matching is that it only works for translation of the template. In case of rotation or size changes, it is ineffective. It also fails in case of only partial views of objects.

15.6.2 Morphological Approach

Morphological approaches can also be used to detect the presence and location of templates. For binary images, using the structuring element as the template and then *opening* the image will result in all locations where the template fits in. For gray images, one may use gray-image morphology. These results are shown for a template in Figure 15.15.

15.6.3 Symbolic

As discussed above, if both models of objects and the unknown object are represented as graphs, then some approach must be used for matching graphical representations. Here we define the basic concepts behind these approaches.

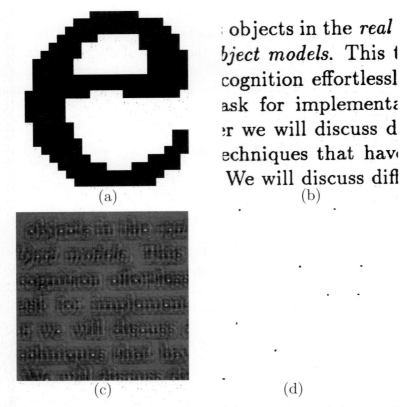

(a)

 objects in the *real
bject models*. This 1
cognition effortlessl
ask for implementa
:r we will discuss d
echniques that hav∢
We will discuss difl

(b)

(c)   (d)

Figure 15.14: A template (a), an image (b), the result of the template matching computations discussed above (c), and the thresholded result to find the match locations (d), $T = 240$.

Graph Isomorphism

Given two graphs (V_1, E_1) and (V_2, E_2), find a 1:1 and onto mapping (an isomorphism) f between V_1 and V_2 such that for $\theta_1, \theta_2 \in V_1, V_2$, $f(\theta_1) = \theta_2$ and for each edge of E_1 connecting any pair of nodes θ_1 and $\theta_1' \in V_1$, there is an edge of E_2 connecting $f(\theta_1)$ and $f(\theta_1')$.

Graph isomorphism can be used only in cases of completely visible objects. If an object is partially visible, or a 2.5-dimensional description is to be matched with a 3-dimensional description, then graph embedding, or subgraph isomorphisms, can be used.

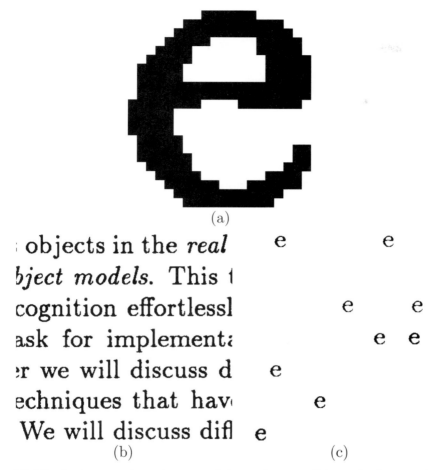

Figure 15.15: A structuring element (a), an image (b), and the result of the morphological opening (c).

Subgraph Isomorphisms

Find isomorphisms between a graph (V_1, E_1) and subgraphs of another graph (V_2, E_2).

A problem with these approaches for matching is that the graph isomorphism is an NP problem. For any reasonable object description, the time required for matching will be prohibitive. Fortunately, we can use more information than that used by graph isomorphism algorithms. This information is available in terms of the properties of nodes. Many heuristics have been proposed to solve the graph matching problem. These heuristics should consider:

- Variability in properties and relations

- Absence of properties or relations

- The fact that a model is *an abstraction* of a class of objects

- The fact that instances may contain extra information.

One way to formulate the similarity is to consider the arcs in the graph as springs connecting two masses at the nodes. The quality of the match is then a function of the goodness of fit of the templates locally and the amount of energy needed to stretch the springs to force the unknown onto the modelence data.

$$
\begin{aligned}
C = & \sum_{d \in R_1} \text{template cost}(d, F(d)) \\
& + \sum_{(d,e) \in R_2} \text{spring cost}(F(d), F(e)) \\
& + \sum_{c \in R_3} \text{missing cost}(c)
\end{aligned}
\tag{15.14}
$$

where $R_1 = \{\text{found in model}\}$, $R_2 = \{\text{found in model } x \text{ found in unknown}\}$, and $R_3 = \{\text{missing in model}\} \cup \{\text{missing in unknown}\}$. This function represents a very general formulation. *Template cost*, *spring cost*, and *missing cost* can take many different forms. Applications will determine the exact form of these functions.

15.6.4   Analogical Methods

A measure of similarity between two curves can be obtained by comparing them on the same frame of reference, as shown in Figure 15.16, and directly measuring the difference between them at every point. Notice that in Figure 15.16 the difference is measured at every point along the x axis. The difference will always be measured along some axis. The total difference is either

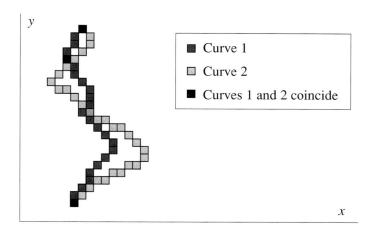

Figure 15.16: Matching of two entities by directly measuring the errors between them.

the sum of absolute errors or the sum of squared errors. If exact registration is not given, some variation of correlation-based methods must be used.

For recognizing objects using three-dimensional models, one may use rendering techniques from computer graphics to find their appearence in an image and then try to compare with the original image to verify the presence of an object. Since the parameters required to render objects are usually unknown, usually one tries to consider some prominent features on three-dimensional models and to detect them and match them to verify the model's instance in an image. This has resulted in development of theories that try to study three-dimensional surface characteristics of objects and their projections to determine *invariants* that can be used in object recognition. Invariants are usually features or characteristics in images that are relatively insensitive to an object's orientation and scene illumination. Such features are very useful in detecting three-dimensional objects from their two-dimensional projections.

Further Reading

Object recognition has been one of the most important topics in machine vision. In one form or another, it has attracted significant attention. Many approaches have been developed for pattern classification. These approaches are very useful in many applications of machine vision. For an excellent introduction to pattern classification, see [70]. Some very good survey papers on object recognition are by Chin and Dyer [57], Binford [34], and Besl and Jain [27].

Many object recognition systems are built upon low-level vision modules which operate upon images to derive depth measurements. These measurements are often incomplete and unreliable and thus adversely affect the performance of higher-level recognition modules. In contrast to this approach, Lowe describes a system in which bottom-up description of images is designed to generate viewpoint-invariant groupings of image features [158]. Brooks' ACRONYM system is a domain-independent model-based interpretation system which uses generalized cylinders for the description of model and scene objects [49, 50]. Some later work along this line was performed under the SUCCESSOR project and is given in [33].

Most object recognition research has considered a small set of objects. If a very large number of objects are to be recognized, the recognition task will be dominated by hypothesis and test approaches. The hypothesis phase will require organization of models indexed by features so that, based on observed features, a small set of likely objects can be selected. Later these selected models may be used to recognize objects by verifying which object from this set is present in the given image. Such approaches are given in Knoll and Jain [143], Ettinger [75], Grimson [93], Lamdan and Wolfson [151].

In many industrial applications, detailed geometric models of objects are available. These models can be used for generating recognition strategies, including feature selection, for three-dimensional objects. CAD-based object recognition is being studied at several places now [34, 99, 32, 221, 178]. An important step in the recognition of three-dimensional objects is to consider their possible two-dimensional projections to determine effective features and recognition strategy. Classification of infinite two-dimensional projection views of objects into topologically equivalent classes, called aspect graphs, was introduced by Koenderink and Van Doorn [144, 145]. Their application to recognition is described by Chakravarty and Freeman [56]. Gigus and Malik [88] developed an algorithm for generating aspect graphs. Recently algorithms have been designed for computing aspect graphs for curved objects also [73, 149, 226].

Ikeuchi and Kanade [120] describe a novel system in which the object and sensor models are automatically compiled into a visual recognition strategy.

The system extracts from the models those features that are useful for recognition and determines the control sequence that must be applied to handle different object appearances. An alternative to this kind of approach is presented by neural network approaches for object recognition. Object recognition is one of the most researched areas in neural networks. Most research in neural networks, however, has addressed only limited two-dimensional objects.

Exercises

15.1 List the major components of an object recognition system. Discuss their role in the recognition task.

15.2 Stereotyping is a phenomenon often criticized in society. Object recognition tasks, however, are dependent on stereotyping. Explain how stereotyping plays an important role in object recognition, particularly its role in relating modelbase and set of features.

15.3 What factors would you consider in selecting an appropriate representation for the modelbase? Discuss the advantages and disadvantages of object-centered and observer-centered representations.

15.4 What is an aspect graph? Develop a generalized aspect graph that is based on image features and their relationships for an object. Where can you use such an aspect graph?

15.5 What is feature space? How can you recognize objects using feature space?

15.6 Compare classical pattern recognition approaches based on Bayesian approaches with neural net approaches by considering the feature space, classification approaches, and object models used by both of these approaches.

15.7 One of the most attractive features of neural nets is their ability to learn. How is their ability to learn used in object recognition? What kind of model is prepared by a neural net? How can you introduce your knowledge about objects in neural nets?

15.8 Where do you use matching in object recognition? What is a symbolic matching approach?

15.9 What is feature indexing? How does it improve object recognition?

15.10 Discuss template matching. In which type of applications would you use template matching? What are the major limitations of template matching? How can you overcome these limitations?

15.11 Sketch the aspect graph of a four-faced trihedral polyhedron with triangular faces.

15.12 A template g is matched with an image f, both shown below, using the normalized cross-correlation method. Find:

    a. The cross-correlation C_{fg}.

    b. $\sum\sum f^2$.

    c. The normalized cross-correlation $M[i,j]$.

$$f = \begin{array}{|c|c|c|c|c|c|c|c|}
\hline
0 & 0 & 0 & 0 & 0 & 0 & 0 & 0 \\\hline
0 & 2 & 4 & 2 & 0 & 0 & 0 & 0 \\\hline
0 & 0 & 2 & 0 & 0 & 0 & 0 & 0 \\\hline
0 & 0 & 2 & 0 & 0 & 0 & 2 & 0 \\\hline
0 & 0 & 0 & 0 & 0 & 0 & 2 & 0 \\\hline
1 & 2 & 1 & 0 & 0 & 2 & 4 & 2 \\\hline
0 & 1 & 0 & 0 & 0 & 0 & 0 & 0 \\\hline
0 & 1 & 0 & 0 & 0 & 0 & 0 & 0 \\\hline
\end{array}
\qquad
g = \begin{array}{|c|c|c|}
\hline
1 & 2 & 1 \\\hline
0 & 1 & 0 \\\hline
0 & 1 & 0 \\\hline
\end{array}$$

Computer Projects

15.1 Implement an object recognition system to recognize objects from their partial views. The objects in an image are from a given set of about 10 objects that are commonly found in an office scene. Select only objects that are more or less two-dimensional (coins, keys, sticky pads, business cards, etc.). Consider the camera to be mounted about 8 feet above the desk. Test your system by considering many random images in which these objects appear in different ways.

15.2 Continuing the above example, now consider that the objects are three-dimensional (mouse, stapler, etc.), and redesign and reimplement a prototype object recognition system. This system should recognize three-dimensional objects from their partial views.

15.3 Now assume that you have a large number of objects in your modelbase. Redesign your system to perform the object recognition task efficiently for a large number of objects.

Appendix A

Mathematical Concepts

Some of the mathematical concepts used in this book are summarized in this appendix. Section A.1 includes some formulas for geometry that are generally useful in machine vision, Section A.2 covers linear spaces, and Section A.3 describes the variational calculus which is used in solving ill-posed problems through regularization.

A.1 Analytic Geometry

Let point $\mathbf{p} = (x, y)$ in two dimensions or $\mathbf{p} = (x, y, z)$ in three dimensions.

The unit vector $\mathbf{u}$ that represents the orientation of a vector $\mathbf{v}$ is the vector of cosines of vector $\mathbf{v}$ with respect to each of the coordinate axes,

$$\mathbf{u} = \frac{1}{\|\mathbf{v}\|} \left(\mathbf{v} \cdot \mathbf{e}_1, \mathbf{v} \cdot \mathbf{e}_2, \ldots, \mathbf{v} \cdot \mathbf{e}_n \right), \tag{A.1}$$

where $\mathbf{e}_i$ is the unit vector for coordinate axis i.

The parametric equation for a line is

$$\mathbf{p}(t) = t\mathbf{u} + \mathbf{p}_0, \tag{A.2}$$

where $\mathbf{u}$ is the unit vector that defines the orientation of the line, $\mathbf{p}_0$ is a point through which the line passes, and $-\infty \le t \le \infty$. This equation can also be used to describe rays (half open lines) and line segments by restricting the domain of t. For $0 \leq t \leq \infty$, Equation A.2 represents the vector starting at point $\mathbf{p}_0$ and pointing in the direction $\mathbf{u}$. For $0 \leq t \leq 1$, Equation A.2 represents the unit line segment between point $\mathbf{p}_0$ and point $\mathbf{p}_1$ given by $\mathbf{p}_1 = \mathbf{p}_0 + \mathbf{u}$.

The equation for the line segment between two points can be written as

$$\mathbf{p}(t) = (1 - t)\mathbf{p}_1 + t\mathbf{p}_2 \tag{A.3}$$

with $0 \leq t \leq 1$.

Three distinct points $\mathbf{p}_0$, $\mathbf{p}_1$, and $\mathbf{p}_2$ define a plane in space. Let $\mathbf{v}_1 = \mathbf{p}_1 - \mathbf{p}_0$ and $\mathbf{v}_2 = \mathbf{p}_2 - \mathbf{p}_0$. If vectors $\mathbf{v}_1$ and $\mathbf{v}_2$ lie in a plane and are not parallel, then the normal to the plane is

$$\mathbf{n} = \mathbf{v}_1 \times \mathbf{v}_2. \tag{A.4}$$

The normal points in the direction of the thumb as the fingers of the right hand sweep from $\mathbf{v}_1$ to $\mathbf{v}_2$. The implicit equation for the plane is the set of points $\mathbf{p}$ that are orthogonal to the plane normal $\mathbf{n}$,

$$\mathbf{n} \cdot (\mathbf{p} - \mathbf{p}_0) = 0, \tag{A.5}$$

where $\mathbf{p}_0$ is a point in the plane and allows the plane to be offset from the origin. If the normal vector $\mathbf{n} = (a, b, c)$, then the plane equation can be written as

$$ax + by + cz + d = 0 \tag{A.6}$$

where $d = -\mathbf{n} \cdot \mathbf{p}_0$ accounts for the displacement of the plane from the origin.

The parametric form of a surface, such as a plane, is an equation of the form $\mathbf{p}(u, v) = (x(u, v), y(u, v), z(u, v))$. The parametric equation for a plane is derived by noting that the origin in the (u, v) domain maps to a point $\mathbf{p}_0 = (x_0, y_0, z_0)$ on the plane, the $\mathbf{u}$-axis in the domain maps to a vector $\mathbf{v}_1$ in the plane, and the $\mathbf{v}$-axis in the domain maps to a vector $\mathbf{v}_2$ in the plane. The parametric equation is

$$\mathbf{p}(u, v) = A \begin{pmatrix} u \\ v \end{pmatrix} + \mathbf{p}_0, \tag{A.7}$$

where the columns of the 3×2 matrix A are the vectors $\mathbf{v}_1$ and $\mathbf{v}_2$.

A.2 Linear Algebra

The notion of a linear space is based on some common assumptions about how physical systems should behave. The power of linear spaces in science and engineering comes from this correspondence between simple mathematical models and real physical systems.

A set of scalars F is a field if the scalars obey the following conditions:

1. If x and y are elements of F, then $x + y$ and xy are elements of F.

2. If x is an element of F, then the additive inverse $-x$ is an element of F.

3. If x is an element of F and $x \neq 0$, then the multiplicative inverse x^{-1} is an element of F.

4. The additive identity 0 and the multiplicative identity 1 are both elements of F.

For example, the set of real numbers with the usual forms of addition and multiplication is a field.

The mathematical model that comes from combining scalars into vectors to represent points in space and many other things is a very powerful model. Let vector $\mathbf{u} = (u_1, u_2, \ldots, u_n)$ and vector $\mathbf{v} = (v_1, v_2, \ldots, v_n)$. Addition of vectors is defined by the scalar addition of the corresponding vector elements,

$$\mathbf{u} + \mathbf{v} = (u_1 + v_1, u_2 + v_2, \ldots, u_n + v_n), \tag{A.8}$$

and multiplication of a vector $\mathbf{v}$ by a scalar a is defined by applying scalar multiplication to the individual elements of the vector:

$$a\mathbf{v} = (av_1, av_2, \ldots, av_n). \tag{A.9}$$

A vector space, also called a linear space, obeys the following conditions:

1. Addition of vectors $\mathbf{u}$ and $\mathbf{v}$ is commutative

$$\mathbf{u} + \mathbf{v} = \mathbf{v} + \mathbf{u}. \tag{A.10}$$

2. Addition of vectors $\mathbf{u}$, $\mathbf{v}$, and $\mathbf{w}$ is associative:

$$(\mathbf{u} + \mathbf{v}) + \mathbf{w} = \mathbf{u} + (\mathbf{v} + \mathbf{w}). \tag{A.11}$$

3. There is a vector denoted by 0 that is the identity element for vector addition:
$$0 + \mathbf{u} = \mathbf{u} + 0 = \mathbf{u}. \tag{A.12}$$

4. For every vector $\mathbf{v}$ there is an additive inverse:
$$\mathbf{v} + (-\mathbf{v}) = 0. \tag{A.13}$$

5. Multiplication of a vector sum by a scalar c is distributed to the individual vectors:
$$c(\mathbf{u} + \mathbf{v}) = c\mathbf{u} + c\mathbf{v}. \tag{A.14}$$

6. Multiplication of a vector by a sum of scalars can be rewritten as the sum of the individual scalar multiplications:
$$(a + b)\mathbf{v} = a\mathbf{v} + b\mathbf{v}. \tag{A.15}$$

7. Multiplication of a vector by a product of scalars is associative:
$$(ab)\mathbf{v} = a(b\mathbf{v}). \tag{A.16}$$

8. There is an identity element for multiplication of a vector by a scalar:
$$1\mathbf{u} = \mathbf{u}. \tag{A.17}$$

The linear space is fundamental to science and engineering because it is the mathematical model for systems that behave linearly, which means that the systems behave in a simple way and are easy to understand and use in design. A linear system S obeys the conditions of superposition and homogeneity:

$$\begin{aligned} S[x + y] &= S[x] + S[y] & \text{(A.18)} \\ S[\alpha x] &= \alpha S[x] & \text{(A.19)} \end{aligned}$$

which say that the response to a sum of inputs is the sum of the responses to the individual inputs, and the response to an input scaled by a constant is the scaled response. The linearity conditions correspond to our intuitive notions about how things should behave. For example, we expect that a light that is twice as bright would make a scene appear to be twice as bright, and we expect that the result of using two lights should be the sum of the results of using each light alone.

A linear combination is a sum of terms multiplied by constant coefficients:

$$a_1b_1 + a_2b_2 + \cdots + a_nb_n. \tag{A.20}$$

An element $\mathbf{v}$ in an n-dimensional vector space V can be represented as a linear combination of n basis vectors:

$$\mathbf{v} = a_1\mathbf{e}_1 + a_1\mathbf{e}_2 + \cdots + a_n\mathbf{e}_n, \tag{A.21}$$

where $\mathbf{e}_i$ is a basis vector and a_i is the corresponding coefficient. A linear transformation is a mapping between vector spaces and can be implemented by matrix multiplication applied to the vector of coefficients that represent an element in a vector space relative to some basis. The natural basis is the set of vectors $\mathbf{e}_1, \mathbf{e}_2, \ldots, \mathbf{e}_n$ with $\mathbf{e}_i = 1$ at position i and zero elsewhere. The coefficients for the natural basis are the coordinates of the vector in the usual Cartesian coordinate system.

Functional analysis extends the notion of vector space to spaces of functions that can be represented by linear combinations of basis functions:

$$f(t) = a_1b_1(t) + a_2b_2(t) + \cdots + a_nb_n(t). \tag{A.22}$$

This is a finite-dimensional vector space, since the basis contains a finite number of basis functions; but there are linear spaces with an infinite number of dimensions that require an infinite number of basis functions, such as Fourier series. Finite-dimensional vector spaces play an important role in statistics, and hence in this book, since a model can be represented by a finite number of parameters that can be estimated by linear regression.

The scalar product supports the notions of the length of a vector and the angle between vectors. The square of the length of a vector is the scalar product of the vector with itself,

$$\|\mathbf{v}\|^2 = \mathbf{v} \cdot \mathbf{v}, \tag{A.23}$$

which is the sum of the squares of the coefficients that represent the vector relative to some basis. The angle between two vectors is the scalar product of the two vectors, normalized by the length of each vector. The angle provides a measure of the difference between two vectors.

Consider two vectors $\mathbf{v}_1$ and $\mathbf{v}_2$ in the usual Euclidean three-dimensional space that are not collinear. Vectors $\mathbf{v}_1$ and $\mathbf{v}_2$ define a plane in space. Any linear combination of $\mathbf{v}_1$ and $\mathbf{v}_2$ is a vector in the plane,

$$\mathbf{v} = a_1\mathbf{v}_1 + a_2\mathbf{v}_2, \tag{A.24}$$

and any point in the plane can be reached with a unique linear combination of the vectors. This plane is a subspace of the three-dimensional Euclidean space, and the vectors $\mathbf{v}_1$ and $\mathbf{v}_2$ form a basis that spans the subspace.

Now consider a third vector $\mathbf{v}_3$ that does not lie in the plane spanned by the vectors $\mathbf{v}_1$ and $\mathbf{v}_2$ as defined in the preceding paragraph. The three vectors $\mathbf{v}_1$, $\mathbf{v}_2$, and $\mathbf{v}_3$ form a basis that spans the entire three-dimensional space. Consider some vector $\mathbf{v}$ that is not coplanar with $\mathbf{v}_1$ and $\mathbf{v}_2$. Vector $\mathbf{v}$ can be written as a linear combination of two vectors,

$$\mathbf{v} = a_1\mathbf{u} + a_2\mathbf{w}, \tag{A.25}$$

such that vector $\mathbf{u}$ lies in the plane spanned by $\mathbf{v}_1$ and $\mathbf{v}_2$, and vector $\mathbf{w}$ is perpendicular to that plane. This concept is fundamental to applications in optimization and regression.

The usual scalar product between vectors $\mathbf{u}$ and $\mathbf{v}$ can be generalized to a quadratic form that includes a weight on each scalar product between corresponding vector elements,

$$a_1 u_1 v_1 + a_2 u_2 v_2 + \cdots + a_n u_n v_n, \tag{A.26}$$

which can be further generalized to a weighted sum of the products between all pairwise combinations of the elements of vectors $\mathbf{u}$ and $\mathbf{v}$:

$$\sum_{i=1}^{n} \sum_{j=1}^{n} a_{ij} u_i v_j. \tag{A.27}$$

This weighted sum can be written in matrix notation as

$$\mathbf{u} A \mathbf{v}^T, \tag{A.28}$$

where the superscript T denotes the transpose of a row vector into a column vector. There is no loss of generality in assuming that matrix A is symmetric.

The concept of eigenvalues and eigenvectors starts simply with the idea that there may be a linear transformation represented by matrix A and some vector $\mathbf{x}$ such that

$$A\mathbf{x} = \lambda\mathbf{x} \qquad (A.29)$$

for some constant λ. Vector $\mathbf{x}$ is an eigenvector of the linear transformation, and λ is its eigenvalue.

Let matrix A be a linear transformation that scales and rotates the natural basis of a vector space so that the axes (basis vectors) correspond in length and orientation with the axes of some ellipsoid that is centered on the origin. Consider a unit vector $\mathbf{u}$ that is already aligned with both an axis of the ellipse and one of the axes of the coordinate system. Since the unit vector is aligned with the axes of the ellipsoid, it will not change orientation under the transformation into the orientation of the ellipsoid, but will change length so that it is scaled to the length of the axis of the ellipsoid; thus,

$$A\mathbf{u} = \lambda\mathbf{u}, \qquad (A.30)$$

where λ is the length of the axis of the ellipsoid. This shows that the eigenvalues and eigenvectors are the length and orientation of the axes of the ellipsoid.

A.3   Variational Calculus

Let $f(x, y)$ be the function that is the solution to a variational problem. The general form of a variational problem is

$$\int \int F(x, y, f, f_x, f_y, f_{xx}, f_{xy}, f_{yy}, \ldots)\, dx\, dy.$$

The solution to the variational problem is given as a partial differential equation, called the Euler equation, which is constructed from a formula involving various partial derivatives of F. The general form for a function of two variables is

$$
\begin{aligned}
F_f - \frac{\partial}{\partial x}F_{f_x} + \frac{\partial^2}{\partial x^2}F_{f_{xx}} - \cdots + (-1)^n \frac{\partial^n}{\partial x^n}F_{f_{x^{(n)}}} \\
- \frac{\partial}{\partial y}F_{f_y} + \frac{\partial^2}{\partial y^2}F_{f_{yy}} - \cdots + (-1)^n \frac{\partial^n}{\partial y^n}F_{f_{y^{(n)}}} = 0 \qquad (A.31)
\end{aligned}
$$

if F does not contain any cross derivatives. In computing the derivatives of the integrand F with respect to a partial derivative of the solution function f, the partial derivative of the solution function is treated as a single variable even though the variable is denoted by a symbol with subscripts. Note how the sign alternates with the order of the derivatives and how the rows of the formula for x and y have the same form. If f is a function of more than two variables, then the Euler equation is extended with an additional sequence of terms for each of the additional variables. If F contains cross derivatives, then there will be additional terms to handle the cross derivatives. If the variational problem requires finding more than one function, then each function will yield another Euler equation.

As an example, consider the problem of determining the surface $z = f(x, y)$ that interpolates a set of data points z_k at locations (x_k, y_k) in a rectangular region R of the image plane, for $k = 1, \ldots, n$. This is an ill-posed problem, since there are an infinite number of functions that can interpolate a set of points. To make the problem well posed, choose the function that is smooth according to the norm

$$\int\int_R \left[\nabla^2 f(x, y)\right]^2 dx\, dy, \tag{A.32}$$

which means choose the function that minimizes Equation A.32 and interpolates the data points z_k at locations (x_k, y_k). Using the calculus of variations, the solution is the biharmonic equation

$$\nabla^4 f(x, y) = 0 \tag{A.33}$$

with boundary conditions

$$f_{yy} = 0 \tag{A.34}$$
$$f_{xxy} = 0 \tag{A.35}$$

along the top and bottom edges of the rectangular domain and boundary conditions

$$f_{xx} = 0 \tag{A.36}$$
$$f_{yyx} = 0 \tag{A.37}$$

along the left and right sides of the rectangular domain. The boundary conditions mean that the interpolated surface does not have to assume any particular value or orientation along the boundary, but should smoothly approach the boundary.

Finding the partial differential equation and boundary conditions for a problem in the variational calculus is only part of the solution. The partial differential equations must be solved by numerical methods.

A.4   Numerical Methods

There are many numerical methods for solving partial differential equations such as the biharmonic Equation A.33 and its boundary conditions. All of the methods involve replacing the partial derivatives with finite difference approximations. There are basically two approaches: (1) replace the partial derivatives in the variational problem, such as Equation A.32, with finite difference approximations to obtain a system of equations that can be solved numerically, or (2) replace the partial derivatives in the partial differential equation derived from the variational problem—Equation A.33, for instance—with finite difference approximations to obtain a system of equations and solve the equations numerically. The numerical methods that solve the variational problem directly can be more efficient, but the iterative methods that solve the partial differential equations are easier to describe and implement.

The simplest finite difference methods solve a partial differential equation at points on a uniform grid. The solution is an array of values for the function $z = f(x, y)$ at grid locations $[i, j]$, with $i = 1, \ldots, n$ and $j = 1, \ldots, m$. The finite difference approximation for the biharmonic evaluated at grid location $[i, j]$ is a linear combination of the function values at neighboring grid locations. Imagine that the n by m grid of function values $f[i, j]$ is unfolded into a long vector $f[k]$, with $k = (i - 1)m + j$. Let N_k be the list of offsets for the grid locations of the neighbors of $f[k]$ in the grid. The biharmonic equation is approximated at each grid location by

$$\nabla^4 f(x, y) \approx a_0 f[k] + \sum_{l \in N_k} a_l f[k + l] = 0 \qquad (A.38)$$

with changes to the coefficients near the boundaries of the domain. For each grid location, this linear equation provides one row in a system of linear equations

$$A f = 0 \qquad (A.39)$$

except at the grid locations where a known value is being interpolated, in which case that row is filled with zeros except at $f[k] = 1$ and the right-hand size is the known value z_k. The system of linear equations can be solved with sparse matrix techniques, or Equation A.38 can be solved for $f[k]$ and this formula repeated over all grid locations that do not have a known value until the solution vector does not change significantly. This is the method of successive approximation and is described in *Numerical Recipes* [197], along with more sophisticated methods for solving partial differential equations and variational problems.

Further Reading

The series of books *Graphics Gems* [89] provide many useful formulas and algorithms for geometry.

Linear spaces are widely used in science and engineering. Lang [153] provides a rigorous introduction to linear spaces. Noble and Daniel [187] cover linear spaces with many practical applications. Naylor and Sell [182] cover functional analysis, which is the extension of vector spaces to spaces of functions, and include many examples from science and engineering.

Many problems can be formulated as optimization problems and solved using the variational calculus. An excellent introduction to the variational calculus is provided by Courant and Hilbert [62, Chap. 4]. The variational calculus produces a partial differential equation that usually must be solved by numerical methods. The book *Numerical Recipes* is an excellent introduction to numerical methods [197].

Appendix B

Statistical Methods

Machine vision could be called statistical geometry, since vision involves estimating geometrical information from image data. This book uses some basic material from statistics, such as the normal distribution and linear regression, which are reproduced in this appendix. The book also uses some new ideas in statistics for formulating measurement algorithms that are robust to unmodeled errors.

B.1 Measurement Errors

There are three types of performance parameters for a sensor or measurement procedure. Resolution or precision is the smallest change in the value that a sensor can report. Repeatability is the variation in repeated measurements of the same quantity. Accuracy is the variation in measurements of a known true value. It is easy to remember the relationship between accuracy and repeatability:

$$\text{Accuracy} = \text{Repeatability} + \text{Calibration}. \tag{B.1}$$

There is a similar relationship between the components of error:

$$\text{Error} = \text{Variance} + \text{Bias}. \tag{B.2}$$

The variance is the error in repeatability for a measurement; the bias is the systematic error due to lack of calibration. For example, suppose that you are measuring the length of a hallway with a yardstick, but instead of a yardstick you accidentally pick up a meter stick. Your measurements would still have the same repeatability, assuming that you were just as careful in laying the measuring stick end to end as you measured the length of the hall. But there would be a systematic bias, a constant proportional to the true length of the hall, due to the incorrect length of the measuring stick. Bias can be removed through careful calibration, but variance (or repeatability) is a characteristic limitation of the measurement method.

The histogram is a useful tool for seeing the distribution, including both variance and bias, of a measurement. The range of measurements on the real line is partitioned into a finite number of intervals called buckets. A one-dimensional integer array of length equal to the number of buckets is used to count the number of occurrences of measurements that fall in the intervals. A plot of the histogram shows the distribution of measurements. The width of the plot is an indication of variance, and the difference between the location of the center of the distribution and the true measurement is an indication of bias.

A measurement y_i of some quantity x can be corrupted by additive error:

$$y_i = x + e_i, \tag{B.3}$$

where e_i is the error in the measurement. If the error were known or if the error could be removed through calibration, then each measurement would be an accurate estimate of the unknown quantity, and no further processing would be necessary. However, the repeatability of a measurement procedure is not perfect, and each measurement will include error drawn from some distribution leading to a distribution of measurements somehow related to the unknown parameter. Statistics is the science of measurement procedures and includes methods for estimating unknown parameters given various assumptions about the characteristics of the errors.

The average of a set of n measurements is

$$\bar{x} = \frac{1}{n} \sum_{i=1}^{n} x_i \tag{B.4}$$

The median is computed by sorting the measurements and choosing the middle element (or averaging the two middle elements if the number of measurements is even). The mode is the location of the peak in the distribution of measurements. The average and median are methods for estimating an unknown parameter from measurements corrupted by additive errors from a symmetric distribution. For example, pixels can be averaged over local neighborhoods, or the pixel values obtained at the same location in a sequence of images can be averaged to reduce the noise in the measurements of gray value.

Several different measurements of the amount of error can be computed. Mean absolute error (MAE) is $\pm\delta_x$, where

$$\delta_x = \frac{1}{n}\sum |x_i - \mu_x| \tag{B.5}$$

and μ_x is the average or median. Root mean square (RMS) error is $\pm\sigma_x$, where

$$\sigma_x^2 = \frac{1}{(n-1)}\sum (x_i - \mu_x)^2 \tag{B.6}$$

and μ_x is the average. Maximum error is $\pm\epsilon_x$, where

$$\epsilon_x = \max_i |x_i - \mu_x| \tag{B.7}$$

and μ_x is the average or median. Note that $\delta \leq \sigma \leq \epsilon$.

B.2   Error Distributions

The binomial distribution models measurement processes with a finite number of outcomes, called events. For example, a fair coin that is flipped once will show heads with probability $1/2$ and tails with probability $1/2$. This set of outcomes is modeled by the polynomial

$$P(x) = 0.5 + 0.5x, \tag{B.8}$$

where the coefficient for each power of x is the probability that heads will occur that many times. If the coin is tossed n times, then the probabilities of various numbers of heads can be determined by expanding the polynomial:

$$P(x; n) = (0.5 + 0.5x)^n. \tag{B.9}$$

The probability that heads will occur i times is the coefficient of x^i in the expanded polynomial. In general, a single measurement with k outcomes can be modeled by a polynomial of order k,

$$P(x) = p_0 + p_1 x + p_2 x^2 + \cdots + p_{k-1} x^{k-1}, \tag{B.10}$$

where p_i is the probability of outcome i and

$$\sum_{i=1}^{k} p_i = 1. \tag{B.11}$$

The cumulative results of various combinations of outcomes after a sequence of n measurements are modeled by raising the polynomial for one measurement to power n,

$$P(x; n) = \left(p_0 + p_1 x + p_2 x^2 + \cdots + p_{k-1} x^{k-1} \right)^n, \tag{B.12}$$

expanding the polynomial, and calculating each coefficient.

The uniform distribution is used to model measurements that are equally likely. For example, if a point is located anywhere with equal probability in the rectangular region defined by points (x_1, y_1) and (x_2, y_2) at the corners, then the probability that the point is at location (x, y) is

$$P(x, y) = \begin{cases} 1/A & \text{if } (x, y) \text{ is in the region} \\ 0 & \text{otherwise} \end{cases} \tag{B.13}$$

where A is the area of the region.

The normal distribution is a useful approximation to the errors in many measurement processes. The normal distribution $N(x; \mu, \sigma^2)$ is

$$N(x; \mu, \sigma^2) = \frac{1}{\sqrt{2\pi}\sigma} e^{\frac{-(x-\mu)^2}{2\sigma^2}} \tag{B.14}$$

where μ is the location of the center of the distribution and σ^2 is the variance. Since bias is usually eliminated by proper calibration, the normal distribution for errors is usually zero mean.

Some measurement processes occasionally produce gross errors, called outliers, in addition to normally distributed errors. Such an error process can be modeled as the mixture of a normal distribution and some unknown, broad-tailed distribution:

$$(1 - \nu)N(x; \mu_1, \sigma_1^2) + \nu B(x; \mu_2, \sigma_2^2), \tag{B.15}$$

where ν represents the odds that an outlier will occur. There is a probability of ν that the measurement will be contaminated with error from the outlier distribution and a probability of $\nu - 1$ that the measurement will be subject to normally distributed errors. Typically, the bias from both error processes is eliminated through proper calibration, and so both error distributions are zero mean:

$$(1 - \nu)N(x; 0, \sigma_1^2) + \nu B(x; 0, \sigma_2^2). \tag{B.16}$$

Since outliers are extreme errors, $\sigma_2 \gg \sigma_1$. As an example, the Cauchy distribution

$$C(x; a, b) = \frac{1}{\pi} \frac{a}{a^2 + (x - b)^2} \tag{B.17}$$

has such broad tails that the variance is infinite.

B.3 Linear Regression

Given n data points and a model with m parameters $a_1, a_2, \ldots, a_m$, the least-squares error in the fit of the model to the data is

$$\chi^2 = \sum_{i=1}^{n} \left(\frac{y_i - y(x_i; a_1, a_2, \ldots, a_m)}{\sigma_i} \right)^2. \tag{B.18}$$

This is weighted least-squares regression, where the weights σ_i are the errors (standard deviations) in the measurements so that less noisy measurements are given more weight.

Often, the error is the same for each measurement, or the individual measurement errors are unknown and assumed to be identical, in which case the least-squares regression problem is to determine the parameters $a_1, a_2, \ldots, a_m$ that minimize

$$\chi^2 = \sum_{i=1}^{n} (y_i - y(x_i; a_1, a_2, \ldots, a_m))^2. \tag{B.19}$$

If the model is linear in the parameters, then the problem is linear regression. A model is linear if it can be represented as a linear combination of basis functions:

$$y(x; a_1, a_2, \ldots, a_m) = a_1 \phi_1(x) + a_2 \phi_2(x) + \cdots + a_m \phi_m(x), \tag{B.20}$$

where the functions $\phi_i(x)$ do not depend on the model parameters. The coefficients $a_1, a_2, a_3, \ldots, a_n$ of the linear combination are the model parameters to be determined through regression. For example, the line

$$y = ax + b \qquad (B.21)$$

is a linear model with parameters a and b.

A linear least-squares regression problem can be reliably solved using standard numerical routines for singular value decomposition [84, pp. 534–539], which also provides a measure of the error in the fitted parameters. Consider fitting a linear model

$$z = a_1 + a_2 x + a_3 y \qquad (B.22)$$

to a set of data points $\{(x_1, y_1, z_1), (x_2, y_2, z_2), \ldots, (x_n, y_n, z_n)\}$. Each data point leads to a constraint

$$z_i \approx a_1 + a_2 x_i + a_3 y_i, \qquad (B.23)$$

and the set of data points leads to a set of constraints that can be written in matrix form:

$$
\begin{bmatrix}
1 & x_1 & y_1 \\
1 & x_2 & y_2 \\
& \vdots & \\
1 & x_n & y_n
\end{bmatrix}
\begin{bmatrix}
a_1 \\
a_2 \\
a_3
\end{bmatrix}
=
\begin{bmatrix}
z_1 \\
z_2 \\
\vdots \\
z_n
\end{bmatrix}. \qquad (B.24)
$$

These equations are usually written in the more compact form

$$AX = B. \qquad (B.25)$$

The A matrix is not square and cannot be inverted directly. One technique is to multiply both sides of Equation B.25 by A^T to form the *normal equations* and solve the normal equations using standard techniques for solving systems of linear equations such as LU decomposition. A better technique is to use singular value decomposition. There are several advantages to using singular value decomposition: (1) it is not necessary to premultiply Equation B.25 to form the normal equations, (2) singular value decomposition handles ill-conditioned systems of equations, and (3) the singular values provided as a by-product of SVD indicate redundancies (unnecessary terms) in the model.

Since singular value decomposition is used frequently for the algorithms in this book and *Numerical Recipes* [197] is an excellent source for good numerical algorithms, the method for solving a linear regression problem using the singular value decomposition routine in *Numerical Recipes* will be presented in detail. (Note that some elements of C programming are described in Appendix C.) The interface for the numerical routine for singular value decomposition is

```
void svdcmp (a, n, m, w, v)
float **a, *w, **v;
int n, m;
```

The routine takes an array a with n rows and m columns and replaces it with the singular value decomposition

$$A = UWV^T. \tag{B.26}$$

The array a is replaced by U in the singular value decomposition, the array w receives the singular values in the diagonal matrix W, and the array v receives the V matrix. Note that n is the number of observations (measurements), while m is the number of parameters in the linear model. It is easy to fill the entries in the array a according to Equation B.24 and call svdcmp to obtain the singular value decomposition. After obtaining the singular value decomposition, use the routine svbksb to solve for the model parameters:

```
void svbksb (u, w, v, n, m, b, x)
float **u, *w, **v, *b, *x;
int n, m;
```

The u, w, and v arrays are the a, w, and v arrays computed by svdcmp. The dimensions n and m are the same as for svdcmp. The b array is the right-hand side of Equation B.24, and the array x is the parameter vector (solution). In other words, the combination of svdcmp and svbksb solve the nonsquare system of linear equations presented as Equation B.24. The code fragment for invoking the routines is

```
float wmin, wmax;
svdcmp (a, n, m, w, v);
wmax = 0.0;
for (j = 1; j <= m; j++)
  if (w[j] > wmax) wmax = w[j];
```

```
wmin = wmax * 1.0e-6;
for (j = 1; j <= m; j++)
  if (w[j] < wmin) w[j] = 0.0;
svbksb (a, w, v, n, m, b, x);
```

Small singular values indicate problems in the regression model. The code provided above sets small singular values to zero, which is one safe way to handle the problem. The constant `1.0e-6` is a typical value but may be different for some applications. If there are small singular values, it is important to carefully analyze the model and determine why the small singular values occur. There may be terms in the model that are unnecessary.

Singular value decomposition is a very reliable algorithm but can fail to produce a good regression estimate for several reasons:

- The wrong model may be used in formulating the regression problem.

- The measurement errors σ_i may be too large.

- The measurement errors may not be from a normal distribution. For example, the errors could be from a broad-tailed distribution.

The probability distribution for χ^2 when the minimum value of the regression norm is obtained is the chi-square distribution with $\nu = n - m$ degrees of freedom. The probability that chi-square should exceed χ^2 by chance is

$$Q = \frac{1}{\Gamma(\frac{\nu}{2})} \int_{\chi^2}^{\infty} e^{-t} t^{\frac{\nu}{2}-1} \, dt. \tag{B.27}$$

The regression error χ^2 should be calculated as part of the regression procedure. The integral can be calculated using numerical methods, but values of Q are provided in statistical tables. If $Q > 0.001$, then the regression fit should probably be rejected. This provides an objective way to evaluate the model fitting procedure. In practice, the value for Q is selected based on knowledge of the application. The value for ν is determined by the number of data points and the order of the model. The corresponding value for χ^2 can be found in statistical tables. If the measured value for χ^2 obtained during regression exceeds the tabulated value, then the algorithm has not succeeded, and the results should be discarded. If excessive values for χ^2 are encountered frequently, then the model is probably wrong or linear regression is not the right approach, and perhaps robust regression should be used instead.

B.4   Nonlinear Regression

If the model is nonlinear in its parameters, the least-squares regression problem is to minimize χ^2 for the nonlinear model. Since the formula for the model is known, both the gradient and Hessian can be calculated. This allows the Levenberg-Marquardt method to be used for solving nonlinear regression problems [197, pp. 540–547].

Seber and Wild [217] is an excellent text on nonlinear regression. In some cases, the arguments to the Levenberg-Marquardt routine in *Numerical Recipes* may not match the intended application, but *Numerical Recipes* provides routines for the Newton-Raphson method, which is discussed in practical texts on nonlinear regression [23].

Further Reading

There are many excellent textbooks on probability and statistics. Any book that describes experimental methods in science and engineering or statistical methods in the social sciences would be sufficient. For example, Beck and Arnold [24] cover linear and nonlinear methods for regression with emphasis on applications in engineering and science. Box, Hunter, and Hunter [43] have written a classic text on experimental methods. Drake [69] has written a basic introduction to probability and statistics, while Papoulis [192] provides more comprehensive coverage. Gaussian error models are used in communications theory [257], which has influenced machine vision and pattern recognition. Vanmarcke [244] and Cressie [63] present probability and statistical methods for spatial data which may be useful as further readings in machine vision. Robust regression methods are summarized in the article by Efron and Tibshirani [72]. Finally, *Numerical Recipes in C* is an excellent source for statistics algorithms [197] and explains regression particularly well.

Appendix C

Programming Techniques

This appendix covers some aspects of C programming that are important for image processing and outlines the design of the image processing routines that were used to produce some of the examples in this book.

C.1 Image Descriptors

Images can be represented in C using image descriptors, which are structures that contain all of the information needed by subroutines to process an image. Image descriptors are an efficient representation and allow an image to be passed to a subroutine through a single subroutine parameter.

An image descriptor contains the following fields:

address	Pointer to the beginning of the image array
height	Number of rows in the image
width	Number of columns in the image
span	Allocated length of each row in the image array

The image descriptor fields should be accessed with macros defined in a header file so that programs are independent of changes in the image representation. A portion of the header file is listed in Program C.1.

Program C.1 The definition of pixels and image descriptors

```
typedef int pixel;

typedef struct _id
{
    pixel *id_address;
    int id_height,
        id_width,
        id_span;
} image_descriptor;

#define address(id_ptr)      ((id_ptr) -> id_address)
#define span(id_ptr)         ((id_ptr) -> id_span)
#define height(id_ptr)       ((id_ptr) -> id_height)
#define width(id_ptr)        ((id_ptr) -> id_width)
```

For convenience, pixels are defined to be 32-bit signed numbers. By redefining the pixel data type and recompiling the programs, the software can be changed to use more common pixel data types such as unsigned bytes. However, image processing algorithms may yield intermediate results that do not conform to the limited range of small integers. More common numerical representations, such as single-precision floating point numbers, may be more suitable for many image processing operations.

The C routine **array** takes a pointer to an image descriptor and returns an array of pointers to the first pixel in each row of the image. A pointer to the array of row pointers is stored in the image descriptor. The routine **array** uses the array stored in the descriptor if it exists; otherwise, it allocates and returns the array of row pointers after storing it in the image descriptor. The storage used by the array of row pointers is reclaimed when the storage used by the image descriptor itself is freed by a call to **free_image**. The array of pointers to the image rows should not be freed explicitly.

The array of image row pointers allows C style array access. As an example of the use of image descriptors, consider the routine listed in Program C.2 for summing the pixels in an image.

Program C.2 A routine for summing the pixels in an image using an array of pointers to the image rows

```
int sum_image(image)
image_descriptor *image;
{
  pixel **image_array = array(image);
  int row, column, sum;

  for (sum = 0, row = 0; row < height(image); row ++)
    for (column = 0; column < width(image); column ++)
      sum += image_array[row][column];
  return (sum);
}
```

Using an array of pointers to the array rows is the easiest method for accessing multidimensional arrays in C and is probably the style that is used most often, but there are situations where it is necessary to write C code that manipulates the pointers directly. Program C.3 sums the pixels in an image without using an array of pointers to the image rows. Pointer arithmetic is coded directly in the program.

Program C.3 A routine for summing the pixels in an image by advancing pointers down the rows and across the columns of the image

```
int sum_image(image)
image_descriptor *image;
{
  pixel *rowptr, *colptr;
  int rowcnt, colcnt, sum;

  for (sum = 0, rowcnt = height(image),
       rowptr = address(image);
       rowcnt > 0;
       rowptr += span(image), rowcnt--)
    for (colcnt = width(image), colptr = rowptr;
         colcnt > 0;
         colcnt--)
```

```
            sum += *(colptr++);
    return (sum);
}
```

The third alternative for accessing the pixels in an image is to code the subscript calculations for accessing each pixel. The formula for accessing the pixel at a specified row and column in an image is

```
address(image)[row * span(image) + column]
```

A program for copying one image to another is listed in Program C.4.

Program C.4 Program to copy one image to another that will work correctly even if the images are not the same size or have different row dimensions

```
copy_image (input, output)
image_descriptor *input, *output;
{
  pixel *input_row_ptr, *output_row_ptr, *inptr, *outptr;
  int rowcnt, colcnt;

  for (rowcnt = min(height(input), height(output)),
       input_row_ptr = address(input),
       output_row_ptr = address(output);
       rowcnt > 0;
       rowcnt--,
       input_row_ptr += span(input),
       output_row_ptr += span(output))
    {
      for (colcnt = min(width(input), width(output)),
           inptr = input_row_ptr,
           outptr = output_row_ptr;
           colcnt > 0;
           colcnt--)
        *(outptr++) = *(inptr++);
    }
}
```

Program C.4 is carefully written to work correctly even if the images are of different sizes or the image spans are different. Defensive coding makes the programs more robust.

The reason for including the span (also called the stride or row dimension) in the image descriptor is that the image may be stored in an array that is wider than the image. For example, Program C.5 returns a subimage of an image. The beauty of image descriptors is that the programs that handle images will work with subimages since the operations using image descriptors are identical.

Program C.5 A routine to return a subimage of an image

```
image_descriptor *make_subimage
                (image, subimage_row, subimage_column,
                 subimage_height, subimage_width)
image_descriptor *image;
int subimage_row, subimage_column,
                                subimage_height, subimage_width;
{
  image_descriptor *subimage;

  subimage = (image_descriptor *)
                          malloc(sizeof(image_descriptor));

  init_image_descriptor (subimage,
                    pixel_address(image, subimage_row,
                                         subimage_column),
                    span(image),
                    subimage_height,
                    subimage_width);
  return (subimage);
}
```

Routines are provided for allocating images and freeing images: `make_image` and `free_image`, respectively.

C.2   Mapping Operators

Some patterns of code occur repeatedly in image processing programs. For example, the code for incrementing pixel pointers and counts to apply a local operator to successive windows of an image will occur without change in every routine that applies local operators to images. These program idioms can be coded in a set of mapping routines that apply an arbitrary operator to an image. The operators are not entirely arbitrary since the number of images must be coded into the argument list. As an example, consider Program C.6, which applies an operator of arbitrary size across successive supports of an input image and places the output into successive pixels of an output image.

Program C.6   Routine to map an operator with arbitrary support across an image and store the results in an output image. The image operator takes a pointer to a window as input and returns a pixel.

```
map_image_support (input, output, function,
                   support_height, support_width)
image_descriptor *input, *output;
int support_height, support_width;
pixel (* function)();
{
  int output_height = height(input) - support_height + 1,
      output_width  = width(input)  - support_width  + 1;
  image_descriptor support_descriptor;
  image_descriptor *support = &support_descriptor;
  int rowcnt, colcnt;
  pixel *input_row_ptr, *output_row_ptr,
        *input_column_ptr, *output_column_ptr;

  init_image_descriptor (support, NULL, span(input),
                         support_height, support_width);

  for (rowcnt = min(output_height, height(output)),
       input_row_ptr = address(input),
       output_row_ptr = address(output);
       rowcnt > 0;
       rowcnt--, input_row_ptr += span(input),
```

```
        output_row_ptr += span(output))
    for (colcnt = min(output_width, width(output)),
         input_column_ptr = input_row_ptr,
         output_column_ptr = output_row_ptr;
         colcnt > 0; colcnt--)
      {
        address(support) = input_column_ptr++;
        *(output_column_ptr++) = (* function)(support);
      }
}
```

As an example of how a mapping function can be used, the following statement uses `map_image_support` listed in Program C.6 and `sum_image` listed in Program C.3 to smooth an image with a 3×3 averaging mask:

```
map_image_support (input, output, sum_image, 3, 3);
```

The statement works because `sum_image` is passed an image descriptor for the 3 by 3 window. Note that the mapping functions may not update the array of row pointers computed by `array`, since this would be inefficient. To use mapping functions, the pointer calculations are coded directly into the routines that implement the local operations.

Mapping functions can be very useful since they allow arbitrarily complicated operators to be applied to images. This is generalized convolution and is used frequently in image processing.

C.3 Image File Formats

There are many image file formats in use, but the formats supported by the *pbmplus* package are portable across different machine architectures. The package handles binary images, gray-level images, and multichannel (color) images and is available on the Internet. The pbmplus package includes C routines for reading and writing image files and utility programs for converting between the pbmplus file formats and other image file formats.

Images stored in pbmplus files are not compressed, but the files can be compressed by programs such as *compress*, which is available on Unix systems, or *gzip* which is available on the Internet.

Further Reading

There are many excellent books on programming in C. The original book on C was written by Kernighan and Ritchie [142]. Harbison and Steele wrote a reference manual on C that is useful for experienced C programmers [104].

The book *Numerical Recipes* [197] describes methods for programming matrix and vector operations, and *Graphics Gems* [89] includes some discussion of programming techniques for image processing.

The Tcl/Tk toolkit is useful for developing programs with interactive interfaces and also provides an extensible command line interpreter [191]. Image processing operations can be implemented as commands that are invoked interactively or through the command line interface. The command language allows machine vision algorithms to be written as scripts, so that repetitive operations can be performed easily. Machine vision applications can be modified by changing the scripts, so it is not necessary to change the image processing programs.

Bibliography

[1] I. E. Abdou and W. K. Pratt. Quantitative design and evaluation of enhancement/thresholding edge detectors. *Proceedings of the IEEE*, 67(5):753–763, May 1979.

[2] J. K. Aggarwal and N. Nandhakumar. On the computation of motion from sequences of images: A review. Technical Report TR-88-2-47, University of Texas, Austin, April 1988.

[3] N. Ahuja and A. L. Abbot. Active stereo: Integrating disparity, vergence, focus, aperture, and calibration for surface estimation. *IEEE Trans. Pattern Analysis and Machine Intelligence*, 15(10):1007–1029, 1993.

[4] J. Aisbett. Optical flow with an intensity-weighted smoothing. *IEEE Trans. Pattern Analysis and Machine Intelligence*, 11(5):512–522, May 1989.

[5] J. Y. Aloimonos. Shape from texture. *Biological Cybernetics*, 58(5):345, 1988.

[6] J. Y. Aloimonos and I. Rigoutsos. Determining the 3-d motion of a rigid planar patch without correspondence, under perspective projection. *Proc. Workshop on Motion*, pages 167–174, 1986.

[7] J. Y. Aloimonos and I. Rigoutsos. Determining the 3-d motion of a rigid surface patch without correspondence, under perspective projection. i. planar surfaces; ii. curved surfaces. *Proc. National Conf. on Artificial Intelligence*, pages 681–688, 1986.

[8] J. Y. Aloimonos, I. Weiss, and A. Bandyopadhyay. Active vision. *Int. J. Computer Vision*, 1:333–356, 1988.

[9] P. Anandan. A computational framework and an algorithm for the measurement of visual motion. *Int. J. Computer Vision*, 2(3):283–310, 1989.

[10] H. Asada and M. Brady. The curvature primal sketch. *IEEE Trans. Pattern Analysis and Machine Intelligence*, 8(1):2–14, January 1986.

[11] N. Ayache and B. Faverjon. Efficient registration of stereo images by matching graph descriptions of edge segments. *Int. J. Computer Vision*, pages 107–131, 1987.

[12] N. Ayache and F. Lustman. Fast and reliable trinocular stereovision. *Proc. First Intl. Conf. Computer Vision*, pages 422–427, 1987.

[13] R. Bajcsy and L. Lieberman. Texture gradient as a depth cue. *Comp. Graph. and Image Proc.*, 5:52–67, 1976.

[14] R. Bajcsy. Active perception. *Proc. IEEE*, 76(8):996–1005, 1988.

[15] H. H. Baker and R. C. Bolles. Generalizing epipolar-plane image analysis on the spatiotemporal surface. *Int. J. Computer Vision*, 3:33–49, 1989.

[16] H. H. Baker and T. O. Binford. Depth from edge and intensity based stereo. *Proc. 7th Int. Joint Conf. AI*, pages 631–636, 1981.

[17] W. H. Baker. *Elements of Photogrammetry*. Ronald Press Company, New York, 1960.

[18] D. H. Ballard. Generalizing the hough transform to detect arbitrary shapes. *Pattern Recognition*, 13(2):111–122, 1981.

[19] D. H. Ballard and C. M. Brown. *Computer Vision*. Prentice-Hall, Englewood Cliffs, New Jersey, 1982.

[20] S. T. Barnard and M. A. Fischler. Computational stereo. *Computing Surveys*, 14(4):553–572, December 1982.

[21] S. T. Barnard and W. B. Thompson. Disparity analysis of images. *IEEE Trans. Pattern Analysis and Machine Intelligence*, 2(4):333–340, July 1980.

[22] R. H. Bartels, J. C. Beatty, and B. A. Barsky. *An Introduction to Splines for Use in Computer Graphics and Geometric Modeling*. Morgan Kaufmann, Los Altos, California, 1987.

[23] D. M. Bates and D. G. Watts. *Nonlinear Regression Analysis and Its Applications*. Wiley, New York, 1988.

[24] J. V. Beck and K. J. Arnold. *Parameter Estimation in Engineering and Science*. Wiley, New York, 1977.

[25] F. Bergholm. Edge focusing. *IEEE Trans. Pattern Analysis and Machine Intelligence*, 9(6):726–741, November 1987.

[26] P. J. Besl. Active, optical range imaging sensors. *Machine Vision and Applications*, 1:127–152, 1988.

[27] P. J. Besl and R. C. Jain. Three-dimensional object recognition. *Computing Surveys*, 17(1):75–145, March 1985.

[28] P. J. Besl and R. C. Jain. Segmentation through variable-order surface fitting. *IEEE Trans. Pattern Analysis and Machine Intelligence*, 10(2):167–192, March 1988.

[29] P. J. Besl and N. D. McKay. A method for registration of 3-d shapes. *IEEE Trans. Pattern Analysis and Machine Intelligence*, 14(2):239–256, February 1992.

[30] J. R. Beveridge et al. Segmenting images using localized histograms and region merging. *Int. J. Computer Vision*, 2(3):311–347, 1989.

[31] J. D. E. Beynon and D. R. Lamb, editors. *Charge-Coupled Devices and Their Applications*. McGraw-Hill Book Company (UK) Limited, London, 1980.

[32] B. Bhanu and C. Ho. CAD-based 3-d object recognition for robot vision. *Computer*, 20(8):19–36, 1987.

[33] T. O. Binford. Spatial understanding: The successor system. *Proc. DARPA Image Understanding Workshop*, pages 12–20, 1989.

[34] T. O. Binford. Survey of model-based image analysis systems. *Int. J. Robotics Research*, 1(1):18–64, 1982.

[35] A. Blake and A. Zisserman. *Visual Reconstruction*. MIT Press, Cambridge, 1987.

[36] D. Blostein and N. Ahuja. Shape from texture: Integrating texture-element extraction and surface estimation. *IEEE Trans. Pattern Analysis and Machine Intelligence*, 11(12):1233–1251, 1989.

[37] R. C. Bolles, H. H. Baker, and D. H. Marimont. Epipolar-plane image analysis: An approach to determining structure from motion. *Int. J. Computer Vision*, 1:7–55, 1987.

[38] R. C. Bolles and M. A. Fischler. A RANSAC-based approach to model fitting and its application to finding cylinders in range data. In *Int. Joint. Conf. Artificial Intelligence*, pages 637–643, 1981.

[39] F. L. Bookstein. *The Measurement of Biological Shape and Shape Change*. Lecture Notes in Biomathematics, 24. Springer, New York, 1978.

[40] F. L. Bookstein. A statistical method for biological shape comparisons. *J. Theoretical Biology*, 107:475–520, 1984.

[41] F. L. Bookstein. Size and shape spaces. *Statistical Science*, 1:181–242, 1986.

[42] F. L. Bookstein. Principal warps: Thin-plate splines and the decomposition of deformations. *IEEE Trans. Pattern Analysis and Machine Intelligence*, 11(6):567–585, June 1989.

[43] G. E. P. Box, W. G. Hunter, and J. S. Hunter. *Statistics for Experimenters: An Introduction to Design, Data Analysis, and Model Building*. Wiley, New York, 1978.

[44] K. L. Boyer and A. C. Kak. Color-encoded structured light for rapid active ranging. *IEEE Trans. Pattern Analysis and Machine Intelligence*, 9(1):14–28, January 1987.

[45] J. M. Brady and H. Asada. Smoothed local symmetries and their implementation. *Int. J. Robotics Research*, 3:36–61, 1984.

[46] M. Brady, J. Ponce, A. Yuille, and H. Asada. Describing surfaces. *Computer Vision, Graphics and Image Processing*, 32(1):1–28, October 1985.

[47] T. J. Broida and R. Chellappa. Estimation of object motion parameters from noisy images. *IEEE Trans. Pattern Analysis and Machine Intelligence*, 8(1):90–99, January 1986.

[48] T. J. Broida and R. Chellappa. Performance bounds for estimating three-dimensional motion parameters from a sequence of noisy images. *J. Opt. Soc. Am. A*, 6(6), June 1989.

[49] R. A. Brooks. Model-based three-dimensional interpretations of two-dimensional images. *IEEE Trans. Pattern Analysis and Machine Intelligence*, 5(2):140–149, 1983.

[50] R. A. Brooks, R. Greiner, and T. O. Binford. The acronym model-based vision system. In *Int. Joint. Conf. Artificial Intelligence*, 1979.

[51] D. C. Brown. Close range camera calibration. *Photogrammetric Engineering*, 37(8):855–866, August 1971.

[52] P. J. Burt. Fast algorithms for estimating local image properties. *Computer Vision, Graphics and Image Processing*, 21:368–382, 1983.

[53] J. F. Canny. A computational approach to edge detection. *IEEE Trans. Pattern Analysis and Machine Intelligence*, 8(6):679–698, November 1986.

[54] P. Cavanaugh. Size and position invariance in the vision system. *Perception*, 7:167–177, 1978.

[55] P. Cavanaugh. Size invariance: Reply to Schwartz. *Perception*, 10:469–474, 1981.

[56] I. Chakravarty and H. Freeman. Characteristic views as a basis for three-dimensional object recognition. *Proc. SPIE Conf. Robot Vision*, 336:37–45, 1982.

[57] R. T. Chin and C. R. Dyer. Model-based recognition in robot vision. *ACM Computing Surveys*, 18(1):67–108, March 1986.

[58] W. F. Clocksin. Perception of surface slant and edge labels from optical flow: A computational approach. *Perception*, 9:253–269, 1980.

[59] W. J. Cody and W. Waite. *Software Manual for the Elementary Functions*. Prentice-Hall, Englewood Cliffs, New Jersey, 1980.

[60] L. Cohen et al. Hierarchical region-based stereo matching. *Proc. Conf. Computer Vision and Pattern Recognition*, pages 416–421, 1989.

[61] R. W. Conners and C. A. Harlow. Equal probability quantizing and texture analysis of radiographic images. *Comp. Graph. and Image Proc.*, 8:447–463, 1978.

[62] R. Courant and D. Hilbert. *Methods of Mathematical Physics*, volume 1. John Wiley & Sons, Interscience Publishers, New York, 1937.

[63] N. A. C. Cressie. *Statistics for Spatial Data*. Wiley, New York, 1991.

[64] G. R. Cross and A. K. Jain. Markov random field texture models. *IEEE Trans. Pattern Analysis and Machine Intelligence*, 5(1):25–39, 1983.

[65] J. L. Crowley and R. Stern. Fast computation of the difference of low-pass transform. *IEEE Trans. Pattern Analysis and Machine Intelligence*, 6:212–222, March 1984.

[66] E. De Micheli, B. Caprile, P. Ottonello, and V. Torre. Localization and noise in edge detection. *IEEE Trans. Pattern Analysis and Machine Intelligence*, 11(10):1106–1117, October 1989.

[67] M. P. do Carmo. *Differential Geometry of Curves and Surfaces*. Prentice-Hall, Englewood Cliffs, New Jersey, 1976.

[68] E. R. Dougherty and C. R. Giardina. *Mathematical Methods for Artificial Intelligence and Autonomous Systems.* Prentice-Hall, Englewood Cliffs, New Jersey, 1988.

[69] A. W. Drake. *Fundamentals of Applied Probability Theory.* McGraw-Hill, New York, 1967.

[70] R. O. Duda and P. Hart. *Pattern Classification and Scene Analysis.* Wiley, New York, 1973.

[71] B. Efron. Computer-intensive methods in statistical regression. *SIAM Review,* 30(3):421–449, September 1988.

[72] B. Efron and R. Tibshirani. Statistical data analysis in the computer age. *Science,* 253:390–395, July 26, 1991.

[73] D. Eggert and K. Bowyer. Computing the orthographic projection aspect graph of solids of revolution. *Proc. IEEE Workshop on Interpretation of 3-D Scenes,* pages 102–108, 1989.

[74] P. Eichel and E. Delp. Sequential edge detection in correlated random fields. In *Proc. Conf. Computer Vision and Pattern Recognition,* pages 14–21, 1985.

[75] G. J. Ettinger. Large hierarchical object recognition in robot vision. *Proc. Conf. Computer Vision and Pattern Recognition,* pages 32–41, 1988.

[76] J.-Q. Fang and T. S. Huang. Solving three-dimensional small-rotation motion equations: Uniqueness, algorithms, and numerical results. *Computer Vision, Graphics and Image Processing,* 26:183–206, 1984.

[77] J.-Q. Fang and T. S. Huang. Some experiments on estimating the 3-d motion parameters of a rigid body from two consecutive image frames. *IEEE Trans. Pattern Analysis and Machine Intelligence,* 6(5):545–554, 1984.

[78] O. Faugeras. *Three-dimensional Computer Vision, a Geometric Viewpoint.* MIT Press, Cambridge, 1993.

[79] C. L. Fennema and W. B. Thompson. Velocity determination in scenes containing several moving objects. *Comp. Graph. and Image Proc.*, 9:301–315, 1979.

[80] M. A. Fischler and R. C. Bolles. Random sample consensus: A paradigm for model fitting with applications to image analysis and automated cartography. *Communications Association of Computing Machinery*, 24(6):381–395, June 1981.

[81] J. D. Foley, A. van Dam, S. K. Feiner, and J. F. Hughes. *Computer Graphics: Principles and Practice*. Addison-Wesley, second edition, 1990.

[82] T. A. Foley. Interpolation and approximation of 3-d and 4-d scattered data. *Computers and Mathematics with Applications*, 13(8):711–740, 1987.

[83] T. A. Foley. Weighted bicubic spline interpolation to rapidly varying data. *ACM Trans. Graphics*, 6:1–18, January 1987.

[84] W. Frei and C. Chen. Fast boundary detection: A generalization and a new algorithm. *IEEE Trans. Computers*, 26(10):988–998, 1977.

[85] S. Geman and D. Geman. Stochastic relaxation, Gibbs distributions, and the Bayesian restoration of images. *IEEE Trans. Pattern Analysis and Machine Intelligence*, 6:721–741, 1984.

[86] J. J. Gibson. Optical motion and transformations as stimuli for visual perception. *Psychological Review*, 64(5):228–295, 1957.

[87] J. J. Gibson. What gives rise to the perception of motion? *Psychological Review*, 75(4):335–346, 1968.

[88] Z. Gigus and J. Malik. Computing the aspect graph for line drawings of polyhedral objects. *IEEE Trans. Pattern Analysis and Machine Intelligence*, 12(2):113–122, 1990.

[89] A. S. Glassner, editor. *Graphics Gems*. Academic Press, New York, 1990.

[90] R. Gonzalez and R. Woods, *Digital Image Processing*. Addison-Wesley, 1992.

[91] R. Gonzalez and R. Woods, *Digital Image Processing*. Addison-Wesley, 1992.

[92] W. E. L. Grimson. *From Images to Surfaces: A Computational Study of the Human Early Vision System*. MIT Press, Cambridge, 1981.

[93] W. E. L. Grimson. Recognition of object families using parameterized models. *Proc. First Int'l. Conf. Computer Vision*, pages 93–101, 1987.

[94] E. Grosso, G. Sandini, and M. Tistarelli. 3-d object recognition using stereo and motion. *IEEE Trans. Systems, Man, and Cybernetics*, 19(6):1465–1476, 1989.

[95] B. Hallert. *Photogrammetry: Basic Principles and General Survey*. McGraw-Hill, New York, 1960.

[96] R. W. Hamming. *Digital Filters*. Prentice-Hall, Englewood Cliffs, New Jersey, second edition, 1983.

[97] F. R. Hampel, E. M. Ronchetti, P. J. Rousseeuw, and W. A. Stahel. *Robust Statistics: An Approach Based on Influence Functions*. Wiley, New York, 1986.

[98] C. Hansen, N. Ayache, and F. Lustman. Towards real-time trinocular stereo. *Proc. Second Int'l. Conf. Computer Vision*, pages 129–133, 1988.

[99] C. Hansen and T. C. Henderson. CAGD-based computer vision. *IEEE Trans. Pattern Analysis and Machine Intelligence*, 11:1181–1193, November 1989.

[100] R. M. Haralick, K. Shanmugam, and I. Dinstein. Textural features for image classification. *IEEE Trans. Systems, Man, and Cybernetics*, 3(6):610–621, 1973.

[101] R. M. Haralick and L. G. Shapiro. Image segmentation techniques. *Computer Vision, Graphics and Image Processing*, 29(1):100–132, 1985.

[102] R. M. Haralick. Digital step edges from zero crossing of second directional derivatives. *IEEE Trans. Pattern Analysis and Machine Intelligence*, 6(1):58–68, January 1984.

[103] R. M. Haralick and L. G. Shapiro. *Computer and Robot Vision*. Addison-Wesley, Reading, Massachusetts, vol. 1, 1992, vol. 2, 1993.

[104] S. P. Harbison and G. L. Steele, Jr. *C: A Reference Manual*. Prentice-Hall, Englewood Cliffs, New Jersey, third edition, 1991.

[105] G. Healey. Using color for geometry-insensitive segmentation. *J. Opt. Soc. Am. A*, 6(6):920–937, 1989.

[106] D. J. Heeger. Optical flow using spatiotemporal filters. *Int. J. Computer Vision*, 1:279–302, 1988.

[107] B. K. P. Horn. *Shape from Shading: A Method for Obtaining the Shape of a Smooth Opaque Object from One View*. PhD thesis, Massachusetts Institute of Technology, 1970.

[108] B. K. P. Horn. Understanding image intensities. *Artificial Intelligence*, 8:201–231, 1977.

[109] B. K. P. Horn. *Robot Vision*. McGraw-Hill, New York, 1986.

[110] B. K. P. Horn. Closed-form solution of absolute orientation using unit quaternions. *J. Opt. Soc. Am. A*, 4(4):629–642, April 1987.

[111] B. K. P. Horn. Relative orientation. *Int. J. Computer Vision*, 4:59–78, January 1990.

[112] B. K. P. Horn and M. J. Brooks. *Shape from Shading*. MIT Press, Cambridge, 1989.

[113] B. K. P. Horn, H. M. Hilden, and S. Negahdaripour. Closed-form solution of absolute orientation using orthonormal matrices. *J. Opt. Soc. Am. A*, 5(7):1127–1135, July 1988.

[114] B. K. P. Horn and B. G. Schunck. Determining optical flow. *Artificial Intelligence*, 17:185–203, 1981.

[115] P. J. Huber. *Robust Statistics*. Wiley, New York, 1981.

[116] M. Hueckel. A local visual operator which recognizes edges and lines. *Journal of the Association of Computing Machinery*, 20:634–646, 1973.

[117] A. Huertas and G. Medioni. Detection of intensity changes with sub-pixel accuracy using laplacian-gaussian masks. *IEEE Trans. Pattern Analysis and Machine Intelligence*, 8(5):651–664, September 1986.

[118] R. A. Hummel. Representations based on zero-crossings in scale-space. In *Proc. Conf. Computer Vision and Pattern Recognition*, pages 204–209, 1986.

[119] V. S. S. Hwang, L. S. Davis, and T. Matsuyama. Hypothesis integration in image understanding systems. *Computer Vision, Graphics and Image Processing*, 36:321–371, 1986.

[120] K. Ikeuchi and T. Kanade. Automatic generation of object recognition programs. *Proceedings of the IEEE*, 76(8):1016–1035, 1988.

[121] A. K. Jain. *Fundamentals of Digital Image Processing*. Prentice-Hall, Englewood Cliffs, New Jersey, 1989.

[122] R. C. Jain. Dynamic scene analysis using pixel-based processes. *Computer*, 12(1):12–18, 1981.

[123] R. C. Jain. Complex logarithmic mapping and the focus of expansion. *SIGGRAPH/SIGART Workshop or Motion: Representation and Control*, pages 42–49, 1983.

[124] R. C. Jain. Direct computation of the focus of expansion. *IEEE Trans. Pattern Analysis and Machine Intelligence*, 5(1):58–64, 1983.

[125] R. C. Jain. Difference and accumulative difference pictures in dynamic scene analysis. *Image and Vision Computing*, 2(2):99–108, May 1984.

[126] R. C. Jain. Segmentation of frame sequences obtained by a moving observer. *IEEE Trans. Pattern Analysis and Machine Intelligence*, 6(5):624–629, 1984.

[127] R. C. Jain, S. L. Bartlett, and N. O'Brien. Motion stereo using ego-motion complex logarithmic mapping. *IEEE Trans. Pattern Analysis and Machine Intelligence*, 9(3):356–369, 1987.

[128] R. C. Jain and A. K. Jain, editors. *Analysis and Interpretation of Range Images*. Springer-Verlag, New York, 1990.

[129] R. C. Jain, D. Militzer, and H.-H. Nagel. Separating non-stationary from stationary scene components in a sequence of real world tv-images. In *Int. Joint. Conf. Artificial Intelligence*, pages 612–618, 1977.

[130] R. C. Jain and H.-H. Nagel. On the analysis of accumulative difference pictures from image sequences of real world scenes. *IEEE Trans. Pattern Analysis and Machine Intelligence*, 1(2):206–214, April 1979.

[131] R. A. Jarvis. A perspective on range finding techniques for computer vision. *IEEE Trans. Pattern Analysis and Machine Intelligence*, 5(2):122–139, 1983.

[132] M. R. M. Jenkin. Tracking three-dimensional moving light displays. *Proc. Workshop Motion: Representation Contr.*, pages 66–70, 1983.

[133] C. Jerian and R. C. Jain. Polynomial methods for structure from motion. *IEEE Trans. Pattern Analysis and Machine Intelligence*, 12(12):1150–1166, 1990.

[134] C. Jerian and R. C. Jain. Structure from motion: A critical analysis of methods. *IEEE Trans. Systems, Man, and Cybernetics*, 1991.

[135] K. Kanatani and T. C. Chou. Shape from texture: General principle. *Proc. Conf. Computer Vision and Pattern Recognition*, pages 578–583, June 1989.

[136] K. Kanatani and T. C. Chou. Shape from texture: General principle. *Artificial Intelligence*, 38(1):1–48, 1989.

[137] H. M. Karara, editor. *Handbook of Non-Topographic Photogrammetry*. American Society of Photogrammetry, 1979.

[138] R. L. Kashyap and R. Chellapa. Decision rules for choice of neighbors in random field models of images. *Comp. Graph. and Image Proc.*, 15:301–318, 1981.

[139] R. Kasturi and R. C. Jain. *Computer Vision: Advances and Applications*. IEEE Computer Society Press, Los Alamitos, California, 1991.

[140] R. Kasturi and R. C. Jain. *Computer Vision: Principles*. IEEE Computer Society Press, Los Alamitos, California, 1991.

[141] J. K. Kearney, W. B. Thompson, and D. L. Boley. Optical flow estimation: An error analysis of gradient-based methods with local optimization. *IEEE Trans. Pattern Analysis and Machine Intelligence*, 9(2):229–244, March 1987.

[142] B. W. Kernighan and D. M. Ritchie. *The C Programming Language*. Prentice-Hall, Englewood Cliffs, New Jersey, second edition, 1988.

[143] T. F. Knoll and R. C. Jain. Recognizing partially visible objects using feature indexed hypotheses. *IEEE J. Robotics and Automation*, 2(1):3–13, March 1986.

[144] J. J. Koenderink and A. J. van Doorn. The singularities of the visual mapping. *Biological Cybernetics*, 24:51–59, 1976.

[145] J. J. Koenderink and A. J. van Doorn. The internal representation of solid shape with respect to vision. *Biological Cybernetics*, 32:211–216, 1979.

[146] T. R. Kohl. *VisionTutor^{TM} Lab Guide*. Amerinex Artificial Intelligence, Inc., Amherst, Massachusetts, 1992.

[147] T. Kohonen. *Content-Addressable Memories*. Springer-Verlag, New York, 1987.

[148] B. Kosko. *Neural Networks and Fuzzy Systems*. Prentice-Hall, Englewood Cliffs, New Jersey, 1992.

[149] D. J. Kriegman and J. Ponce. Computing exact aspect graphs of curved objects: Solids of revolution. *Proc. IEEE Workshop on Interpretation of 3-D Scenes*, pages 116–122, 1989.

[150] E. P. Krotkov. *Exploratory Visual Sensing for Determining Spatial Layout with an Agile Stereo Camera System*. PhD thesis, University of Pennsylvania, 1987.

[151] Y. Lamdan, J. T. Schwartz, and H. J. Wolfson. Object recognition by affine invariant matching. *Proc. Conf. Computer Vision and Pattern Recognition*, pages 335–344, 1988.

[152] P. Lancaster and K. Šalkauskas. *Curve and Surface Fitting*. Academic Press, New York, 1986.

[153] S. Lang. *Linear Algebra*. Addison-Wesley, second edition, 1971.

[154] R. K. Lenz and R. Y. Tsai. Techniques for calibration of the scale factor and image center for high accuracy 3-d machine vision metrology. *IEEE Trans. Pattern Analysis and Machine Intelligence*, 10(5):713–720, September 1988.

[155] M. Levine. *Vision in Man and Machine*. McGraw-Hill, New York, 1985.

[156] S.-P. Liou and R. C. Jain. Motion detection in spatio-temporal space. *Computer Vision, Graphics and Image Processing*, 45:227–250, 1989.

[157] H. C. Longuet-Higgins and K. Prazdny. The interpretation of a moving retinal image. *Proc. R. Soc. Lond. B*, 208:385–397, 1980.

[158] D. G. Lowe. Three-dimensional object recognition from single two-dimensional images. *Artificial Intelligence*, 31(3):355–395, 1987.

[159] Y. Lu and R. C. Jain. Behavior of edges in scale space. *IEEE Trans. Pattern Analysis and Machine Intelligence*, 11(3):337–356, 1989.

[160] Y. Lu and R. C. Jain. Reasoning about edges in scale space. *IEEE Trans. Pattern Analysis and Machine Intelligence*, 14(4):450–468, 1992.

[161] S. B. Marapane and M. M. Trivedi. Region-based stereo analysis for robotic applications. *IEEE Trans. Systems, Man, and Cybernetics*, 19(6):1447–1464, 1989.

[162] S. B. Marapane and M. M. Trivedi. Multi-primitive hierarchical (mph) stereo analysis. *IEEE Trans. Pattern Analysis and Machine Intelligence*, 16(3):227–240, 1994.

[163] D. Marr. *Vision: A Computational Investigation into the Human Representation and Processing of Visual Information*. W. H. Freeman and Company, San Francisco, 1982.

[164] D. Marr and E. C. Hildreth. Theory of edge detection. *Proc. R. Soc. Lond. B*, 207:187–217, 1980.

[165] T. Matsuyama. Expert systems for image processing: Knowledge-based composition of image analysis processes. *Computer Vision, Graphics and Image Processing*, 48:22–49, 1989.

[166] J. E. W. Mayhew and J. P. Frisby. Psychophysical and computational studies towards a theory of human stereopsis. *Artificial Intelligence*, 17:349–385, 1981.

[167] G. Medioni and R. Nevatia. Segment-based stereo matching. *Computer Vision, Graphics and Image Processing*, 31:2–18, 1985.

[168] B. T. Mitchell and A. M. Gillies. A model-based computer vision system for recognizing handwritten ZIP codes. *Machine Vision and Applications*, 2(4):231–243, 1989.

[169] A. Mitiche, S. Seida, and J. K. Aggarwal. Using constancy of distance to estimate position and displacement in space. *IEEE Trans. Pattern Analysis and Machine Intelligence*, 10(4):594–599, July 1988.

[170] F. H. Moffitt and E. M. Mikhail. *Photogrammetry*. Harper & Row, Cambridge, third edition, 1980.

[171] J. L. Mohammed, R. A. Hummel, and S. W. Zucker. A gradient projection algorithm for relaxation methods. *IEEE Trans. Pattern Analysis and Machine Intelligence*, 5(3):330–332, May 1983.

[172] H. P. Moravec. Towards automatic visual obstacle avoidance. *Proc. 5th Int. Joint Conf. Artificial Intell.*, 2:584, August 1977.

[173] D. W. Murray and B. F. Buxton. Scene segmentation from visual motion using global optimization. *IEEE Trans. Pattern Analysis and Machine Intelligence*, 9(2):220–228, March 1987.

[174] H.-H. Nagel. Formation of an object concept by analysis of systematic time variations in the optically perceptible environment. *Comp. Graph. and Image Proc.*, 7:149–194, 1978.

[175] H.-H. Nagel. On the estimation of optical flow: Relations between different approaches and some new results. *Artificial Intelligence*, 33:299–324, 1987.

[176] H.-H. Nagel and W. Enkelmann. An investigation of smoothness constraints for the estimation of displacement vector fields from image sequences. *IEEE Trans. Pattern Analysis and Machine Intelligence*, (5):565–593, September 1986.

[177] V. S. Nalwa. *A Guided Tour of Computer Vision*. Addison-Wesley, Reading, Massachusetts, 1993.

[178] N. Narasimhamurthi and R. C. Jain. Computer-aided, design-based object recognition: Incorporating metric and topological information. *Proc. SPIE Digital and Optical Shape Representation and Pattern Recognition*, 938:436–443, 1988.

[179] S. K. Nayar, K. Ikeuchi, and T. Kanade. Shape from interreflections. *Proc. 3rd Intl. Conf. Computer Vision*, pages 2–11, 1990.

[180] S. K. Nayar and Y. Nakagawa. Shape from focus. *IEEE Trans. Pattern Analysis and Machine Intelligence*, 16(8):824–831, 1994.

[181] S. K. Nayar and K. Ikeuchi. Photometric sampling: A method for determining shape and reflectance of surfaces. In H. Freeman, editor, *Machine Vision for Inspection and Measurement*. Academic Press, New York, 1989.

[182] A. W. Naylor and G. R. Sell. *Linear Operator Theory in Engineering and Science*. Springer-Verlag, New York, 1982.

[183] A. M. Nazif and M. D. Levine. Low level image segmentation: An expert system. *IEEE Trans. Pattern Analysis and Machine Intelligence*, 6(5):555–577, 1984.

[184] S. Negahdaripour and B. K. P. Horn. Direct passive navigation. *IEEE Trans. Pattern Analysis and Machine Intelligence*, 9(1):168–176, January 1987.

[185] R. Nevatia. Depth measurement by motion stereo. *Comp. Graph. and Image Proc.*, 5:203–214, 1976.

[186] R. Nevatia. *Machine Perception.* Prentice-Hall, Englewood Cliffs, New Jersey, 1982.

[187] B. Noble and J. W. Daniel. *Applied Linear Algebra.* Prentice-Hall, Englewood Cliffs, New Jersey, second edition, 1977.

[188] L. O'Gorman and R. Kasturi. *Document Image Analysis.* IEEE Computer Society Press, Los Alamitos, California, 1995.

[189] R. Ohlander, K. Price, and D. R. Reddy. Picture segmentation using a recursive region splitting method. *Comp. Graph. and Image Proc.*, 8(3):313–355, 1978.

[190] A. V. Oppenheim and R. W. Schafer. *Digital Signal Processing.* Prentice-Hall, Englewood Cliffs, New Jersey, 1975.

[191] J. K. Ousterhout. *Tcl and the Tk Toolkit.* Addison-Wesley, 1994.

[192] A. Papoulis. *Probability, Random Variables, and Stochastic Processes.* McGraw-Hill, New York, 1965.

[193] T. Pavlidis. Curve fitting as a pattern recognition problem. In *Proc. Int. Conf. on Pattern Recognition*, pages 853–859, 1982.

[194] A. Pentland. Fractal-based description of natural scenes. *IEEE Trans. Pattern Analysis and Machine Intelligence*, 6(6):661–674, November 1984.

[195] W. K. Pratt. *Digital Image Processing.* Wiley, New York, 1978.

[196] W. K. Pratt. *Digital Image Processing.* Wiley, New York, second edition, 1991.

[197] W. H. Press, B. P. Flannery, S. A. Teukolsky, and W. T. Vetterling. *Numerical Recipes in C: The Art of Scientific Computing.* Cambridge University Press, 1988.

[198] J. M. S. Prewitt. Object enhancement and extraction. In B. S. Lipkin and A. Rosenfeld, editors, *Picture Processing and Psychopictorics.* Academic Press, New York, 1970.

[199] L. R. Rabiner and B. Gold. *Theory and Application of Digital Signal Processing.* Prentice-Hall, Englewood Cliffs, New Jersey, 1975.

[200] A. Ravishankar Rao. *Taxonomy for Texture Description and Identification.* Springer-Verlag, New York, 1990.

[201] E. Rich. *Artificial Intelligence.* McGraw-Hill, New York, 1983.

[202] L. G. Roberts. Machine perception of three-dimensional solids. In J. K. Aggarwal, R. O. Duda, and A. Rosenfeld, editors, *Computer Methods in Image Analysis.* IEEE Computer Society, Las Alamitos, California, 1977.

[203] I. Rock. *The Logic of Perception.* MIT Press, Cambridge, 1983.

[204] D. F. Rogers and J. A. Adams. *Mathematical Elements for Computer Graphics.* McGraw-Hill, New York, 1976.

[205] A. Rosenfeld and A. C. Kak. *Digital Picture Processing.* Academic Press, New York, 1976.

[206] A. Rosenfeld and A. C. Kak. *Digital Picture Processing.* Academic Press, New York, second edition, 1982. Two volumes.

[207] P. J. Rousseeuw and A. M. Leroy. *Robust Regression and Outlier Detection.* Wiley, New York, 1987.

[208] F. F. Sabins. *Remote Sensing: Principles and Interpretation.* W. H. Freeman, New York, second edition, 1987.

[209] V. Salari and I. K. Sethi. Feature point correspondence in the presence of occlusion. *IEEE Trans. Pattern Analysis and Machine Intelligence*, 12(1):87–91, 1990.

[210] H. Samet. *The Design and Analysis of Spatial Data Structures*. Addison-Wesley, Reading, Massachusetts, 1990.

[211] R. J. Schalkoff. *Digital Image Processing and Computer Vision*. Wiley, New York, 1989.

[212] B. G. Schunck. Robust estimation of image flow. In *Proceedings SPIE Sensor Fusion II, Human and Machine Strategies*, vol. 1198, pp. 116–127, 1989.

[213] E. L. Schwartz. Computational anatomy and functional architecture of striate cortex: A spatial mapping approach to perceptual coding. *Vision Research*, 20:645–669, 1980.

[214] E. L. Schwartz. A quantitative model of the functional architecture of human striate cortex with application to vision illusion and cortical texture analysis. *Biological Cybernetics*, 37:63–76, 1980.

[215] E. L. Schwartz. Cortical anatomy, size invariance, and spatial frequency analysis. *Perception*, 10:455–468, 1981.

[216] E. L. Schwartz. Columnar architecture and computational anatomy in primatevisual cortex: Segmentation and feature extraction via spatial frequency coded difference mapping. *Biological Cybernetics*, 42:157–168, 1982.

[217] G. A. F. Seber and C. J. Wild. *Nonlinear Regression*. Wiley, New York, 1989.

[218] R. Sedgewick. *Algorithms*. Addison-Wesley, second edition, 1988.

[219] I. K. Sethi and R. C. Jain. Finding trajectories of feature points in a monocular image sequence. *IEEE Trans. Pattern Analysis and Machine Intelligence*, 9(1):56–73, 1987.

[220] M. Shah, A. Sood, and R. C. Jain. Pulse and staircase edge models. *Computer Vision, Graphics and Image Processing*, 34:321–343, 1986.

[221] L. G. Shapiro. A CAD-model-based system for object localization. *Proc. SPIE Digital and Optical Shape Representation and Pattern Recognition*, 938:408–418, 1988.

[222] A. Singh. An estimation-theoretic framework of image-flow computation. *Proc. Third Int. Conf. Computer Vision*, pages 168–177, 1990.

[223] S. S. Sinha and B. G. Schunck. A two stage algorithm for discontinuity-preserving surface reconstruction. *IEEE Trans. Pattern Analysis and Machine Intelligence*, 14(1):36–55, January 1992.

[224] C. C. Slama, editor. *Manual of Photogrammetry*. American Society of Photogrammetry, Falls Church, Virginia, fourth edition, 1980.

[225] I. Sobel. Camera models and machine perception. *Stanford AI Memo 121*, May 1970.

[226] T. Sripradisvarakul and R. C. Jain. Generating aspect graphs for curved objects. *Proc. IEEE Workshop on Interpretation of 3-D Scenes*, pages 109–115, 1989.

[227] L. H. Staib and J. S. Duncan. Boundary finding with parametrically deformable models. *IEEE Trans. Pattern Analysis and Machine Intelligence*, 14(11):1061–1075, 1992.

[228] G. Stockman. Object recognition and localization via pose clustering. *Computer Vision, Graphics, and Image Processing*, 40:361–387, 1987.

[229] M. Subbarao. Interpretation of image flow: Rigid curved surfaces in motion. *Int. J. Computer Vision*, 2(1):77–96, 1988.

[230] S. L. Tanimoto and A. Klinger. *Structured Computer Vision, Machine Perception through Hierarchical Computational Structures*. Academic Press, New York, 1980.

[231] S. Tehrani, T. E. Weymouth, and B. G. Schunck. Interpolating cubic spline contours by minimizing second derivative discontinuity. In *International Conference on Computer Vision*, pages 713–716, 1990.

[232] D. Terzopoulos. Multilevel computational processes for visual surface reconstruction. *Computer Vision, Graphics and Image Processing*, 24:52–96, 1983.

[233] D. Terzopoulos. Image analysis using multigrid relaxation methods. *IEEE Trans. Pattern Analysis and Machine Intelligence*, 8(2):129–139, March 1986.

[234] R. Y. Tsai. A versatile camera calibration technique for high-accuracy 3d machine vision metrology using off-the-shelf cameras and lenses. *IEEE Trans. Robotics and Automation*, 3(4), August 1987.

[235] R. Y. Tsai and T. S. Huang. Uniqueness and estimation of three-dimensional motion parameters of rigid objects with curved surfaces. In *Proc. Conf. Pattern Recognition and Image Processing*, pages 112–118, 1982.

[236] R. Y. Tsai and T. S. Huang. Uniqueness and estimation of 3-d motion parameters and surface structures of rigid objects. In S. Ullman and W. Richards, editors, *Image Understanding 1984*. Ablex Publishing, Norwood, New Jersey, 1984.

[237] R. Y. Tsai and T. S. Huang. Uniqueness and estimation of three-dimensional motion parameters of rigid objects with curved surfaces. *IEEE Trans. Pattern Analysis and Machine Intelligence*, 6(1):13–27, January 1984.

[238] J. K. Tsotsos et al. A framework for visual motion understanding. *IEEE Trans. Pattern Analysis and Machine Intelligence*, 2(6):563–573, 1980.

[239] M. Tuceryan and A. K. Jain. Texture analysis. In C. H. Chen, L. F. Pau, and P. S. P. Wang, editors, *Handbook of Pattern Recognition and Computer Vision*, pages 235–276. World Scientific Publishing Company, 1993.

[240] J. R. Ullman. Discrete optimization by relational constraint satisfaction. *IEEE Trans. Pattern Analysis and Machine Intelligence*, 4(5):544–551, September 1982.

[241] S. Ullman. *The Interpretation of Visual Motion*. MIT Press, Cambridge, 1979.

[242] S. Ullman. Maximizing rigidity: The incremental recovery of 3-d structure from rigid and nonrigid motion. *Perception*, 13:255–274, 1984.

[243] L. Van-Ban and D. T. Lee. Out-of-roundness problem revisited. *IEEE Trans. Pattern Analysis and Machine Intelligence*, 13(3):217–223, March 1991.

[244] E. Vanmarke. *Random Fields: Analysis and Synthesis*. MIT Press, Cambridge, 1983.

[245] G. Wahba. *Spline Models for Observational Data*. Society for Industrial and Applied Mathematics, Philadelphia, Pennsylvania, 1990.

[246] B. A. Wandell. The synthesis and analysis of color images. *IEEE Trans. Pattern Analysis and Machine Intelligence*, 9(1):2–13, January 1987.

[247] A. B. Watson and A. J. Ahumada, Jr. Model of human visual-motion sensing. *J. Opt. Soc. Am. A*, 2(2):322–342, 1985.

[248] H. Wechsler. *Computational Vision*. Academic Press, Boston, 1990.

[249] J. Weng, T. S. Huang, and N. Ahuja. Motion and structure from line correspondences: Closed-form solution, uniqueness, and optimization. *IEEE Trans. Pattern Analysis and Machine Intelligence*, 14(3):318–336, March 1992.

[250] P. M. Will and K. S. Pennington. Grid coding: A novel technique for image processing. *Proc. IEEE*, 60(6):669–680, 1972.

[251] P. Winston. *Artificial Intelligence*. Addison-Wesley, Reading, Massachussetts, third edition, 1992.

[252] A. P. Witkin. Recovering surface shape and orientation from texture. *Artificial Intelligence*, 17:17–45, 1981.

[253] A. P. Witkin. Scale-space filtering. In *Int. Joint. Conf. Artificial Intelligence*, pages 1019–1022, August 1983.

[254] P. R. Wolf. *Elements of Photogrammetry.* McGraw-Hill, New York, second edition, 1983.

[255] L. B. Wolff. Accurate measurement of orientation from stereo using line correspondence. *Proc. Conf. Computer Vision and Pattern Recognition*, pages 410–415, 1989.

[256] R. J. Woodham. Analysing images of curved surfaces. *Artificial Intelligence*, 17:117–140, 1981.

[257] J. M. Wozencraft and I. M. Jacobs. *Principles of Communication Engineering.* Wiley, New York, 1965.

[258] Y. Yakimovsky. Boundary and object detection in real world images. *Journal of the Association of Computing Machinery*, 23:599–618, 1976.

[259] M. Young. *Optics and Lasers Including Fibers and Integrated Optics.* Springer-Verlag, Berlin, second revised edition, 1984.

[260] A. L. Yuille and T. Poggio. Fingerprint theorems for zero crossings. *J. Opt. Soc. Am. A*, 2:683–692, May 1985.

[261] A. L. Yuille and T. Poggio. Scaling theorems for zero-crossings. *IEEE Trans. Pattern Analysis and Machine Intelligence*, 8:15–25, January 1986.

[262] Z. Zhang, O. D. Faugeras, and N. Ayache. Analysis of a sequence of stereo scenes containing multiple moving objects using rigidity constraints. *Proc. Second Int. Conference on Computer Vision*, pages 177–186, 1988.

Index